When Science and Christianity Meet

May 2016

WHEN SCIENCE & CHRISTIANITY MEET

Edited by

David C. Lindberg and Ronald L. Numbers

The University of Chicago Press
Chicago and London

The University of Chicago Press, Chicago 60637
The University of Chicago Press, Ltd., London

Paperback edition 2008
Printed in the United States of America

17 16 15 14 13 12 11 10 09 08 2 3 4 5 6

ISBN-13: 978-0-226-48214-9 (cloth)
ISBN-13: 978-0-226-48216-3 (paper)
ISBN-10: 0-226-48214-6 (cloth)
ISBN-10: 0-226-48216-2 (paper)

Library of Congress Cataloging-in-Publication Data

When science and Christianity meet / edited by David C. Lindberg and Ronald L. Numbers.
p. cm.
Includes bibliographical references and index.
ISBN 0-226-48214-6 (alk. paper)
1. Religion and science—History. I. Lindberg, David C. II. Numbers, Ronald L.

BL245 .B35 2003
261.5'5—dc21

2002155569

♾ The paper used in this publication meets the minimum requirements of the American National Standard for Information Sciences—Permanence of Paper for Printed Library Materials, ASNI Z39.48-1992.

To Susan E. Abrams, loyal friend, editor par excellence, and indefatigable supporter of scholarship on the history of science.

CONTENTS

ILLUSTRATIONS

ACKNOWLEDGMENTS

This book is the product of years of reflection by the editors on problems of communicating to university undergraduates and the educated public the fruits of historical scholarship on the relationship between science and Christianity. Rather than attempting to write a synthetic survey, we looked for contributors who would join us in addressing discrete episodes or themes in the long history of Christianity's engagement with science—and do so in a manner accessible to a general audience.

A conference in March 1999, after the essays were drafted and circulated, brought together the eleven contributors to the volume with a roughly equal number of potential users of the book—teachers and students—for the sole purpose of ensuring that the authors had achieved the level of accuracy and accessibility to which we aspired. Jim Miller, senior program associate with the Program of Dialogue between Science and Religion at the American Association for the Advancement of Science, first proposed such a get-together. The Center for Theology and the Natural Sciences in Berkeley hosted the conference, which also received support from the Counterbalance Foundation of Seattle and the Center for Research in Science at Azusa Pacific University. In addition to our authors, the participants included:

- Kate Grayson Boisvert, Los Medanos College
- Gary Ferngren, Oregon State University

- Rebecca J. Flietstra, Point Loma Nazarene University
- Nathan G. Hale Jr., Piedmont, California
- Peter M. J. Hess, Center for Theology and the Natural Sciences
- David M. Knight, University of Durham (United Kingdom)
- James Miller, American Association for the Advancement of Science
- Elizabeth Moore, University of Oregon
- Edward T. Oakes, S.J., Regis University
- Ronald P. Olowin, Saint Mary's College
- Richard Olson, Harvey Mudd College
- Mark Railey, Center for Research in Science, Azusa Pacific University
- Frank M. Turner, Yale University
- Adrian Wyard, Counterbalance Foundation.

For financial and organizational support that made the conference possible, we are indebted to Jim Miller, Peter Hess, Mark Railey, and Adrian Wyard. We also wish to thank Nicolaas Rupke and two anonymous readers of the book manuscript for many helpful suggestions; Michael Koplow for an extraordinary job of copyediting; Jennifer Howard for seeing the manuscript through to completion; Stephen Wald for preparing the index; and Carson Burrington for assistance with the illustrations. To our wives, Greta Lindberg and Karen Steudel Numbers, we express gratitude for their good-natured fortitude as they have endured our long-standing preoccupation with the relationship between science and Christianity. Finally, dedication of this book to Susan E. Abrams is an expression of gratitude for her long-standing friendship, editorial advice, and staunch support of this volume since its conception.

INTRODUCTION

Scholars have long debated how best to characterize the relationship between science and Christianity. No generalization has proved more seductive and tenacious than that of "conflict." Indeed, the two most widely read books on the subject carry the titles *History of the Conflict between Religion and Science* (1874) and *A History of the Warfare of Science with Theology in Christendom* (1896). In the former, the New York chemist-historian John William Draper lashed out at the Roman Catholic church for its alleged centuries-long hostility toward science, arguing that the Vatican's antipathy toward science had left its hands "steeped in blood." (No scientist, to our knowledge, ever lost his life because of his scientific views, though the Italian Inquisition did incinerate the sixteenth-century Copernican Giordano Bruno for his heretical *theological* notions.) In the latter, a monumental two-volume work, the American historian Andrew Dickson White depicted the engagement of Christianity and science as a series of "battles" between narrow-minded, dogmatic theologians and truth-seeking men of science.[1]

Not surprisingly, many Christians took offense at such negative and largely unwarranted characterizations. They pointed out that Christian Europe gave birth to modern science and that a large majority of contributors to science were professing Christians. (Even today, according to a recent survey, approximately 40 percent of American scientists profess belief in "a God to whom one may pray in expectation of receiving an answer"; countless others accept a less personal deity.)[2] Some Christian apologists have gone so far as to reframe the relationship between science and Christianity as an essentially harmonious engagement, arguing that science could have developed only in a culture, such as Christendom, where belief in an orderly cosmos, created and regulated by a divine being, was widely held.[3] However, such sweeping claims misrepresent the scientific achievements of ancient Greeks and medieval Muslims and exhibit a thorough misunderstanding of the history of early Western science. It is true, of course, that Christianity has aided science in many specific ways. For example, in a recent history of solar observatories in cathedrals the distinguished Berkeley historian John Heilbron concluded that "the Roman Catholic church gave more financial and social support to the study of astronomy for over six centuries, from the recovery of ancient learning during the late Middle Ages into the Enlightenment, than any other, and, probably, all other, institutions." Heilbron's assertion, though counterintuitive to many readers, rests on sound historical evidence.[4]

By the last quarter of the twentieth century many historians of science and Christianity were growing increasingly uncomfortable with the triumphalist narratives of both the warriors and the harmonizers. Largely setting aside the question of who was "right" or "wrong" by the standards of present-day science or theology, these historians attempted to evaluate the historical involvement of science and Christianity in terms of the values and knowledge of the historical actors themselves. In short, they laid aside apologetic and polemical goals, choosing to understand rather than to judge, to the point where it became nearly impossible to discern their personal religious sentiments from their writings on the subject of science and religion.

In 1981 we brought together a group of such scholars to reassess the history of science and Christianity. This conference resulted in a volume entitled *God and Nature: Historical Essays on the Encounter between Christianity and Science* (1986). As we noted in the introduction to that volume, virtually every chapter portrayed

> a complex and diverse interaction that defies reduction to simple "conflict" or "harmony." Although instances of controversy are not hard to find, it cannot be said that scientists and theologians—much less science and Christianity—engaged in protracted warfare. Likewise, although cases of mutual support are numerous, it would be a travesty to maintain that Christianity and science have been perennial allies. Some Christian beliefs and practices seem to have encouraged scientific investigation, others to have discouraged it; the interaction varied with time, place, and person.[5]

Despite our effort to provide balanced, nonpartisan accounts that tilted toward neither science nor Christianity—and our willingness to admit conflict whenever it occurred—the book's rejection of the warfare thesis left some readers with the false impression that we sided with the harmonist school of historians.[6]

During the past fifteen years or so a number of histories of science and religion have continued to undermine the notion that the historical relationship between science and Christianity could be reduced to simple generalities. The most influential of these works has been John Hedley Brooke's *Science and Religion: Some Historical Perspectives* (1991). In this book Brooke synthesized and expanded on the revisionist scholarship of the previous generation, arguing for what might be called "the complexity thesis." Avoiding simplistic formulas of conflict and harmony, he reveled in the rich diversity of interplay between science and Christianity. He described a thoroughly entangled relationship, with religious beliefs not only providing "presupposition, sanction, even motivation for science" but also regulating "discussions of method" and playing "a selective role in the evaluation of rival theories."[7]

As in history generally, increasing suspicion of so-called master narratives and growing attention to the diversity of human experience has called even mid-level generalizations—about particular nations, denominations, and classes—into question. Take just one example: in 1979 James R. Moore published a major study of Protestant reactions to Darwinism in North America and Great Britain, in which he argued that Calvinist theology, such as that typically associated with Presbyterians and Congregationalists, made it easier to embrace unadulterated Darwinism than did more liberal strains of theology. Recently, however, the geographer and historian of science David N. Livingstone has shown that even Calvinist responses to evolution varied markedly from one locale to another. In Belfast, Ireland, roiled by John Tyndall's declaration that all "religious theories, schemes and systems which embrace notions of cosmogony . . . must . . . submit to the control of science, and relinquish all thought of controlling it," Calvinists stoutly resisted Darwinism, while in Princeton, New Jersey, a Presbyterian stronghold inclined to harmonize science and Christianity, Calvinists tended to be much more welcoming of Darwin's new theory. In other words, geography trumped theology.[8]

Lately the very terms "science" and "Christianity" have come under scrutiny.[9] We often speak of what science states or what Christianity claims, as though science and Christianity were existing things, capable of speaking and claiming. But we must never forget that it is people who do the believing, the speaking, the teaching, and the battling—even when speaking with authority on behalf of the community of scientists and Christians. And when human beings are involved, so are human agendas and interests. To complicate matters historically, science did not appear as a cultural category with its current meaning until about two hundred years ago, and the term "scientist" is even more recent. And to make things even worse, no universally accepted definition of "science" exists. Does the term embrace the "science" of the Church of Christ, Scientist? Does it include parapsychology and creation science, as well as astronomy and zoology? Are molecular biologists and nuclear physicists more representative of science than social and behavioral scientists? And who gets to decide?

The term "Christianity" presents its own problems. We may speak of it as a single entity, but in fact it comes in many varieties, equipped with a host of different theologies and therefore with the potential for many different responses to the same scientific claim. Moreover, those of us who write about the history of science and Christianity have been guilty of concentrating almost all of our attention on America and Western Europe, despite the fact that Christendom has always been solidly represented in eastern Europe, western Asia, and Africa. This volume may appear to deal with "Christianity" in general, but except for Saint Augustine, who resided in fifth-century North Africa, it says virtually nothing about the attitudes toward or contributions to science of Christians who lived outside of the North Atlantic community (with the exception of Italy); and it limits itself to the Christian theologies that have been historically dominant in America and Western Europe.[10]

In this volume, dubbed "Son of *God and Nature*" by an irreverent colleague, we present a collection of twelve case histories that illustrate a variety of encounters between Christianity and science. In referring to these essays as "case histories," we are declaring that they do not offer a comprehensive account of the historical relations between science and Christianity. Rather they are discrete historical studies, covering a dozen of the most notorious, most interesting, or most instructive instances of encounter between these two powerful cultural forces. Authors were chosen without regard for personal religious belief or practice (though we believe that they run the gamut from agnostics to Christian evangelicals), but rather for their ability and willingness to address a nonspecialist audience while measuring up to the highest standards of historical scholarship. The charge to our authors was to tell the story of their assigned case "like it was," in all of its particularity. The goal was to examine the interaction between science and Christianity in a dozen concrete, historically situated, local settings. As readers will discover, every case is unique, reflecting different historical actors, different agendas, different social and intellectual contexts, leading in the end to a rich variety of different outcomes.

Do the stories contained in this volume reveal a pattern? Yes, if a pattern may be vague, imprecise, and variable and yet be considered a pattern. Our authors address the varied relationships between two powerful cultural traditions attempting, sometimes, to occupy the same intellectual and social ground. Points of conflict were not as numerous as rumor would have it, but historically many of them were judged critically important. Occasionally those conflicts that genuinely existed assumed the form of aggressive hostility, but the norm was interaction of a far more complicated sort, falling somewhere along the spectrum that separates the harmonious from the bellicose, with peaceful coexistence at its midpoint. Tensions, as often as not, were internal to the minds and hearts of individual people, striving to preserve both their loyalty to the principles of Christian theology and their commitment to the basics of scientific knowledge. Because most participants considered peace preferable to warfare, it was customary to search for avenues of compromise and accommodation. On occasion, one of the parties emerged a winner and the other a loser, but more often a workable peace was crafted.

If the stories told in this volume are representative, historical study does not reveal science and Christianity locked in deadly combat; nor does it disclose an interaction of unfailing support and mutual compatibility. The relationship between science and Christianity proves to be much more intricate and interesting than these traditional alternatives allow, richly varied and nuanced, thoroughly human, and imbued with the same complexity that we find in other areas of human experience.

1

The Medieval Church Encounters the Classical Tradition: Saint Augustine, Roger Bacon, and the Handmaiden Metaphor

David C. Lindberg

The Problem

According to widespread popular belief, the period of European history known as the Middle Ages or medieval period (roughly the years 450–1450) was a time of barbarism, ignorance, and superstition. The epithet "Dark Ages" often applied to it nicely captures this opinion. As for the ills that threatened literacy, learning, and especially science during the Middle Ages, blame is most often laid at the feet of the Christian church, which is alleged to have placed religious authority above personal experience and rational activity, thereby snuffing out the faint sparks of scientific and other forms of intellectual creativity that had survived the barbarian invasions of late antiquity.[1]

But this is a caricature, the acceptance of which has proved an obstacle to an understanding of the Middle Ages as they really were. It is true that the early centuries of the medieval period, like those of late antiquity, saw a great deal of political and social turmoil. It is also true that literacy and learning, in this early period, were in a state of decline. But an account that fails to acknowledge differences among geographical regions and change over time cannot do justice to the complex medieval reality. An accurate account will reveal that learning grew from small beginnings in the early Middle Ages to become a thriving industry in the later Middle Ages; that important scientific achievements emerged during this period; and that the church and its theology maintained a relationship to the natural sciences far too complicated to be captured by simple black-and-white categories such as adversaries or allies. Unquestionably, some portions of the classical tradition gave rise to suspicion, hostility, and even ecclesiastical condemnation. However, such cases were exceptional; far more commonly, critical reflection about the nature of the world was tolerated and even encouraged. In their quest to understand the world in which they lived, medieval scholars employed all of the resources at their disposal, including inherited scientific ideas, personal observation, rational inference, and religious tradition. And they did so with as much integrity as one finds today in the average university professor and with far less interference from the church than the caricature of the Middle Ages would suggest.

By way of developing and defending these claims, I propose to concentrate on two historical figures who have contributed mightily to the image of the Middle Ages: Saint Augustine of Hippo (354–430), the early church father who did more to determine medieval Christian attitudes toward pagan science than any other person, and Roger Bacon (ca. 1220–ca. 1292), the most notorious scientific figure of the Middle Ages, widely acclaimed for his rejection of authority and his campaign on behalf of mathematical and what he called "experimental" science. (I employ the term "pagan" without pejorative intent, to mean simply non-Judeo-Christian.) I do not claim, of course, that the lives of Augustine and Bacon present us with the whole story of medieval encounter between science and religion, but I do believe that an examination of their careers will reveal the basic contours of that story.[2]

The Middle Ages and the Classical Tradition

Several preliminaries must first occupy us. About 850 years separate Augustine and Bacon. What are the chronological divisions associated with this long period of European history? There were no catastrophes or achievements so decisive or conspicuous that we can use them as chronological markers, and the boundaries are therefore intrinsically fuzzy. But in round numbers the declining years of the Roman Empire run from about A.D. 180 to 450. Church historians know this as the "patristic period"—an era during which Christian doctrine was codified by a series of church councils and influential church fathers. The characteristics that strike us as distinctively medieval emerged gradually in the course of the fifth century. The early medieval period is customarily dated from about 450 to 1000. This was followed by a period of European recovery, 1000–1200, and the high or later Middle Ages, roughly 1200–1450. The story recounted in this essay runs from the closing decades of the patristic period to the first seventy-five years of the high Middle Ages.

Was there, in fact, any science worthy of the name during this long period? Certainly many of the ingredients of what we now regard as science were present: languages for describing nature, methods for exploring it, factual and theoretical claims that emerged from such explorations, and criteria for judging the truth or validity of the claims thus made. Moreover, it is clear that pieces of the resulting medieval knowledge were for all practical purposes identical to what is now taken to be genuine science (planetary astronomy and geometrical optics are good examples).

But patristic and medieval approaches to nature also differed from ours in significant ways. Knowledge about the world of nature was then an integral part of the larger philosophical enterprise—a characteristic that modern scientists would find alien. Theology and religion were regarded as legitimate participants in the investigation and formulation of truths about the natural world far more frequently than they are today.[3] Observational evidence, though regularly employed in the validation of theoretical claims during the medieval period, had a profile considerably lower than in modern science. The motivation for pursuing science and the institutions where that pursuit took place were quite different from the modern ones. The governmental support that drives big science today would have been inconceivable during the patristic and medieval periods. And the mechanisms now available for disseminating scientific knowledge are far more efficient than were those operating in a culture that antedated the printing press and electronic media.

Given these similarities and differences, are we justified in calling this patristic and medieval effort "science"? This question is a matter of dispute among historians of science. Some prefer the cautious expression "natural knowledge." Others speak of "natural philosophy," in order to call attention to the integral relationship in that earlier era between the pursuit of natural knowledge and the pursuit of other forms of understanding. And some boldly use the expression "science" or "natural science," declaring thereby that the objects of their scholarship, although not identical to modern science, are the ancestors of modern scientific disciplines and practices and therefore are entitled to claim the family name. This seems to me a pointless debate. The important thing is to agree on what we are talking about and to employ terminology that facilitates communication on that subject. In the following pages, I will employ all three of the aforementioned competing locutions indiscriminately, as synonyms. I will also employ expressions denoting specific branches of the pursuit of natural knowledge, such as "mathematical science," "astronomy," "cosmology," "optics," "meteorology," and "medicine." The reader should understand that at no point do I wish to maintain *identity* between the patristic and medieval enterprises thus named and their modern descendants.

Augustine and Bacon encountered the natural sciences as elements of the classical tradition; and if we wish to understand their attitudes toward the natural sciences, we must look briefly at the whole of which these sciences were a part. The classical tradition consisted of the accumulated learning of ancient Greece, transmitted vertically through time and horizontally across geographical, cultural, and linguistic boundaries, adjusting itself in the process to new cultural and linguistic circumstances and undergoing significant modifications. The classical tradition included poetry, drama, history, political theory and ethics, metaphysics or theology, and the natural sciences. It also included the rules of effective reasoning, writing, and arguing. Prominent within the portion of the classical tradition devoted to nature were the writings of the philosophers Plato and Aristotle, members of the Stoic and Epicurean philosophical schools, the mathematician Euclid, the astronomer Ptolemy, the physician Galen, their followers, and their critics—writings that addressed topics ranging from medicine and the mathematical sciences to meteorology, cosmology, and the relationship of all this to the gods. It is critical to bear in mind that these were pagan writings, produced outside the Christian fold, sometimes inconsistent with Christian doctrine, and potentially the objects of hostility from a Christian audience.[4]

The transmission and fate of the classical tradition is a subject to which we could easily devote a book-length analysis. But the short version is this: As Rome extended its power over the Mediterranean basin in the centuries after 200 B.C., broad cultural contact between Greeks and Romans (encouraged by widespread bilingualism among the Roman upper classes) introduced a thin, popularized version of the classical tradition into Roman education and Roman culture. A few Greek works were translated into Latin; but as bilingualism and the conditions that had favored scholarship diminished in the declining years of the Roman Empire (after about A.D. 180), Roman audiences (initially pagan but gradually becoming Christian) were increasingly limited to pieces of the classical tradition that had been explained, epitomized, or otherwise appropriated by Latin authors.[5] The Western church fathers of the patristic period and Christian authors of the early Middle Ages were forced to rely on this derivative, Latinized (but still philosophically vigorous) version of the classical tradition.

Meanwhile, a far richer, more complete version of the scientific portions of the classical tradition followed a roundabout itinerary that allowed it to burst onto the scene in Christian Europe in the twelfth century.[6] This version, which included many original Greek sources, was first carried eastward into western and central Asia, where (after the rise of Islam, generally dated to A.D. 622) it was translated into Arabic and assimilated by Muslim intellectuals. It moved across north Africa to Spain with the expansion of the Islamic Empire (seventh and eighth centuries). Finally, as a result of the reconquest of Spain by Christian armies, these original sources, along with the extensive Arabic literature inspired by them, were translated from Arabic into Latin (primarily in the twelfth century and first half of the thirteenth) and entered at last into the mainstream of medieval Christian culture. About the same time, many of the same materials were translated into Latin from original Greek versions to which Western Europeans had gained access.[7]

The Early Church and the Classical Tradition

The process of assimilation, however, was fraught with difficulties. The classical tradition, owing to its pagan origin, clashed with Christian doctrine on fundamental issues, including the nature and identity of the divine being, the problem of good and evil, the relationship between creator and creation, and the sources of religious authority. The early church fathers (who, we must recall, had access only to the thinner version of the classical tradition) found much to fear in it.[8]

The church father who has come to symbolize this fear was Tertullian (fl. 195–215), a highly educated critic of the classical tradition, who converted to Christianity after completion of his own superb classical education. Tertullian wrote extensively against heresy, attacking the classical tradition as its incubator. He lashed out at logic and dialectic (the art of constructing logical arguments) and specifically at "wretched Aristotle," who "invented dialectic . . . , the art of constructing and destroying, elusive in its claims, contrived in its conjectures, harsh in argumentation, prolific in contentions, a nuisance even to itself." And his often-quoted warning against curiosity ("No curiosity is required of us after Christ Jesus, no investigation after the Gospel") is regularly interpreted as an expression of the opinion that the Christian requires no knowledge beyond that which biblical revelation furnishes.[9] Not only is this a caricature of Tertullian's true position, but it is also not representative of patristic attitudes (although this has proved no obstacle to its wide dissemination).[10]

This attitude imputed to Tertullian is at an extreme end of a broad spectrum of patristic opinion. If the pagan learning embodied in the classical tradition appeared dangerous, it also proved indispensable, and the level of hostility expressed by Tertullian in his moments of rhetorical overkill was the exception rather than the rule. Total repudiation of the classical tradition by the church fathers was, as a practical matter, impossible. Many had been educated in the classical tradition before converting to Christianity and had acquired habits of rational inquiry that could not have been easily tossed aside. Moreover, the tools of rational discourse and some of the assumptions of Greek philosophy were required for the development of Christian doctrine and defense of the faith against its detractors. And finally, there is the simple fact that much of the content of the classical tradition was theologically benign. It would have been absurd for educated Christians to repudiate the intellectual riches of the classical tradition in everything from botany to physics and medicine to metallurgy—thereby dooming themselves to a state of barbaric ignorance. From the fact that Christians were wary of theological dangers in Greco-Roman philosophy and religion it does not follow that they were prepared to renounce all aspects of the larger Greco-Roman culture that (we must never forget) was also their culture.

Consequently, many of the church fathers expressed at least limited approval of the classical tradition. For example, the second- and third-century writers Athenagoras, Clement, and Origen all found Greek philosophy a useful tool in the defense of Christianity. Athenagoras marshaled the authority of Plato, Aristotle, and the Stoics in favor of monotheism. Clement attacked the earliest Greek philosophers for their atheism. But he also acknowledged that certain philosophers and poets bore testimony to the truth, and that within the philosophical tradition there is "a slender spark, capable of being fanned into flame, a trace of wisdom and an impulse from God."[11] Tertullian himself viewed Christian religion as the fulfillment of Greek rationality, and he both advocated and engaged in philosophical activity.

Augustine and the Natural Sciences

The church father who most influentially defined the proper attitude of medieval Christians toward pagan learning was Augustine (fig. 1.1). A leading teacher of rhetoric, subsequently (in 386) a convert to Christianity at the age of thirty-two, and eventually (after 395) bishop of Hippo in North Africa, Augustine was a prolific writer of books on theological and philosophical topics (more than a hundred of which survive).[12] Many of these works contain passages that suggest quite a cautious, or even a negative, attitude toward pagan learning. In his *Confessions,* Augustine warned against the dangers of curiosity: "Besides that lust of the flesh which lies in the gratification of all senses and pleasures, . . . there pertains to the soul, through the same senses of the body, a certain vain and curious longing, cloaked under the name of knowledge and learning." In the same treatise, Augustine expressed regret for the effort he had devoted to mastering the liberal arts (including logic, geometry, and arithmetic)—effort, he wrote, that "served not to my use, but rather to my destruction."[13]

But it would be a mistake to infer from such fragments that Augustine re-

Figure 1.1. A medieval representation of Saint Augustine, from a manuscript copy of his *City of God* in the Biblioteca Medicea-Laurenziana in Florence.

nounced rational activity in general or the classical tradition in particular. That he opposed false or heretical reasoning and the philosophical systems that gave rise to it is not in doubt, and he was skeptical of any large-scale investment in the classical tradition. However, rational activity properly grounded in a life of faith and applied to appropriate objects (especially the articles of the faith and their rational underpinnings) was indispensable. Heretical reasoning about the Trinity, he pointed out, "is to be shunned and detested, not because it is reasoning, but because it is false reasoning. . . . Therefore, just as you would be ill advised to avoid all speaking because some speaking is false, so you must not avoid all reasoning because some reasoning is false."[14]

What legitimacy, then, did Augustine attach to rational activity directed toward objects having limited or negligible religious relevance? And in particular, what was his attitude toward the rational and empirical investigation of the material world in which we live? Certainly Augustine placed low priority on such investigations. In his *Literal Commentary on Genesis,* he noted that scholars frequently present long discussions of the form and shape of the heaven, matters that "the sacred writers," in their profound wisdom, "have omitted." "Such subjects," he continues, "are of no profit for those who seek eternal happiness, and, what is worse, they take up very precious time that ought to be given to what is spiritually beneficial."[15] Augustine elaborated in his *Enchiridion* (a handbook of basic Christian doctrine), cautioning that we should not be alarmed if Christians are ignorant of the natural knowledge contained in the classical tradition. It is sufficient for them to understand that God is the only cause of created things. In *On Christian Doctrine,* he argued that within pagan learning, "aside from the history of things both past and present, teachings which concern the corporeal senses, including the experience and theory of the useful mechanical arts, and the sciences of disputation and of numbers, I consider nothing to be useful."[16] Augustine proposed the compilation of handbooks to provide Christians with all that they need to know of each discipline:

> I think it might be possible . . . to collect . . . and record . . . explanations of whatever unfamiliar geographical locations, animals, herbs and trees, stones and metals are mentioned in the Scripture. The same thing could be done with numbers so that the rationale only of those numbers mentioned in Holy Scripture is explained.[17]

In the opinion of this most influential theologian of early Christendom, natural philosophy was of very modest religious utility.

But modest religious utility, it turns out, was no cause for dismissal. In his *Literal Commentary on Genesis,* Augustine made clear that although scriptural knowledge is vastly superior to knowledge gained through the senses, the latter is inestimably superior to ignorance. Moreover, he worried that Christians, naïvely interpreting Scripture, might express absurd opinions on cosmological issues, thus provoking ridicule among better-informed pagans and bringing the Christian faith into disrepute. "Even non-Christians," he wrote, know

> something about the earth, the heavens, and the other elements of this world, about the motion and orbit of the stars and even their size and

> relative positions, about the predictable eclipses of the sun and moon, the cycles of the years and the seasons, about the kinds of animals, shrubs, stones, and so forth. . . . Now, it is a disgraceful and dangerous thing for an infidel to hear a Christian, presumably giving the meaning of Holy Scripture, talking nonsense on these topics; and we should take all means to prevent such an embarrassing situation, in which people show up vast ignorance in a Christian and laugh it to scorn.[18]

So we see that there were contexts in which Augustine's attitude toward pagan works on natural philosophy was relatively favorable. In *On Christian Doctrine,* he admonished: "If those who are called philosophers, especially the Platonists, have said things which are indeed true and are well accommodated to our faith, they should not be feared; rather, what they have said should be taken from them as from unjust possessors and converted to our use." Pagan learning contains not only "superstitious imaginings," but also "liberal disciplines more suited to the uses of truth." A good Christian, he concluded, "should understand that wherever he may find truth, it is his Lord's."[19]

In Augustine's view, then, natural knowledge was not to be loved, but to be used. "We should use this world and not enjoy it," he wrote in *On Christian Doctrine,* "so that . . . by means of corporeal and temporal things we may comprehend the eternal and spiritual."[20] The material and temporal must be compelled to serve the spiritual and eternal. Natural philosophy, pagan in origin, is legitimized—indeed, sanctified—by the service it performs for the faith—especially as a source, when allegorically interpreted, of moral and theological truths and for the assistance it lends to the interpretation of Scripture. The natural sciences must be pressed into service as the handmaidens of theology and religion.

Did Augustine practice what he preached? Did he, in fact, put the natural sciences to work on behalf of theology and religion? Although such a project was not his primary mission, whenever the natural sciences impinged on, or could be mobilized in support of, his episcopal, pastoral, or theological responsibilities, Augustine put them to work. There is no clearer example than his *Literal Commentary on Genesis,* where he made copious use of the natural sciences contained in the classical tradition to explicate the Creation story as found in the first three chapters of the Book of Genesis and other biblical passages that address (or appear to address) scientific questions. Here we encounter Greco-Roman ideas about lightning, thunder, clouds, wind, rain, dew, snow, frost, storms, tides, plants and animals, matter and form, the four elements, the doctrine of natural place, seasons, time, the calendar, the planets, planetary motion, the phases of the Moon, astrological influence, the soul, sensation, sound, light and shade, and number theory. For all of his worry about overvaluing the natural philosophy of the classical tradition, Augustine applied it with a vengeance to biblical interpretation.

Several examples will illustrate. Psalm 103:2 (in Vulgate; 104:2 in the King James and Revised Standard Versions) poses a cosmological problem by virtue of its reference to the "stretching out" of the material heavens (containing the fixed stars) "like a skin"—a passage that could be taken to teach the idea of flat heavens and thus to be inconsistent with the spherical cosmology of the classical tradition. Fearful that among people educated in the classical tradition, Scripture might fall into disrepute if the "flat heavens" theory were presented to them as its true meaning, Augustine decided that he must address this (otherwise unimportant) question. He began by arguing that cosmological matters such as this were omitted from Scripture because of their irrelevance to the attainment of salvation. Nevertheless, if it were to be demonstrated that the heavens are spherical, the Christian would be obliged to demonstrate that the scriptural claim is not inconsistent with this account. And this is an easy demonstration to perform, since we all know that a skin can be stretched into a spherical form, as in an inflated leather ball.[21]

If a solution to this cosmological problem was easily found, a physical problem associated with Aristotle's theory of the elements posed a more serious challenge to the interpretation of Scripture. According to Aristotle, four elements (in combination) are the constituents of all other material things: earth, water, air, and fire. Earth and water, Aristotle held, are heavy and therefore have a natural tendency to descend. Fire and air are light and naturally ascend. Earth is the heavier of the two heavy elements; fire the lighter of the two light elements. Therefore, if the ideal arrangement defined by the natural tendencies of the four elements were ever to be achieved, we would have the spherical body of the earth at the center of the universe, surrounded by spherical shells (moving from bottom to top) of water, air, and fire (see fig. 1.2). Continuous change in the universe, including transmutation of the elements, prevents this ideal arrangement from being completely achieved, but it is useful nonetheless as an account of the natural places of the four elements.

The problem that Augustine had to confront was that according to Genesis 1:6, the "firmament" (the celestial sphere to which, it was believed, the stars were affixed) separated certain waters located below it from other waters situated above it. The waters below the firmament were undoubtedly the oceans and rivers on the face of the earth, but how was it possible for other waters to be above the firmament—and, therefore, also above the spheres of air and fire? A parallel problem was posed by Psalm 135:6 (in Vulgate; 136:6 in the King James and Revised Standard Versions), where the psalmist refers to the spreading of the earth above the waters. Augustine's solution was to argue, first, for a possible

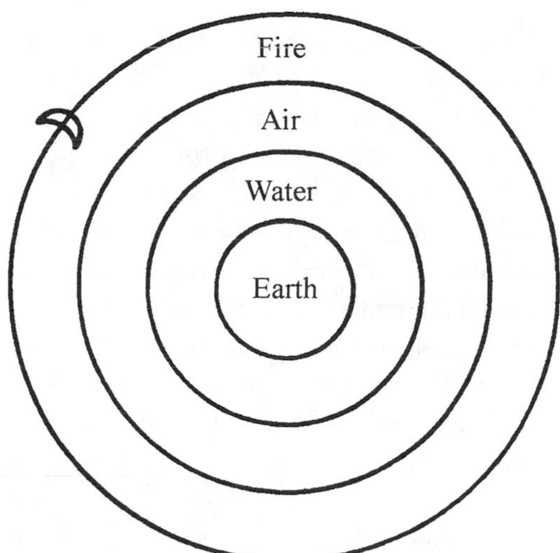

Figure 1.2. The natural place of everything in the terrestrial (sublunar) region is illustrated by this ideal arrangement of the elements, which would be achieved if each were to be situated in its natural place. Note the spherical Earth, accepted by all educated medieval people. Beyond the sphere of fire is the sphere of the Moon, and beyond that the celestial region.

figurative interpretation of at least the passage from the Psalms. However, he was determined also to take the literal sense seriously. In the case of earth placed above water, the psalmist may simply have been referring to promontories or roofs over caverns, where we all acknowledge that earth can be positioned above water.

As for the placement of water above the firmament, Augustine offers an extended restatement and affirmation of the Aristotelian theory of the elements and their natural places. That air belongs above water is evident from an experiment with a jar submerged upside down in water:

> The jar seems to be empty, but it is obvious that it is full of air when it is thus placed with its mouth down into water. Then, finding no outlet in the higher part of the vessel, and being unable by nature to break through the waters below and make its way from beneath, the air fills all the vessel, withstands the water, and does not allow it to enter. But place the jar so that the mouth is not downwards but to the side, and the water will flow in below while the air escapes above.[22]

How then is it possible for water to be situated above the firmament? Why doesn't its weight cause it to descend toward the natural place defined by Aristotle? Augustine's solution is to argue that water in the form of water vapor (as in clouds) is known to reside above air. And if this is true, it is possible that even finer droplets of water can be situated above the firmament without in any way violating the principles of Aristotelian physics.[23]

Augustine's concern in these first two examples was the interpretation of Scripture in ways consistent with the cosmology and physics of the classical tradition. As a final example, I call attention to a case of interaction with the natural sciences provoked by Augustine's theological and pastoral concerns. It is impossible to find a serious natural philosopher before the seventeenth century who doubted the reality of astrological influence on the terrestrial region. Lunar influence on the tides and solar influence on climate and the seasons were irrefutable; and there seemed no reason to suppose that similar, though less obvious, influences from the other planets were any less genuine. But such traditional beliefs provoked immediate theological opposition when some interpreters inferred from them that the human will must then be constrained by celestial forces. The moment freedom of the will was threatened, so were the Christian doctrines of human responsibility, sin, and salvation.

Augustine believed in the reality of celestial forces, but rejected their influence on the mind on account of the fatalistic implications of such a view:

> In what pertains to fate, let us be loyal to the true faith and wholeheartedly reject all subtleties of astrologers and their so-called scientific observations . . ., which they fancy established by their theories. With such talk they try to undermine even the foundations of our belief in prayer, and with headstrong impiety they treat evil-doing that is justly reprehensible as if God were to blame as the Maker of the stars, and not man as the author of his own sins.[24]

Christians, he warned, should have nothing to do with astrologers. But to discredit astrology, he did not merely point to its theological dangers; he also advanced scientific arguments. His most powerful argument, presented in both *The City of God* and *Literal Commentary on Genesis,* was based on the empirical observation that twins, conceived simultaneously and born only moments apart, frequently experience radically different fates. If people are unwilling to credit this argument, Augustine urges, let them go out into the world and observe for themselves.

What are we to make of Augustine's attitude toward the classical natural sciences? Three points are worth making. First, it is important to appreciate his deep ambivalence: of only indirect relevance to salvation and the Christian's terrestrial pilgrimage, the natural sciences could be of no more than secondary or tertiary importance in the estimation of a theologian such as Augustine. Nevertheless, Augustine judged them indispensable for scriptural interpretation and defense of the faith. Second, despite this ambivalence, it was Augustine who most influentially articulated the handmaiden formula as a rationale for pursuing the natural sciences—a rationale that would govern attitudes toward the natural sciences well beyond the end of the Middle Ages. Although of no intrinsic value, in his view, the natural sciences of the classical tradition acquired value extrinsically insofar as they proved useful handmaidens to theology and the church. Third, Augustine put the natural sciences to work in his role as pastor, theologian, and biblical interpreter, demonstrating by example what the handmaiden formula prescribed in words.

From Augustine to Bacon

Nearly eight hundred years separated Augustine's death from Roger Bacon's birth. It is not the object of this paper to discuss these centuries in detail, but a brief account will bridge the gap and help us to position Bacon properly in the scientific and religious history of the Middle Ages.[25]

One of the most characteristic features of the early Middle Ages was the practice of monasticism, which established itself in Western Christendom during Augustine's lifetime. Monasteries were institutions (separate for men and women) to which people retreated in order to separate themselves from the world in order to pursue holiness and spirituality. They tended to be rural and were, at least in theory, self-sufficient. Monks and nuns committed themselves to lives of manual labor, contemplation, and worship and (within the dominant Benedictine order) took vows of poverty, chastity, and obedience; also of residency for life in the monastery in which their vows were taken. Because of the requirement that every monk and nun be literate enough to read the Bible and other devotional literature, monasteries developed their own schools to teach reading and writing; they also established *scriptoria,* where the necessary books were produced by copyists (fig. 1.3). These two features of monastic life were a vital contribution to the preservation of literacy and learning through a very dangerous period of European history.

Education in monastic schools was based primarily on biblical and devotional sources. But pagan literature of the classical tradition, though not abundant, was available and put to use when it contributed to biblical interpretation or some other manifestly religious purpose. The natural sciences were seldom prominent, but their handmaiden status was recognized and contributed to the preservation of a thin version of the classical scientific tradition within the monasteries. And on occasion we find outstanding scholars, educated in monasteries or associated with monastic communities, making original contributions to the natural sciences or contributing conspicuously to their preservation and dissemination.[26]

Figure 1.3. A medieval scribe copies a book. From a thirteenth-century manuscript in the Bodleian Library, Oxford (MS Bodley 602, fol. 36r).

Europe experienced a period of extraordinary political, social, and economic renewal in the eleventh and twelfth centuries. The return of something resembling centralized government, capable of establishing secure borders and reducing the level of internal violence, led to rapid population growth, reurbanization, and economic development, which in turn created opportunities for educated people. Education shifted its center of gravity from rural monasteries to urban cathedral schools, which grew rapidly in number and size. The purpose of education in the cathedral schools, like that in the monasteries, was religious; but the curriculum of the cathedral schools was based on a broader conception of the range of subjects that were religiously useful and might therefore be pressed into service as faithful handmaidens of religion and theology.

For their knowledge of the classical tradition, early medieval scholars had been largely dependent on the thin, derivative version that had survived the slow disintegration of the Roman Empire in Latin sources. Now, in the twelfth century, Western Christendom gained access to the original Greek heritage through translations either of the Greek originals or more frequently of Arabic versions (accompanied by original Arabic treatises that extended the classical achievement). Translation of this thick version of the classical tradition into Latin became a serious enterprise from the end of the eleventh century, accelerating in the twelfth, and tapering off in the thirteenth. Greek-to-Latin translations made in southern Italy and Arabic-to-Latin translations rendered primarily in Spain (sometimes through the intermediary of Hebrew or Spanish) gave Western scholars access for the first time in centuries to the full range of classical sources.

Finally, a new institution, dedicated to higher learning, evolved just in time to offer a home to the newly translated literature. The University of Bologna emerged about 1150. The University of Paris followed in 1200, Oxford University in 1220, and another two dozen universities before the middle of the fifteenth century. In these universities, students who enrolled in the faculty of arts could obtain an advanced education in logic, the humanities, and the natural sciences. After completing the arts curriculum, students could proceed to graduate faculties in law, medicine, or theology. Here the classical tradition became securely institutionalized for the first time in history.

Roger Bacon and the New Learning

It was against the background of these extraordinary educational developments that Roger Bacon (fig. 1.4) was born into a prosperous English family some time between 1214 and 1220. We know nothing of his youth and early schooling, but he obtained his advanced education at the Universities of Oxford and Paris. In the 1240s he taught at the University of Paris, where he was one of the first professors to lecture on the scientific works of Aristotle. At some point he expanded his scope to include mathematical and what he called "experimental" science. And he joined the Franciscan order, one of a pair of newly founded mendicant orders.[27] We do not know what attracted Bacon to the Franciscans; perhaps it was the order's reputation for learning and holiness or Bacon's conviction that membership in the order would in some way advance his scholarly program. However, Bacon soon found himself at odds with his Franciscan superiors and brothers, and he seems to have been placed for a time under close supervision in

Figure 1.4. One of the few images of Roger Bacon, dating from the twentieth century (courtesy of the Friar Bacon Pub in Oxford). Possibly a less-than-perfect representation of the thirteenth-century friar.

the Parisian convent. His fame rests on a trio of treatises written at the pope's request, in which he laid out an ambitious plan for the reform of learning in Christendom, based on knowledge of the natural sciences contained in newly translated works of the classical tradition.[28] He died about 1292.

The campaign on which Bacon thus embarked was both ambitious and delicate. He was attempting to steer a middle course between two equally unacceptable dangers. On the one hand, the newly available materials presented themselves as an autonomous body of secular knowledge, having no religious function, prepared to serve no master but human rational capacities—in short, an expression of the "vain curiosity" against which Augustine had so urgently warned. On the other hand, Bacon feared that those who wielded power within Christendom would overreact to the dangers of the new learning and repudiate an inestimable treasure. This is, in broad outline, the same problem that we have already seen in the patristic period: how to deal with an intellectual heritage that was simultaneously dangerous and indispensable, how to salvage it and put it to work without condoning its intolerable theological consequences, how to discipline it without emasculating it. Theologians of the patristic and early medieval periods had developed working arrangements that made it possible to live with, and indeed gain the benefits of, the classical tradition. In Bacon's day the problem remained, but the circumstances had changed: the theological dangers of the classical tradition in the fuller version available in the thirteenth century (with Aristotelian philosophy as its centerpiece) were both more numerous and more intractable, while the utility of this version of the classical tradition (enriched, as it was, by nearly the full Aristotelian corpus, as well as contributions of late antiquity and medieval Islam) was also a great deal more obvious.[29]

What exactly were the theological dangers? A number of specific Aristotelian claims trampled on critically important Christian doctrines. Aristotle's description of an eternal universe, which had neither beginning nor end, was radically opposed to the Christian doctrines of Creation and final judgment. The universe described by Aristotle was deterministic, governed by causal chains rigidly prescribed by the natures of things. This determinism, combined with Aristotle's description of a deity who never intervened in the operation of the cosmos, clashed with the Christian doctrines of divine omnipotence, providence, and miracles. And Aristotle's theory of the soul, as generally understood, left no room for personal immortality.

But the overriding menace may have been the rationalistic outlook in which Aristotle's philosophy came packaged. A broad mastery of the Aristotelian works made available through the translations of the twelfth century (far more complete than the modest collection of logical works available during the early Middle Ages) left the distinct impression that the only road to truth was Aristotelian demonstration, based on sense experience and rational inference. The new learning contained in the thick version of the classical tradition thus entered Christian Europe in the twelfth and thirteenth centuries under the banner of extreme rationalism. If one observed closely the more radical defenders of the new learning, it might appear that even theological doctrines would be compelled to pass the test of syllogistic demonstration. Human reason seemed poised to shove biblical revelation aside and become the sole arbiter of truth.[30]

We might expect, given the circumstances, that the religious authorities would have moved with single-minded zeal to suppress the new learning. But in fact the response was quite mixed and contested. For example, the teaching of Aristotle's works on the natural sciences was forbidden in the faculty of arts at the University of Paris in 1210 and 1215, though the ban did not extend beyond the borders of Paris. And before the end of the century the bishop of Paris would issue a condemnation of a long list of propositions drawn from Aristotelian natural philosophy and other works in the classical tradition.[31] But this is only a small part of the picture. The new learning also had its defenders—scholars (including churchmen) who recognized its overwhelming explanatory power in every area of human knowledge. Aristotle's natural philosophy offered a convincing framework and a powerful methodology for thinking about cosmology, meteorology, matter theory, motion, light, sensation, psychology (the theory of the soul), and biological phenomena of all kinds. The mathematical astronomy of Ptolemy and the tradition spawned by it were able to predict planetary positions with considerable accuracy. The geometrical optics of Euclid, Ptolemy, and their Islamic followers dealt convincingly with the reflection and refraction of light and the principal phenomena of visual perception. And medical treatises offered anatomical and physiological knowledge, along with descriptions of diseases and therapies. For educated people in general, but especially for professors and students in the medieval universities, the new learning was simply too attractive and too persuasive to surrender.

We do not know exactly when or how Bacon became involved in attempts to promote the new learning; in the 1240s he was lecturing on Aristotle at the University of Paris, in the 1250s he was writing on mathematical and "experimental" science, and by the 1260s he had embarked on his campaign in defense of the natural sciences and other portions of the classical tradition in its thick version. Although by this time, Aristotle's "books on nature" had been admitted into the arts curriculum at Paris, Bacon understood the position of the natural sciences more generally to be still precarious. The mathematical sciences and what he called "experimental" science were widely ignored, and even the fate of Aristotle's works (as events were soon to reveal) remained undecided. Bacon's aim was to head off any further threat—either from those who undervalued the new learning (including religious authorities, who had the power to suppress it) or from champions of the new learning, prepared to carry its radical rationalistic program to the point of compromising Christian doctrine. His goal was to demonstrate that the pagan learning of the classical tradition was a vital resource, capable of offering essential services to theology and the church; and moreover that it posed no insuperable religious threat, that suitably disciplined and purged of error, it would serve as a faithful handmaiden of religion and the church. This was Augustine's handmaiden formula, rearticulated at a moment in history when there was a great deal more learning, some of it potentially unruly, to be brought into servitude to religion.[32]

Bacon's campaign on behalf of the new learning found its fullest and most systematic expression in what is known as his *Opus maius (Larger work),* written in the 1260s at the request of Pope Clement IV. The central theme of this book, drawn from Augustine's *On Christian Doctrine,* is that there exists one perfect, all-encompassing wisdom, the ultimate source of which is Scripture. Philosophy is not an autonomous intellectual enterprise but, properly practiced, the unfolding and expounding of scriptural truth. Indeed, Bacon goes so far as to maintain that for something to be against Scripture, it is sufficient for that something merely to be independent of Scripture: "Whatever is not connected to [Scripture] is proved to be against it and is to be abhorred by Christians."[33]

The criterion of legitimacy that Bacon here adopts would appear, at first glance, to be so narrow as to exclude major portions of the classical tradition. However, Bacon proceeds to claim, if not to demonstrate, that almost everything qualifies for handmaiden status. He reminds his readers of the passages in *On Christian Doctrine* in which Augustine claims that all truth, wherever we may find it, is God's truth and, consequently, admonishes them to snatch philosophical truth from pagans as from unlawful possessors. But this is to sanctify all true philosophical claims: philosophical truths are automatically Christian truths, and the entire world of learning, insofar as it achieves truth, also achieves legitimacy. Bacon is quite explicit about this, claiming that philosophy as a whole "is worthy and belongs to sacred truth."[34] But to give this claim credibility, Bacon must elaborate on the benefits for Christendom of the various philosophical disciplines. He devotes hundreds of pages to the effort, unsystematically spewing forth arguments of mixed quality, interrupted by frequent digressions. We can do no more than sample the results.

Bacon begins with the easy disciplines: the study of grammar and foreign languages. Most obviously we need both grammatical knowledge and command of foreign languages if we are to understand the Bible, which was not written in Latin, but translated from Hebrew and Greek. Moreover, the biblical text has been corrupted and can be corrected only by those with linguistic knowledge. Foreign languages (chiefly Greek and Arabic) are also required for successful translation of *scientific* works from other tongues. Christendom requires foreign languages for the sake of commercial transactions and diplomatic relations with other nations. Without foreign languages, missionaries will achieve no success in the conversion of Jews, Muslims, and other non-Christians; nor will they be equipped to subdue those who refuse to convert. Finally, words have extraordinary powers by which miracles and other great deeds are performed.

> How many tyrants and evildoers have been confounded by powerful words, and convinced more than through wars! And not only by the words of the saints or the faithful, but even by the words of philosophers have they been so stricken that they were compelled to obey the truth.[35]

Bacon explains such remarkable manifestations of verbal power by noting that superior things naturally have power over their inferiors, that the human rational soul is superior to all created things except angels, and that the principal activity of the rational soul is to produce words. It follows that words (especially if expressed under the appropriate constellations, with "great desire and fixed intention" by a rational soul "unstained by sin") can perform mighty acts.[36]

The case for mathematics and the mathematical sciences is not as easily made, and Bacon devotes more than three hundred pages (in the standard modern edition of his works) to the project. His overarching claim is that mathematics is important because all other knowledge depends on it. Bacon offers many "proofs" of this fundamental, preparatory role of mathematics. Mathematics, he argues, was the first "science" to emerge: discovered by the sons of Adam, it passed through Noah and Abraham to the Egyptians. It has proved itself quasi-innate and the most easily learned of the sciences—accessible, Bacon assures his reader, even to the clergy. It is a source of examples for all other subject matters. Only mathematics offers true demonstration; therefore, any subject that aspires to certainty must be founded on it. And more. Bacon sums up the case by concluding that "there are four great sciences, without which the other sciences cannot be grasped. . . . Of these, mathematics is the gate and key. . . . And . . . a command of mathematics prepares the mind and elevates it to confirmed knowledge of all things."[37]

But the suggestion that mathematics be valued for what it can do for the various branches of philosophy is only indirectly an argument for its *religious* utility. Do the mathematical sciences serve theology and the church directly? Bacon devotes the last 225 of this section's 300 pages to demonstrating that indeed they do. Astronomy, among the most prominent of the mathematical sciences, is one of his favorite examples (fig. 1.5). Any subject that investigates the heavens, he argues, inevitably has theological relevance. Moreover, the vast size of the universe (revealed by astronomers) fills us with awe and evokes praise of the creator. Theological treatises are filled with questions about the number, shape, and nature of the heavens and about planetary models and motions—matters on which astronomers can provide instruction. An understanding of chronology, which is dependent on astronomy, is essential for the construction of a sacred history of the world, from the Creation to the Antichrist. Chronology promises also to assist calculation of the dates of the Creation, of Noah's exit from the Ark, and of the Israelites' arrival in the Sinai desert.[38]

However, Bacon was worried about the traditional association (Greek in origin) of mathematics with magic, divination, and deterministic astrology, which had, in his opinion, brought mathematics into disrepute.[39] Bacon attempts to rescue it by distinguishing between "true" and "false" mathematics. The false kind, associated with magic, is deterministic, claiming to "judge infallibly concerning all future events." Practitioners of this false mathematics employ "charms, incantations, conjurations, superstitious sacrifices, and various frauds," and ascribe Christ's miracles to similar practices. The true mathematics, by contrast, is nondeterministic and places Christ's miracles in a separate category, out of reach of the analysis applicable to other remarkable phenomena.[40]

Moreover, practitioners of the true mathematics share an understanding of a cosmology (of which Bacon was one of the major defenders) based on the universal radiation of force. According to this cosmology, everything in the universe radiates its force or image in all directions, so that everything influences everything else. The cosmos is thus an intricate network of forces responsible for

Figure 1.5. A medieval astronomer makes observations with an astrolabe for calendric purposes. From a thirteenth-century manuscript in the Bibliothèque de l'Arsenal in Paris (MS 1186, fol. 1v).

phenomena as diverse as light, the radiation of heat, lunar influence on the tides, and the efficacy of prayer. In such a world, astrological influence (of a nondeterministic variety) is ubiquitous. For example, it influences the human body and its organs, as well as human temperament, and helps to explain the longevity of the Old Testament patriarchs. It explains why northern peoples have manners different from those of southern peoples. It even explains the emergence of various sects throughout human history, provides arguments for the superiority of Christianity, and promises to foretell the coming of the Antichrist.[41]

Worries about the Antichrist (foretold in various biblical passages and a topic of lively interest in Bacon's day) are a recurrent theme in Bacon's defense of the classical tradition in general and the mathematical sciences in particular.[42] He is convinced that the Antichrist and his minions will appear armed with astrological power:

> I write of these matters not only out of the quest for wisdom, but also because of the dangers that confront and will continue to confront Christians and the church of God on account of unbelievers, especially the Antichrist, who will make use of the power of learning [here Bacon has in mind especially the understanding of astrological forces] and will turn everything to evil use. By means of words . . . and stellar effects, also strong desire to do evil combined with sure purpose and extreme audacity, he himself will bring ill fortune and cast a spell not only on individual people but also on cities and regions. And by this grand means he will accomplish whatever he wishes without war, and men will obey him like beasts.[43]

The church, if it is not suitably prepared, will be vulnerable to the Antichrist's evil purposes. A knowledge of the mathematical sciences, Bacon believes, is an essential part of that preparation.

Bacon has much more to say about the religious utility of philosophy in general and natural philosophy in particular. He presents a lengthy account of geographical knowledge required for missionary activity and other religious purposes. Well over 150 pages of the modern printed edition of the *Opus maius* are devoted to an original exposition of the science of light and vision, based on newly translated Greek and Arabic sources. This material, justified like the rest by its religious utility, frequently circulated independently of the rest of the *Opus maius*, deeply influencing optical thought for the next 350 years. And he offers a plea for the pursuit of "experimental" science—a methodology for the acquisition and certification of knowledge, as well as a body of practical knowledge and lore allegedly acquired by experimental or experiential means.[44]

Besides having its individual merits, practically every discipline, it turns out, is capable of contributing to biblical interpretation. Relying heavily on Augustine, Bacon finds that history, logic, grammar, mathematics, metaphysics, medicine, the natural sciences (including agriculture), and the principles of navigation must all be grasped if we are "to avoid complete ignorance of what the Scriptures wish to convey when they introduce certain figures of speech drawn from these arts." For example, we will be unable to grasp the meaning of the biblical admonition to be "wise as serpents and innocent as doves" (Matthew 10:16) if we are ignorant of the characteristics of serpents and doves. And we will never understand the psalmist's meaning when he writes "Preserve me, O Lord, as the pupil of your eye" (Psalm 17:8) unless we know enough about the preservation of the pupil (what we call the "crystalline lens") to judge how we might be similarly preserved.[45]

But the relationship is reciprocal. If the natural sciences contribute to our interpretation of the Bible, the Bible contains truths that will clarify our understanding of the world of nature. A particularly good example is the theory of the rainbow. All attempts to explain the rainbow, Bacon argues, have failed to identify its "final cause" (that is, its purpose)—an explanatory element that Aristotle regarded as a necessary part of any successful explanation. But Scripture makes the purpose of the rainbow clear when it tells us of God's promise to Noah, after the Flood, that he would never again destroy the earth in that manner and, moreover, that the rainbow would be a sign of this covenant: "I set my bow in the cloud, and it shall be a sign of the covenant between me and the earth" (Genesis 9:13). And from this clearly identified final cause, Bacon claims, we can infer other causal factors in the production of the rainbow—specifically (through an intricate chain of inferences), that the rainbow is produced by an infinity of reflections and refractions in innumerable drops of falling water.[46] Bacon's claim to have been led from the biblical account of the purpose of the rainbow to an understanding of the role of reflection and refraction in its production may stretch credibility. But the important thing is what Bacon employed this example to prove—namely, that embedded in Scripture are truths applicable to the natural sciences. In this case, as in innumerable others, Bacon claims, religion offers invaluable instruction to her handmaidens; the handmaiden relationship between science and religion is mutually beneficial.

Conclusion

Let us be clear about the purpose of this essay. It has not been my aim to praise Augustine or Bacon for great scientific achievements or for the anticipation of any aspect of modern science. Augustine lived near the end of the patristic period, Bacon during the high Middle Ages, and it should surprise nobody to learn that their conceptions of nature were representative of the eras to which they belonged. Each was a significant actor in the intellectual life of his age, helping to determine attitudes toward the classical scientific tradition. Each had a continuing influence that extended well beyond his lifetime. But the notion that great people truly transcend the eras that produce them is the stuff of mythology and fiction. So this essay does not aim to measure Augustine and Bacon by modern scientific or religious yardsticks, but merely to explore the ways in which the scientific beliefs and attitudes (patristic or medieval in character) of these two influential men interacted with the Christian context in which they lived.

What may be the most surprising conclusion (in the light of the mythology that has grown up around Bacon) is the close resemblance between the two. Bacon, celebrated for his alleged repudiation of authority and anticipation of modern scientific methodology, was in fact thoroughly Augustinian, and thus a solid traditionalist, in his attitude toward the value and legitimacy of the natural sciences. Augustine had no use for the natural sciences as ends in themselves, but he accepted them and even esteemed them (if their high profile in his *Literal Commentary on Genesis* is any measure) as valuable, if sometimes problematic, handmaidens. Roger Bacon shared these opinions. He feared that the problematic aspects of the classical tradition would lead to its neglect, or even repudiation, by medieval Christendom—an eventuality that he was determined to head off. His *Opus maius* is an extended plea for the reliability and religious utility of the handmaiden named "natural science." As for his attitude toward the possible autonomy of the natural sciences, he noted that "one science is mistress of the others—namely, theology, for which the others are integral necessities and which cannot achieve its ends without them. And it lays claim to their virtues and subordinates them to its nod and command." And in case this wasn't clear enough, he continued: "Every investigation of mankind that is not directed toward salvation is totally blind and leads finally to the darkness of hell."[47]

It does not follow, of course, that the differences between Augustine and Bacon were insignificant. Most fundamentally, the two men were born in different eras and worked in different religious and intellectual contexts. Augustine was a theologian, living in an era when pagan philosophies such as neo-Platonism were living alternatives to Christianity and therefore a threat of the utmost gravity. Bacon, by contrast, was a philosopher, living in a Christian culture that felt no serious threat to its hegemony, but where the fate of the classical sciences was still in doubt. It should come as no surprise, therefore, that Augustine was more worried about the threat to theology posed by classical natural philosophy than the threat to classical natural philosophy posed by Christian theology; whereas in Bacon's hierarchy of worries the order appears to have been reversed.

Moreover, Bacon lived at a moment in history when a far more complete version of the classical tradition was available than in Augustine's day, poised (depending on your point of view) either to threaten the faith or to serve it. Whereas Augustine thought small pieces of the classical tradition would prove religiously useful, Bacon was prepared to argue that almost everything could be of service. Bacon proceeded to demonstrate how the handmaiden formula could be stretched to justify a remarkable array of natural sciences, some of which seemed (at first glance) to have little or no utility for theology or religion. The remarkable success with which he and such contemporaries as Albert the Great (Albertus Magnus) and Thomas Aquinas[48] wielded the handmaiden formula helped to overcome opposition to the new learning, create for it an institutional home within the medieval universities, mendicant orders, and an occasional medieval court, and ultimately anchor it firmly within European intellectual life. Not for the first time in history, a traditionalist strategy served as a vehicle for fundamental change.

How did the natural sciences fare as handmaidens? The answer depends, of course, on time, place, local circumstance, and the specific scientific claims in question. In technical disciplines, such as the mathematical sciences, handmaiden status had little influence on the course, the general shape, or the specific content of the science. For example, the mathematical models of medieval planetary astronomy depended for both their general form and their specific detail on astronomical models inherited from Greek and Islamic predecessors, designed to account for collections of planetary observations, both ancient and medieval. And in Bacon's grand synthesis (partly mathematical) of Greek and Islamic theories of light and vision, nobody has identified any theoretical claim whose content was influenced by Bacon's membership in a Christian culture.

In disciplines that impinged on broader issues of causation or worldview, handmaiden status clearly made a difference. Medieval accounts of the origin of the cosmos were, of course, largely framed by the Genesis story. Opinions about the number and nature of celestial spheres lying beyond the planetary spheres were influenced by the Genesis account of waters both above and below the firmament. In the later Middle Ages, certain distinguished natural philosophers were moved to develop theories of motion consistent with the certainty (guaranteed by the Christian doctrine of divine omnipotence) that God, had he wished, could have endowed the cosmos as a whole with a rotational motion. And for a final example, a substantial current of antiastrological sentiment during the Middle Ages took its inspiration from Augustine and other theologians who were opposed to astrology because, in its deterministic form, it threatened the ideas of human free will and responsibility.[49]

What we see during the Latin Middle Ages, then, is a complicated interplay between scientific and religious beliefs. University professors and other university-educated Europeans placed high value on both the Christian and classical traditions. They were committed to the central doctrines of Christian theology and accepted the authority of the church to determine what these doctrines were. They also recognized the extraordinary explanatory power of the classical natural sciences, and many hitched their reputations and their careers to the further development and dissemination of these sciences. It should come as no surprise that when such people discovered areas of conflict between the religious and the scientific traditions, they looked for ways of easing the tensions. Compromise, clarification, reinterpretation, revision, the identification of outright error—all played a part in the achievement of peace. None of this occurred without a struggle, of course, and a certain amount of skirmishing took place then and has continued to the present. But by the end of the Middle Ages, the classical sciences had accommodated themselves to Christian doctrine and the needs of a Christian culture, while Christian theology had taken its form, its method, and some of its content from the classical tradition. The classical sciences had entered the mainstream of European culture—a position from which they have never been dislodged.

Acknowledgments

I gratefully acknowledge helpful comments on this paper by Michael Shank and participants in the "Case Histories" conference in Berkeley, March 1999, especially Gary B. Ferngren.

2

Galileo, the Church, and the Cosmos

David C. Lindberg

Galileo's struggle with the Roman Catholic church over the arrangement of what we now call the "solar system" was a multifaceted event. It is well known, of course, that it was a clash of ideas—between scientific claims fervently held by a small band of scientific reformers on the one hand and opposing theological doctrines supported by centuries of church tradition on the other. Unfortunately, this version of the story has proved itself vulnerable to simplistic black-and-white elaborations. In the overheated rhetoric of the likes of Andrew Dickson White and many popular writers, this has become a tale of combat to the death between the voices of scientific freedom and the forces of theological intolerance—and, as a welcome bonus, an opportunity to bash the Catholic church.[1] As such, it may make good drama, but it is seriously deficient as history.

Other writers, with the same concentration on cosmological and theological ideas but with greater respect for the historical record, have published nuanced studies that explore the quality of Galileo's cosmological arguments and the nature of the theological currents then at work in the Catholic church. Here we begin to see important aspects of the struggle between Galileo and the leadership of the church. At stake were not only the meaning of certain biblical passages that addressed (or appeared to address) cosmological issues, but also the larger question of who had the right to determine cosmological (and other scientific) truth. Tension over such matters has had a long history within Christendom, and it was an important factor in the Galileo affair.

But if we were to stop here the story would be incomplete, for (as sketched above) it omits the human dimension. We must never forget that, strictly speaking, ideas cannot clash and theoretical claims cannot, of themselves, engage in combat. It is people who fight over theoretical and methodological claims, people who clash over ideological issues. And when people are involved, human interests and local circumstances are inevitably present as well. My purpose in the present essay, therefore, is to bring the story down to Earth and reveal it as a concrete historical event, situated in time and space—influenced, without a doubt, by cosmological and theological beliefs, but also powerfully shaped by local circumstances and the interests of an important group of historical actors.[2]

Galileo Galilei (fig. 2.1) was born on 15 February 1564 in the city of Pisa—some fifty miles from Florence, under whose control it had fallen in 1406. The eldest son of a court musician of considerable talent, Galileo was educated first in Pisa, later in a monastery school in the hills outside Florence. In 1581 he took up medical studies at the University of Pisa, but a year later he abandoned them; subsequently he studied mathematics privately with a tutor and in the long run

Figure 2.1. Galileo Galilei (1624).

undertook an extended program of self-education. He never earned a university degree, but he eventually became a university professor. In 1589 Galileo assumed teaching duties in mathematics at the University of Pisa, thanks to the intervention of Guidobaldo del Monte, a distinguished mathematician and an influential member of the nobility. Three years later Galileo moved to the University of Padua (in the Venetian Republic), where he remained until 1610, when he resigned in order to return to Florence as philosopher and chief mathematician to the grand duke of Tuscany.[3] Galileo first became seriously involved in the heliocentric debate just before this return, in the years 1609–10.

Heliocentrism

What was this heliocentric debate? Western cosmologies, since at least the fourth century B.C., had been geocentric: they had featured a spherical earth in the center of the universe and assigned planetary status to the Sun, which circled the earth with a daily motion.[4] This was the view of the Greek philosopher Aristotle (d. 322 B.C.), the Greek astronomer Ptolemy (fl. A.D. 150), and their many medieval commentators. Heliocentric (Sun-centered) systems were not unheard of, but they survived in late antiquity and the Middle Ages merely as curiosities. However, in 1543 Nicolaus Copernicus, a church official and accomplished astronomer from northern Poland, published a book, *On the Revolutions of the Celestial Orbs,* in which he took the heliocentric system (now fully equipped with mathematical models capable of predicting planetary positions) and defended it as a true description of the universe.

Copernicus's book was a highly technical astronomical text, dominated by detailed geometrical models for all of the planets; and we must make a brief foray into these geometrical details if we are to understand what was at stake in the heliocentric debates. The fundamental idea underlying Copernicus's models was borrowed from Ptolemy—namely, that two or more uniform circular motions can be combined to produce a nonuniform composite. If we set aside various complexities, Ptolemy's basic geocentric model can be represented as follows (fig. 2.2): a given planet P moves uniformly around a small circle called an "epicycle"; meanwhile, the center of this epicycle moves uniformly around a large circle called a "deferent" (or carrying circle), the center of which is at point C. These two uniform circular motions, when combined and viewed from the fixed earth, are meant to replicate the observed behavior of P as viewed against the background of the fixed stars. However, to produce models that actually worked with quantitative precision, Ptolemy found it necessary to complicate the geometry in two ways. First, he was willing to shift the deferent slightly, so

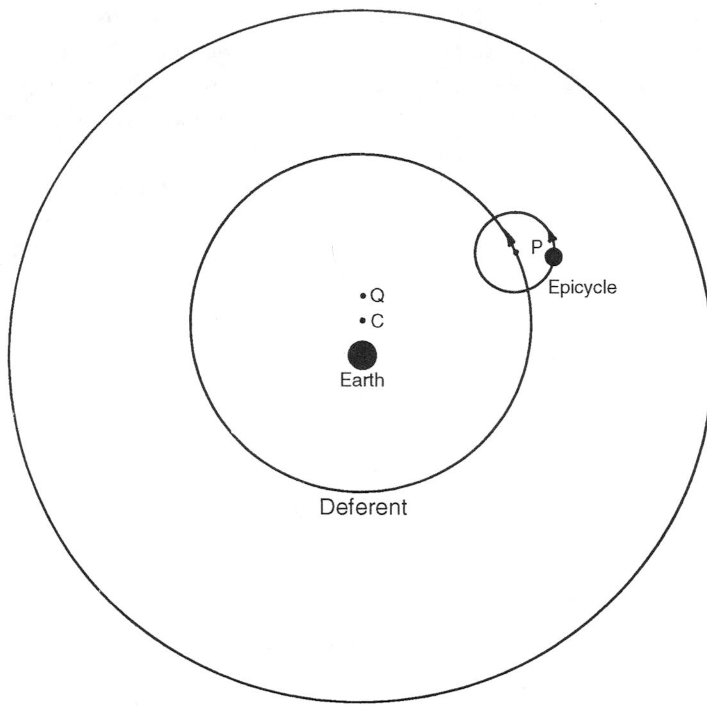

Figure 2.2. Ptolemaic geocentric model for a typical planet. The various elements in this and other geometrical diagrams in this article are not drawn to scale.

that its geometrical center no longer coincided with the geometrical center of the earth (which remained in the center of the universe). Second, he allowed the center of the epicycle to move about the deferent with a motion that swept out equal angles in equal times as viewed *not* from the center of the deferent (as a straightforward understanding of the expression "uniform motion" would seem to require), but from a noncentral point, the "equant point" Q. This device, while retaining uniformity of angular motion about the equant point, actually allowed the center of the epicycle to move with variable speed around the circumference of the deferent.

Copernicus's models (fig. 2.3) differed from those of Ptolemy in a number of minor respects, but two major ones deserve mention. First and most obviously, the central cosmological object in the Copernican system was the Sun; in the Ptolemaic, the earth. Second, Copernicus banished the equant from his heliocentric model in favor of uncompromised uniformity of motion. He required the center of an epicycle to sweep out equal angles in equal times as viewed from the *center* of its associated deferent. Like the earth in the Ptolemaic

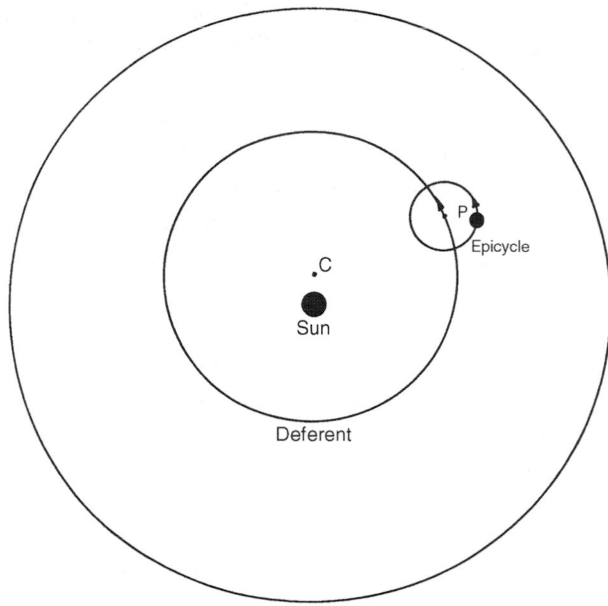

Figure 2.3. Copernican heliocentric model for a typical planet. Epicycles in the heliocentric model are much smaller than their counterparts in the geocentric model, and no equant points are allowed. The center of the Sun coincides with the center of the stellar sphere, but not with the center, C, of the planetary deferent.

models, the Sun is at the center of the fixed stellar sphere but not at the center of the deferent.[5]

Because Copernicus's book was highly technical, written for a small audience of mathematically proficient astronomers, it was little known and less read. Contrary to legend, its publication created no public stir. But the book *did* secure an audience among astronomers, many of whom employed it for calculating planetary positions, while denying its claim to cosmological truth.

Why did those astronomers who first mastered Copernicus's *Revolutions* refuse to accept the truth of heliocentrism? Because the evidence that could be marshaled in the middle of the sixteenth century in support of the heliocentric model as physically true was not convincing. No observation, taken by itself, could prove that the Sun rested and the earth moved. Predictions using the new system (in the form given it in the *Revolutions*) were no more accurate than those offered by the old.[6] The advantages of heliocentrism perceptible at the time were limited. First, elimination of the Ptolemaic equant meant that all motions were uniform about their centers—a feature of the heliocentric system viewed by some as a return to the proper foundational principles of astronomy. Second, Copernicus's lunar theory and his theory of the precession of the equinoxes were recognized as technically superior to Ptolemy's.

Third, the heliocentric system had a number of advantages that fall into the elusive realm of what we might call simplicity, order, coherence, or intelligibility. For example, the Ptolemaic system offered no obvious principle by which to determine the true sequence of the planetary orbits. Copernicus, working within a heliocentric framework, was able to argue that the planets are arranged in space, from innermost to outermost, according to their increasing periods of revolution about the Sun (fig. 2.4). He could also explain retrograde motion of the planets—the odd reversal of direction exhibited by the planets in their slow motion through the fixed stars. Predicted but unexplained by Ptolemy's geocentric astronomy, retrograde motion was revealed by heliocentrism to be a simple case of optical illusion, owing to the fact that we observe those planets from a moving platform (see fig. 2.5). Finally, the Ptolemaic model for every planet except the Moon (considered in this model to be the planet closest to the earth) contained some annual element: either the planet circled its epicycle annually, or the

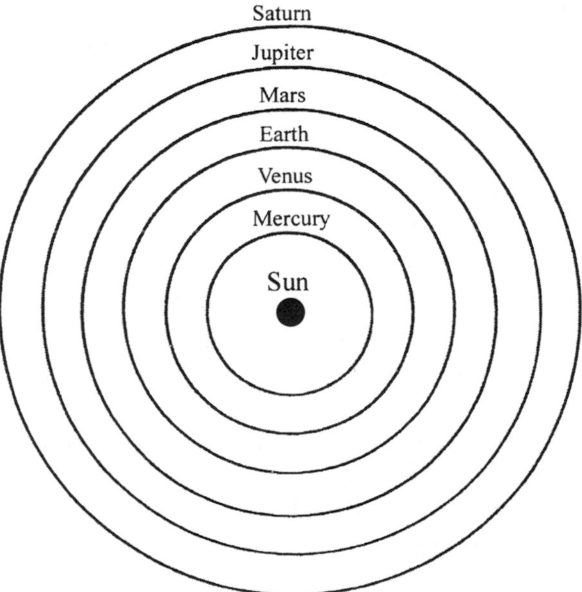

Figure 2.4. The order of the planets based on the heliocentric system (approximate orbital periods in parentheses). Proceeding outward from the Sun: Mercury (88 days), Venus (225 days), Earth (365 days), Mars (687 days), Jupiter (12 years), Saturn (30 years).

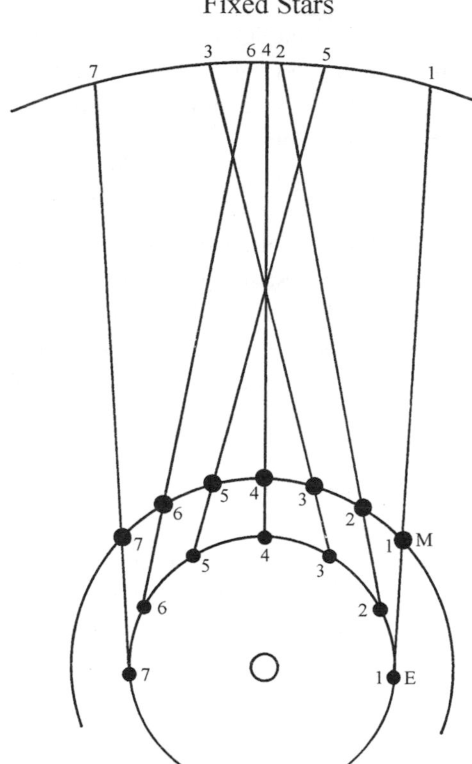

Figure 2.5. Retrograde motion of Mars explained on the heliocentric model. The position of Mars is observed at uniform intervals against the background of the fixed stars. Although both Mars (M) and the Earth (E) (from which Mars is observed) move with uniform motion, the line of sight in relation to the fixed stars (as indicated by the sequence of numbers at the top of the drawing) swings from right to left, then from left to right, then once again from right to left.

center of the epicycle circled the deferent annually. Inexplicable within a geocentric cosmology, these elements could be explained by a heliocentrist as the result of a simple perceptual error—the mistaken attribution to the observed bodies of motion actually possessed by the observer. Significant though such heliocentric advantages may seem in retrospect, none of them was widely regarded in the sixteenth century as decisive evidence of the truth of the heliocentric model. It was understood that such virtues as uniformity, simplicity, order, coherence, and intelligibility do not guarantee the truth of a theory; and before the end of the century respectable alternatives were available, to which one might be rationally committed.[7]

Alongside these advantages of the heliocentric model were several powerful disadvantages. First, putting the earth in motion represented a massive violation of everyday common sense. Second, removal of the earth from the center of the cosmos represented a destructive attack on Aristotle's physics—the only comprehensive system of physics in existence—and therefore represented a serious violation of scientific common sense. Third, to put the earth in motion was to put it into the heavens, thereby destroying the dichotomy between the heavens and the earth, which had served as a fundamental cosmological premise wherever Aristotelian philosophy prevailed for the previous two millennia. Fourth, the absence of stellar parallax (apparent change in the relative positions of a pair of stars, expected if those stars were viewed from a planet conceived to be moving through an orbit with a diameter of ten million miles) offered powerful empirical evidence against heliocentrism (fig. 2.6).[8] As a result, few people in the half century after publication of the *Revolutions* took the system seriously as a description of physical reality. Those astronomers and natural philosophers who rejected heliocentrism did so *not* because of blind conservatism or religious intolerance, but because of their commitment to widely held scientific principles and theories. Indeed, the first serious critics were young astronomers in the German universities, who perceived the simplicity and intelligibility of the heliocentric theory and used it for calculations, but regarded it as physically impossible. They understood that simplicity and intelligibility do not guarantee truth.[9]

From the Catholic church, a key player in our story, there was scarcely a stir. Copernicus had been talked into publishing his book by various friends, includ-

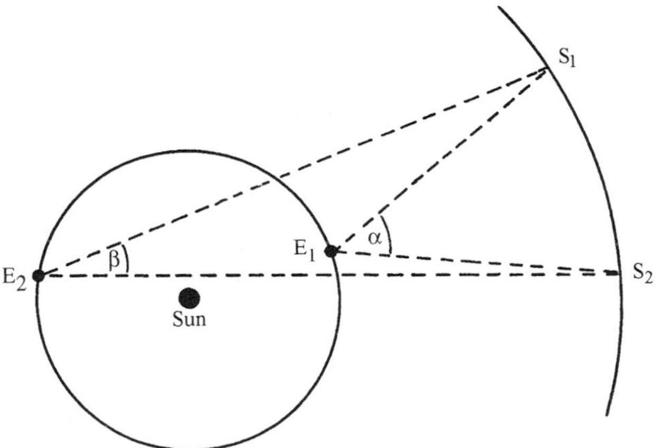

Figure 2.6. Stellar parallax on the heliocentric model. Suppose that the angular separation (α) between the two stars, S_1 and S_2, is measured from Earth in position E_1. Wait about six months, until Earth is in position E_2 (now judged, according to contemporary conceptions of the size of Earth's orbit, to be approximately ten million miles farther from the two stars). The angular separation (β) between the same two stars should now be substantially smaller. But in fact there is no measurable difference. It follows either that the stars are so unimaginably far away that ten million miles is infinitesimal by comparison or that the Earth is not in orbit around the Sun, but fixed in the center of the stellar sphere.

ing ecclesiastical officials. He had dedicated the *Revolutions* to the pope. And, except for one or two people, nobody judged his ideas dangerous—foolish, perhaps, but hardly a threat.

Galileo and Heliocentrism

The event that triggered Galileo's involvement in cosmological issues was the invention of the telescope by a Dutchman in the summer of 1608.[10] Galileo learned of it about a year later from his friend Paolo Sarpi and quickly figured out how to produce an eight-power version, later a twenty-power instrument. Equally important was his decision to turn the telescope to the heavens, converting a toy into an instrument of scientific investigation.

Galileo observed, successively, the fixed stars, the Moon, the planets, and eventually the Sun.[11] He saw previously invisible stars, thereby multiplying the population of the heavens roughly tenfold. He perceived craters, mountains, and valleys on the Moon (fig. 2.7). He discovered that the face of Venus passed through a complete set of phases, from thin crescent to full circular disk. Galileo observed four "little stars" accompanying Jupiter in its motion through the starry sphere, tagging along like dogs on a leash—stars that Galileo would

Figure 2.7. Galileo's sketch of the Moon as seen through the telescope. From Galileo's *Starry Messenger,* in *Le Opere di Galileo Galilei,* national ed., vol. 3, pt. 1 (Florence: Barbèra, 1892), p. 66.

identify, in his published account of these telescopic observations, as satellites. Saturn appeared to have "ears"—interpreted by Galileo as a pair of satellites accompanying the planet "on its flanks," but ultimately understood (after Galileo's death) as rings.[12] And finally, Galileo observed spots moving about on the surface of the Sun; he was not the first to do so and subsequently became involved in a bitter priority dispute over sunspots with the astronomer Christopher Scheiner.

Galileo apparently counted himself a Copernican from the mid-1590s; but it was the publication of an account of his telescopic observations in two small books, the *Starry Messenger* (1610) and *Letters on Sunspots* (1613), that propelled him into a serious heliocentric campaign. Galileo's telescopic observations certainly did not demonstrate the truth of the heliocentric model. However, they did, when deployed in his arguments, undermine some of the more powerful objections against heliocentric cosmology—a far cry from proving that heliocentric cosmology is true.

Consider several examples. What was the cosmological relevance of the discovery that the lunar surface is mountainous and pocked with craters? One of the most potent arguments against the mobility of the earth was the implausibility of the claim that this great earthen rock that we inhabit, the very symbol of stability, was sailing at breakneck speed through the heavens. But Galileo, armed with his lunar observations, argued that the Moon resembles the earth in topography (in the *Starry Messenger* he compares the lunar surface to that of Bohemia)[13] and therefore in substance; and the Moon sails through the heavens in all cosmologies. Consequently, it is illegitimate to deny the earth planetary status on grounds of its rocky substance. If this seems to us like a convincing proof, that is because we were convinced before we encountered the proof; to those who had grown up in a culture where the stability and centrality of the earth were reinforced every day in dozens of ways, the argument deriving from the appearance of the lunar surface would have appeared a great deal more problematic.

The argument from the satellites of Jupiter is of the same general type. In the reigning geocentric model of Aristotle and Ptolemy, the Moon is simply the planet situated closest to the central earth. When Copernicus removed the earth from the center, the Moon went along with it, for even the most basic lunar observations revealed that the Moon always has the same apparent diameter, from which it follows that the Moon has an earth-centered motion. The Moon thus obtained the status of satellite. Not only was it the only satellite in the planetary system (an ugly ad hoc feature of the heliocentric model, according to its critics), but it alone, among all the celestial bodies, circled about a center other than the stable center of the cosmos. However, if one accepted Galileo's discovery of Jupiter's satellites, then satellites and a center of celestial motions other than the center of the universe had to be acknowledged on any cosmology. Once again, an objection against heliocentric cosmology had been seriously weakened.

The phases of Venus gave rise to quite a different sort of argument. The standard geocentric model for the motion of Venus, devised by Ptolemy, predicted that Venus would never appear as a full circular disk (fig. 2.8).[14] If one believed Galileo's claim that the phases of Venus varied from thin crescent to full circle, the Ptolemaic model in its existing state was doomed.[15] However, it does not follow that *geocentrism* was doomed, for the geocentric model could be modified to accommodate Galileo's discovery. Indeed, the geocentric model already proposed and defended by the great Danish observational astronomer Tycho Brahe (1546–1601) predicted precisely what Galileo had discovered (fig. 2.9).[16] In short, although the observed phases of Venus were inconsistent with Ptolemy's specific model for Venus (and therefore probably offered rhetorical advantages to Galileo and other heliocentrists), they did not, if the matter were examined carefully, really tip the scales one way or the other in the geocentric-heliocentric debate. Galileo (eager to gain rhetorical advantage wherever he could) framed his argument in the *Letters on Sunspots* in such a way as to suggest that the phases of Venus, as he described them, defeated the only available geocentric model and consequently established the validity of Copernicus's heliocentric alternative "with absolute necessity."[17]

So Galileo had arguments, rather than proof. And in marshaling these arguments, he deployed his remarkable rhetorical gifts ("unexcelled in the annals of science," according to one modern historian of rhetoric).[18] Galileo's aim was not to write carefully reasoned scholarly papers of the sort astronomers and cosmologists now write, but to influence public opinion and win the cosmological

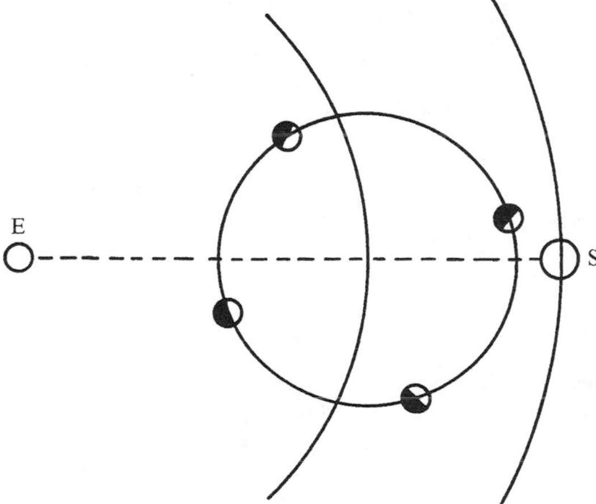

Figure 2.8. Ptolemy's model for Venus. Venus is shown in four different positions on its epicycle, the center of which is always situated on the line connecting Earth (E) and the Sun (S) (needed to guarantee that Venus will always appear close to the Sun). It is apparent that the illuminated half of Venus always directly faces the Sun and, therefore, faces mostly away from Earth. It follows that if this model describes the real world, terrestrial observers will never see more than a thin crescent of Venus.

debate in the public arena.[19] His arguments were as notable for their emotional power as for their logical power. The *Starry Messenger,* with which Galileo's campaign began, was written in Latin and therefore clearly intended for the highly educated. However, the audience that Galileo had in mind was much broader than Copernicus's audience of astronomical specialists; this book (like all of Galileo's cosmological publications) contained no mathematics and no astronomical technicalities and was meant for anybody capable of reading Latin.

Opening Confrontations

But why did Galileo run into trouble? Why did he encounter opposition, whereas seventy years earlier heliocentrism had caused no stir? Because in those seventy years the climate of opinion within the Catholic church had changed. This changed climate, to put it in simplest terms, was a result of the Protestant Reformation—a broad reform movement within the Catholic church, which reached its climax in the first half of the sixteenth century and culminated in the

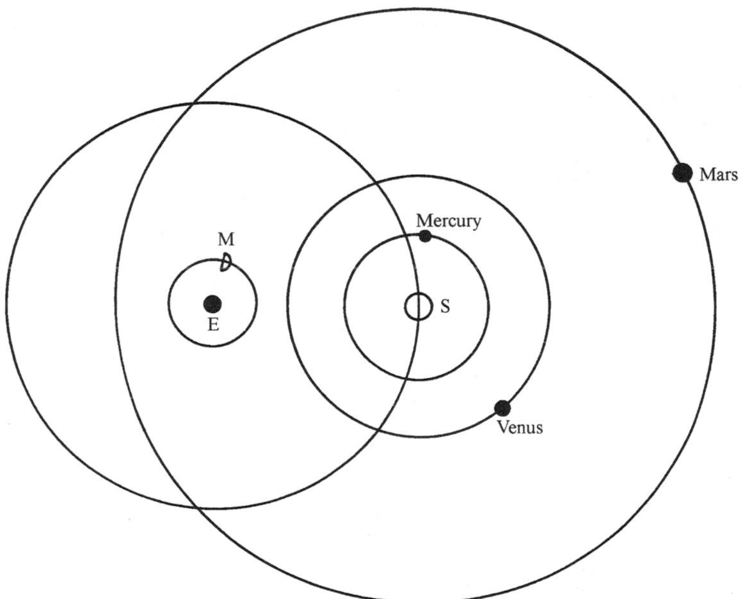

Figure 2.9. The geo-heliocentric cosmology of Tycho Brahe. Earth (E) is situated at the center of the stellar sphere (not shown). The Sun (S) and Moon (M) orbit Earth. The remaining planets are satellites of the Sun, accompanying it in its annual course through the fixed stars, while orbiting it, each with its characteristic period. Jupiter and Saturn, not shown, also have Sun-centered motions.

splitting of Western Europe roughly into Protestant and Catholic halves.[20] Naturally, the leaders of the Catholic church reacted to such a catastrophic event: having just lost half of Europe as a result of what could be construed as a relaxed policy toward dissent and controversy, they turned cautious. Driven by principles that emerged from the Council of Trent (1545–63), they took giant steps toward converting the church into a centralized, authoritarian bureaucracy capable of enforcing orthodox belief. This church bureaucracy was a great deal more worried about controversy than the medieval church had been. And it took a much stricter view of biblical interpretation, moving (in the years after Trent) toward literalism and refusing to embrace any interpretation not sanctioned by church tradition or the church fathers. A decree on the interpretation of Scripture that emerged from the council reads:

> The Council decrees that, in matters of faith and morals . . ., no one, relying on his own judgment and distorting the Sacred Scriptures according to his own conceptions, shall dare to interpret them contrary to that sense which Holy Mother Church, to whom it belongs to judge their true sense and meaning, has held and does hold, or even contrary to the unanimous agreement of the Fathers.[21]

This emphatic statement was a repudiation of the Protestant notion that Scripture stands alone as the proper authority for Christian belief and practice, in no way dependent on church tradition.

Galileo's troubles began soon after publication of the *Starry Messenger*.[22] In 1611, he made a visit to Rome to plead the case for his telescopic discoveries in person. The Jesuits at the Collegio Romano confirmed his telescopic observations (but not the heliocentric interpretation that he gave them) and treated him as a celebrity.[23] However, back in Florence Galileo's attempt to press the case for heliocentrism was beginning to run into opposition. Cosmology had always been a moderately sensitive issue, and the Giordano Bruno affair ten years earlier had perhaps increased the danger.[24] By posing a threat to the literal sense of certain scriptural passages, heliocentrism became a concern around which opposition to Galileo among conservative Florentine Dominicans could coalesce. Three years later, controversy over sunspots between Galileo and the Jesuit astronomer Christopher Scheiner resulted in a cooling of relations between Galileo and his Jesuit friends and supporters.

About the same time, Galileo got caught up in a debate regarding the possible mobility of the earth that had begun during a breakfast at the Medici court (reported to Galileo by his friend and student, Benedetto Castelli). Particularly worrisome for Galileo and his cosmological campaign was the extent to which the participants in this debate had been prepared to decide the issue by appeal to Scripture. There was clearly a need, Galileo perceived, for a sophisticated discussion of the principles of interpretation (or exegesis) applicable to the allegedly scientific content in Scripture. Galileo sought to supply this discussion in an open letter to Castelli—a handwritten letter that was soon circulating in multiple copies. Galileo argued in this letter that the sole purpose of the Bible was to persuade readers "of those articles and propositions which are necessary for . . . salvation and surpass all human reason." When the biblical text oversteps those limits, addressing matters that are within reach of sensory experience and rational knowledge, God does not expect these God-given capacities to be abandoned. In order to be widely understood, the scriptural writers accommodated themselves to popular conceptions; consequently, in matters of scientific dispute, the interpreter need not be limited to the "apparent meaning of the words."[25] It follows that theologians, before committing themselves to an interpretation of such passages, would be well advised to examine the demonstrative arguments of scientists or natural philosophers.

Galileo had embarked on a dangerous course. In this letter (subsequently expanded and published as the *Letter to the Grand Duchess Christina*), Galileo articulated principles of biblical interpretation emanating from Augustine (354–430) and long accepted within organized Christendom.[26] Nevertheless, advice from a layman on such matters was not welcomed by the theologians and other officials who staffed the church bureaucracy, particularly when that advice ran (in its willingness to push aside the opinion of the church fathers, as well as the literal sense of Scripture) so obviously against prevailing exegetical opinion as it had developed within the church since Trent.

Galileo's letter provoked a strong reaction, still largely local. The bishop of Fiesole (a hill town just outside Florence) called for the jailing of Copernicus.[27] The bishop was unaware that Copernicus had been dead for seventy years; he knew only that this Copernicus was a menace and should be kept off the streets. A little later, an elderly member of the Dominican order, Niccolò Lorini, secured a copy of Galileo's letter to Castelli and sent it to the Inquisition (the bureau in Rome responsible for ensuring correct belief and dealing with matters of heresy, also known as the "Holy Office"). In his cover letter, Lorini accused Galileo of adopting rash and possibly heretical principles of biblical exegesis.[28]

Galileo had no knowledge of this specific accusation, but he was aware that trouble was brewing and decided that he must go to Rome once more to make his case personally within the halls of power. He was convinced that he had decisive scientific arguments, and he naïvely supposed that such arguments would carry him to victory over the geocentric opposition. Moreover, memories of that triumphant visit to Rome a few years earlier encouraged optimism. Once in Rome (late 1615 and early 1616), Galileo cut quite a figure, arguing his case with passion wherever opportunity presented itself. A certain Antonio Querengo has left a vivid account of Galileo's persuasive efforts:

> He discourses often amid fifteen or twenty guests who make hot assaults upon him, now in one house, now in another. But he is so well prepared that he laughs them off; and although the novelty of his opinion leaves people unpersuaded, yet he reveals the futility of most of the arguments with which his opponents try to defeat him. Monday, . . . in the house of Federico Ghislieri, he achieved wonderful feats; and what I liked most was that, before answering the opposing arguments, he amplified and strengthened them with new grounds that appeared invincible, so that, in subsequently demolishing them, he made his opponents look all the more ridiculous.[29]

The Florentine ambassador to Rome, whose obligation it was to protect Galileo, was not pleased. Reporting to the grand duke of Tuscany, he wrote that Galileo "is vehement and stubborn and very worked up in this matter; and it is impossible, when he is around, to escape from his hands. And this business is not a joke, but may become of great consequence, and the man [Galileo] is here under our protection and responsibility."[30]

Galileo received plenty of attention, but he did not convince the people who counted (though he certainly had his supporters). His arrogant, impetuous style seems, on balance, to have been more effective in stirring up trouble and making enemies than (as he had hoped) in calming the waters. Indeed, near the end of February 1616, while Galileo was still in Rome, the Inquisition, finally acting on the charges made several years earlier by Lorini, censured two propositions that embodied the essentials of the heliocentric system: that the Sun is at rest in the center of the universe, and that the earth is in motion around it. The consultors to the Inquisition declared these propositions to be "philosophically absurd" and either "formally heretical" (the former proposition) or "at least erroneous in faith" (the latter). The published decree placed Copernicus's *Revolutions* on the Index of Prohibited Books "until corrected."[31] There it remained until 1835.

What lay behind this decision of the Inquisition? In simplest and specific terms, the question was whether or not the heliocentric hypothesis is consistent with traditional church teachings (based on a literal reading of biblical passages that appear to address, directly or indirectly, cosmological matters). The larger issue that lay behind this question was that of epistemological authority: are cosmological truth-claims dependent on science or on theology—on conclusions drawn from reason and sense experience or on the content of biblical revelation as interpreted by the fathers of the church? And if both are to be taken into account, how are competing cosmological and theological claims to be adjudicated? Reason and sense, Galileo claimed, taught the mobility of the earth. But revelation literally interpreted, said the leading theologians, clearly taught its fixity. Half a dozen passages from the Bible were appealed to by the theologians. Two of them speak to the issue with particular clarity:

> Ecclesiastes 1:5: The Sun rises and the Sun goes down, and hastens to the place where it rises.
>
> Psalm 93:1: . . . the world [i.e., the earth] is established; it shall never be moved.

These are manifestly expressions of geocentric cosmology. A third passage also yields easily to a geocentric interpretation:

> Joshua 10:12–13: Then spoke Joshua to the Lord in the day when the Lord gave the Amorites over to the men of Israel; and he said in the sight of Israel, "Sun, stand thou still at Gibeon, and thou Moon in the valley of Aijalon." And the Sun stood still, and the Moon stayed, until the nation took vengeance on their enemies.

Note that Joshua commanded the Sun to stop, in violation of what was obviously its natural state of motion. Were the cosmos heliocentric, Joshua would have been obliged to command the earth to cease rotation on its axis.

Added to these biblical texts was a further epistemological consideration. It was widely held by leading theologians, both Catholic and Protestant, that the natural sciences were in principle incapable of determining the true world system with certainty. Only the Creator knows what lies behind the celestial motions; the human intellect has no access to this divine knowledge. We cannot climb up into the heavens to find out what is really going on; and the kind of evidence we can gather from our terrestrial vantage point simply does not settle the matter. The same lesson was taught by the Ptolemaic astronomical tradition, which had been dominated by strong "instrumentalist" tendencies—treating the mathematical models for the planets as mere mathematical instruments, designed to predict observed planetary positions but not to describe celestial reality.

Given these arguments, what would the balance sheet have looked like to members of the Inquisition? On the side of geocentric cosmology were clear biblical statements to the effect that the Sun moves, while the earth is fixed. In support of heliocentric cosmology were no scientific proofs, but scientific opinions and arguments, asserted within a climate that cast doubt on the ability of the human intellect ever to discover the true cosmological system. Thus it was not (as members of the Inquisition saw it) divinely inspired biblical certainties against convincing scientific demonstrations, but biblical certainties against improbable scientific conjectures. From the church's perspective, no choice could have been easier.

It is tempting, from a modern perspective, to propose that the leading theologians of the church ought to have modified their interpretation of the relevant biblical texts in order to get into step with scientific opinion. But we must keep in mind that the position adopted by the Inquisition *was* in step with the majority, if not the latest, scientific opinion. And it would have been a most remarkable event had its members taken elaborate measures to abandon their own deeply held principles of biblical interpretation, as well as the traditional cosmological opinions of the church fathers, while simultaneously rejecting the majority opinion of qualified astronomers.

Let us return briefly to the course of events. The Inquisition formally censured heliocentrism in 1616, declaring it false and heretical. But Galileo faced no personal danger. He was charged with no offense; he was not declared a heretic. He was simply summoned by Cardinal Roberto Bellarmino, representing the Inquisition, and informed that heliocentrism had been declared false and heretical and was not to be held or defended. A surviving affidavit issued by Bellarmino states that Galileo had

> been notified of the declarations made by the Holy Father and published by the Sacred Congregation of the Index, whose content is that the doctrine attributed to Copernicus (that the earth moves around the sun and the sun stands at the center of the world without moving from east to west) is contrary to Holy Scripture and therefore cannot be defended or held.[32]

However, the Galileo file of the Inquisition contains another document that has figured prominently in the story of the Galileo affair. This document, a memorandum dated 26 February 1616, asserts that the meeting between Galileo and Bellarmino was also attended by the commissary general of the Inquisition, Angelo Segizzi, who ordered Galileo "to abandon completely the . . . opinion that the sun stands still at the center of the world and the earth moves, and henceforth not to hold, teach, or defend it in any way, either orally or in writing."[33] The importance of this document, if authentic, is that the order issued by Segizzi, forbidding all discussion of heliocentrism, went subtly but significantly beyond the command issued by Bellarmino, which merely enjoined Galileo not to "defend or hold" (as true) the heliocentric hypothesis. But is the document authentic? Probably, although the contrary opinion has also been vigorously defended.[34] Either way, it found its way into the Inquisition's file on Galileo; there it remained, waiting to be used against Galileo in the trial of 1633.

The decree of 1616 brought Galileo's public campaign on behalf of heliocentrism to a halt. Galileo returned to Florence, where he turned, of necessity, to the pursuit of other scientific interests. Toward the end of 1618 three comets passed through the European skies in quick succession. These comets caused considerable excitement and elicited a certain amount of commentary on the nature of comets, including attempts to extract from the phenomena of comets conclusive arguments against heliocentrism. Galileo was drawn into controversy with a Jesuit mathematics professor, Orazio Grassi, who had written on the subject; and the two were soon attacking each other in pseudonymous treatises and through intermediaries. The controversy culminated in Galileo's publication of a treatise, *The Assayer* (1623)—among other things, a bitter attack on Grassi, in which Galileo accused him of rude behavior, fraud, and intellectual theft.[35] *The Assayer* set forth the foundations of a mechanistic account of nature and later emerged as a landmark in the development of seventeenth-century science; but its importance for Galileo and heliocentrism lay in the poisoning of the waters between Galileo and the Jesuits, with whom Galileo had managed, until now, to maintain friendly relations.

Dialogue Concerning the Two Chief World Systems

Paul V, pope during the first decades of Galileo's heliocentric campaign, died in 1621. After the two-year papacy of Gregory XV, Maffeo Barberini (descended from a wealthy Florentine family) ascended to the papal throne as Urban VIII. Urban was considered an intellectual, a man of vision, and a moderate on the subject of heliocentrism. Moreover, he was an admirer and friend of Galileo. Three years earlier he had written a poem honoring Galileo for some of his telescopic discoveries; and just six weeks before his election to the papacy, he had thanked Galileo for the latter's congratulatory letter sent in acknowledgment of a doctorate received by the future pope's nephew. "I remain much obliged to Your Lordship," Barberini wrote, "for your continued affection towards me and mine"; and he assured Galileo "that you will find in me a very ready disposition to serve you out of respect for what you so merit and for the gratitude I owe you."[36]

The election of his friend Barberini to the papacy must have seemed to

Galileo an incredible stroke of luck. Galileo quickly concluded that he must exploit these changed circumstances in an attempt to revive the battle for heliocentrism. He requested an audience with the pope. In the course of six meetings, the two got around to the subject of cosmology. Urban made clear his belief that humans were, in principle, incapable of achieving certainty regarding cosmological matters. To develop a model that would make accurate astronomical predictions, he argued, was not to prove the truth of that model; and on that point, of course, he was perfectly correct, both in his terms and in ours. According to a report from one of Urban's confidants of a conversation between Urban and Galileo (perhaps during one of these audiences), Urban argued:

> Let us grant you that all of your demonstrations are sound and that it is entirely possible for things to stand as you say. But now tell us, do you really maintain that God could not have wished or known how to move the heavens and the stars in some other way? [Urban assumes that Galileo concedes the point.] Then you will have to concede to us that God can, conceivably, have arranged things in an entirely different manner, while yet bringing about the effects that we see. And if this possibility exists, which might still preserve in their literal truth the sayings of Scripture, it is not for us mortals to try to force those holy words to mean what to us, from here, may appear to be the situation.[37]

Nonetheless, from his discussions with the pope, Galileo came to understand that he was free to write about heliocentrism, so long as he treated it as mere hypothesis. Although Galileo was forbidden by the decree of 1616 to defend heliocentrism as true, there could be no objection to a treatise that explored the pros and cons of the heliocentric model.

Galileo set to work, completing his *Dialogue on the Two Chief World Systems* in 1629, after several delays owing to illness. His goal, no doubt, was to treat heliocentrism as a hypothesis, in accordance with Urban's admonition, but in the process to demonstrate unmistakably that it was the *best* hypothesis available. As the book emerged from his pen, however, it had become nothing less than a powerful argument on behalf of the indubitable *truth* of heliocentrism; no reader could have understood it otherwise. Nor did one have to read between the lines to perceive this as Galileo's purpose, for in the *Dialogue* itself he repeatedly claimed to have demonstrated the "truth" of his conclusions.[38]

In the *Dialogue,* Galileo skillfully refuted the standard objections against heliocentrism—the claim, for example, that on a moving earth a heavy object dropped from a tower would not strike the ground at the foot of the tower. On the positive side of the ledger, he pointed to such advantages of heliocentrism as its superior ability to explain retrograde motion. Crucial to his positive case was an argument from the tides, which claimed that the only adequate explanation of the tides was to see them as the sloshing of water in the great sea basins owing to the double motion of the earth (annual about the Sun and daily on its axis). At the close of the four days of dialogue, after bombarding his readers with arguments in favor of heliocentrism, Galileo had his mouthpiece, Salviati, disclaim belief in the truth of heliocentrism, portraying it as a fancy, a chimera, or a paradox, intended only to display the physical arguments on behalf of heliocentrism without any claim of physical truth. "I do not ask and have not asked from others," Salviati states, "an assent which I myself do not give to this fancy, and I could very easily regard it as a most unreal chimera and a most solemn paradox."[39] Simplicio (who had staunchly defended traditional Aristotelian geocentric cosmology throughout the dialogue) responds:

> I confess that your idea [of explaining the tides on the heliocentric model] seems to me much more ingenious than any others I have heard, but I do not thereby regard it as true and convincing. Indeed, I always keep before my mind's eye a very firm doctrine, which I once learned from a man of great knowledge and eminence, and before which one must give pause. From it I know what you would answer if . . . you are asked whether God with His infinite power and wisdom could give to the element water the back and forth motion we see in it by some means other than by moving the containing basin; I say you will answer that He would have the power and the knowledge to do this in many ways, some of them even inconceivable by our intellect. Thus, I immediately conclude that in view of this it would be excessively bold if someone should want to limit and compel divine power and wisdom to a particular fancy of his.[40]

The close resemblance between this speech, emanating from a "man of great knowledge and eminence," and the pope's argument, quoted above, is unmistakable. That Galileo put it into the mouth of the slow-witted Aristotelian laughingstock of the dialogue did not escape Urban's notice.

Galileo submitted the book through the appropriate channels for licensing by the church. Father Niccolò Riccardi, whose responsibility it was to make the final decision on suitability for publication, understood the delicacy of the subject matter and worried about Galileo's manuscript. He knew that Urban had granted Galileo permission to publish a book on the subject of cosmology, but he wasn't certain that this was the book the pope had had in mind. After several long delays and various revisions, including a new preface and conclusion expressly written to make clear that heliocentrism was no more than a hypothesis, Riccardi granted the official imprimatur.[41] The book, written in Italian (thereby violating contemporary scholarly custom in an attempt to reach the broadest possible lay readership), was printed in Florence and appeared in bookshops in February 1632.

The Trial and Recantation

To appreciate the reception of Galileo's *Dialogue* at the papal court, we need to understand the tense circumstances then prevailing in Rome, which had become a hotbed of fear and suspicion.[42] Europe was midway through the Thirty Years' War; the power of the papacy was threatened by the Spanish, who controlled half of the Italian peninsula; and the pope himself had recently come under heavy criticism for adopting positions of political expediency apparently favorable to the Protestant king Gustavus Adolphus of Sweden. On top of everything else, a horoscopic prediction of Urban's imminent death began to circulate, falsely rumored to have come from Galileo's hand. Under the circumstances, whatever equanimity Urban may have possessed dissolved into irritability, mistrust, and authoritarian behavior; it was not a propitious moment for the appearance of a book by this same Galileo on the theologically sensitive topic of cosmology.

Urban quickly discovered (perhaps with help from Galileo's enemies) his argument in the mouth of Simplicio. He became convinced that Galileo had brazenly ignored his admonition about the hypothetical character of heliocentrism and, moreover, by choosing to have his argument voiced by the inept Simplicio, had knowingly made the pope an object of ridicule. Such flagrant insubordination could not go unpunished. A letter from the Florentine ambassador, Francesco Niccolini, to his superior in Florence reveals the pope's state of mind:

> Yesterday I did not have time to report to Your Most Illustrious Lordship what had transpired (in a very emotional atmosphere) between myself and the Pope in regard to Mr. Galilei's work. . . .
> I too am beginning to believe, as Your Most Illustrious Lordship well expresses it, that the sky is about to fall. While we [Niccolini and Urban] were discussing those delicate subjects of the Holy Office, His Holiness exploded into great anger, and suddenly he told me that even our Galilei had dared entering where he should not have, into the most serious and dangerous subjects which could be stirred up at this time. I replied that

> Mr. Galilei had not published without the approval of his ministers and that for that purpose I myself had obtained and sent the prefaces to your city. He answered, with the same outburst of rage, that he had been deceived by Galileo.

Niccolini added later in the letter, "Thus I had an unpleasant meeting, and I feel the Pope could not have a worse disposition toward our poor Mr. Galilei." In another letter, written six days later, Niccolini reported: "the Pope believes that the Faith is facing many dangers and that we are not dealing with mathematical subjects here but with Holy Scripture, religion, and Faith."[43]

It was inevitable, under the circumstances, that the judicial machinery of the Inquisition would be set in motion. Printing of Galileo's *Dialogue* was immediately suspended. The pope appointed a special commission to investigate the case, charging it to weigh "every smallest detail, word for word, since one is dealing with the most perverse subject one could ever come across."[44] An alarmed Galileo, informed by friends of developments in Rome, sought to ameliorate the situation through the intervention of the Florentine ambassador and others favorable to his cause. The latter group included some of the very officials involved in the proceedings against Galileo, for the papal court could not claim perfect uniformity of thought. In the end Galileo (now nearly sixty-nine years old and in poor health) was compelled to make the long trip (about two hundred miles) from Florence to Rome for interrogation and trial. Traveling in a coach provided by his patron, the grand duke of Tuscany, Galileo left Florence toward the end of January 1633 and (after a period of quarantine at the border between Tuscany and the Papal States, required by fear of plague) arrived in Rome three weeks later. There he was comfortably housed, first in the residence of the Florentine ambassador, subsequently (during the actual trial) in a suite of rooms provided by the Inquisition, attended by his own servant.[45]

As the case unfolded in the course of the trial, it proved to be only indirectly about biblical interpretation and cosmological theories (as was the decision of 1616 condemning heliocentrism).[46] The trial of 1633 was about disobedience and flagrant insubordination: the issues dealt with in the decree of 1616 were not reexamined; its conclusions were merely reasserted.[47] The memorandum of 26 February 1616 was discovered in the Inquisition's file on Galileo, and its broader prohibition was used to seal the case against Galileo—to prove beyond the shadow of a doubt that he had disobeyed orders issued by the Inquisition. Even without this disputed memorandum, however, Galileo's disobedience, and therefore his guilt, would surely have been clear to everybody.

In the course of the trial, which began in the middle of April, Galileo was given the opportunity to make a statement. He revealed that since the previous interrogation he had been thinking continuously about the charge in the 1616 memorandum that had forbidden him to "hold, teach, or defend" heliocentrism in any manner whatsoever. "It occurred to me," he added,

> to reread my published *Dialogue,* which I had not seen for three years, in order carefully to ascertain whether, contrary to my most sincere intention, there had inadvertently fallen from my pen anything from which a reader or the authorities might infer . . . some sign of disobedience on my part. . . . I was able to obtain a copy of my book. . . . And, because I had not seen it for so long, it seemed to me like a new writing and by another author. I freely confess that it seemed to me composed in such a way that a reader ignorant of my real purpose might have reason to think that the arguments presented for the false side, which I really intended to refute, were expressed in such a way as to . . . compel conviction . . . rather than to be easily refuted.[48]

It was a nice try, but it would not do the job. As for Segizzi's injunction, described in the memorandum of 1616, Galileo claimed to have no recollection of it and therefore to have had no intention to deceive the Inquisition when he failed to mention it in the course of his application for the licensing of his book.[49]

The case dragged on into the second half of June. On 21 June, Galileo was summoned and asked "whether he holds or has held, and for how long, that the Sun is the center of the world and the earth is not the center of the world but moves with a daily motion." In his now famous recantation, Galileo replied:

> A long time ago, . . . before the decision of the Holy Congregation of the Index, and before I was issued that injunction, I was undecided and regarded the two opinions, those of Ptolemy and Copernicus, as disputable, because either the one or the other could be true in nature. But after the above-mentioned decision, . . . all my uncertainty vanished, and I held, as I still hold, as very true and undoubted, Ptolemy's opinion, namely, the stability of the earth and the motion of the sun.[50]

Sentence was passed the next day:

> We say, pronounce, sentence, and declare that you, the above-mentioned Galileo, because of the things deduced in the trial and confessed by you . . . , have rendered yourself . . . vehemently suspected of heresy, namely, of having held and believed a doctrine that is false and contrary to the divine and Holy Scripture: that the sun is the center of the world and does not move from east to west, that the earth moves and is not the center of the

> world, and that one may hold and defend as probable an opinion after it has been declared and defined as contrary to Holy Scripture.

The sentencing document proceeded to condemn Galileo "to formal imprisonment in this Holy Office at our pleasure. As a salutary penance we impose on you to recite the seven penitential Psalms once a week for the next three years. And we reserve the authority to moderate, change, or condone wholly or in part the above-mentioned penalties and penances."[51] Galileo's imprisonment was subsequently commuted to house arrest.

As a final event in this tragedy, Galileo was required to "abjure" his error—that is, to renounce it under oath. Kneeling in the Dominican convent of Santa Maria sopra Minerva, he said:

> I, Galileo, son of the late Vincenzio Galilei of Florence, . . . having before my eyes and touching with my hands the Holy Gospels, swear that I have always believed, I believe now, and with God's help will believe in the future all that the Holy Catholic and Apostolic Church holds, preaches, and teaches. However, whereas, after having been judicially instructed with injunction by the Holy Office to abandon completely the false opinion that the sun is the center of the world and does not move, and that the earth is not the center of the world and does move, and not to hold, defend, or teach this false doctrine in any way whatever, orally or in writing; and after having been notified that this doctrine is contrary to Holy Scripture, I wrote and published a book in which I treat of this already condemned doctrine and adduce very effective reasons in its favor, without refuting them in any way; therefore, I have been judged vehemently suspected of heresy. . . .
>
> Therefore, desiring to remove from the minds of Your Eminences and every faithful Christian this vehement suspicion, rightly conceived against me, with a sincere heart and unfeigned faith I abjure, curse, and detest the above-mentioned errors and heresies, and in general each and every other error, heresy, and sect contrary to the Holy Church; and I swear that in the future I will never again say or assert, orally or in writing, anything which might cause a similar suspicion about me; on the contrary, if I should come to know any heretic or anyone suspected of heresy, I will denounce him to this Holy Office.[52]

The abjuration was a formality, of course. It is foolish to accuse Galileo, as some have done, of selling out in the battle for freedom of thought and expression. No such battle existed in early seventeenth-century Italy; and if there had been one, it is doubtful that Galileo would have joined it. The significance of Galileo's condemnation and abjuration is simply the humiliation of Italy's greatest living scientist and the end of his campaign on behalf of heliocentrism.

For the remaining nine years of his life, Galileo was under house arrest, comfortably situated in his rented villa just outside Florence, with few restrictions on who could come and go. He turned his attention to other scientific problems, principally mechanics and the strength of materials, preparing the manuscript that was to be published in the Netherlands as the *Discourse on Two New Sciences* (1638). As punishment for his defense of heliocentrism, Galileo suffered neither torture nor imprisonment. But he lost his freedom of movement and, perhaps most important of all, his voice on the subject of cosmology.

Conclusions

Is this merely a story of heroic and not-so-heroic deeds leading to a tragic end, or can it teach us something about the historical relations of science and religion? There are, indeed, important lessons to be learned from the Galileo affair, but not the ones customarily drawn.

First, the Galileo affair is often presented as a simple ideological conflict: scientific rationalism versus religious authority. But as we have seen, it had an enormous human and political dimension as well. Science and religion as such cannot interact, but scientists and theologians can. Theoretical and methodological positions come down to earth and enter the real world only insofar as they are defended by humans; and when flesh and blood make an appearance, we are apt to find that personal interest and political ambition are as important as ideological stance. There were old scores to settle, egos to stroke, and careers to be made. Dominican-Jesuit politics clearly figured in the drama.[53] Galileo's friendship with members of the pro-Spanish faction at the papal court may also have played a role, as did his alienation of the Jesuits and his apparent betrayal of the pope. Galileo's personality was a consistent and important factor; indeed, it seems clear that had he played his cards differently, with more attention to diplomacy, Galileo might well have carried out a significant campaign on behalf of heliocentrism without condemnation.

It follows, in the second place, that the outcome of the Galileo affair was powerfully influenced by local circumstances. The Galileo affair was not merely about universal or global aspects of science and religion, or universal beliefs of scientists and religious leaders, but also (as we have seen) about the local circumstances impinging on individual historical actors—fears, rivalries, ambitions, personalities, political context, and socioeconomic circumstance. Historical events are situated in time and space; they are contingent and local, and our analysis must respond to this reality.

Third, the Galileo affair is consistently and simplistically portrayed as a battle between science and Christianity—an episode in the long warfare of science and theology. This is how Andrew Dickson White viewed the matter in the nineteenth century, and his interpretation has become part of the Western cultural heritage. But consider the complex reality. Every one of the combatants, whether church official or disciple of Galileo, called himself a Christian; and all, without exception, acknowledged the authority of the Bible. Many on both sides of the struggle, including Galileo, were theologically informed, capable of articulating carefully reasoned theological positions. Similarly, all of the principal combatants, from Urban VIII to Galileo and his supporters (even Grand Duchess Christina, on the sidelines), possessed informed, rationally defensible, and strongly held cosmological beliefs.

But the complexity does not end here. Among the clergy, differences of opinion regarding principles of biblical interpretation were tolerated; and some clergy, adopting Galileo's exegetical principles, counted themselves among his vocal supporters. Meanwhile, among people with special expertise in astronomy and cosmology, heliocentrism (viewed as an account of cosmological reality) remained a minority opinion. It follows that conflict was located as much *within* the church (between opposing theories of biblical interpretation) and *within* science (between alternative cosmologies) as *between* science and the church. This analysis is an attempt not to downplay the magnitude or seriousness of the conflict over heliocentrism, but simply to map it. And the point is that it is impossible to identify clearly defined battle lines falling along a divide separating heliocentric scientists, prepared to overlook the Bible or interpret it allegorically, from geocentric theologians or clergy, committed to church tradition and biblical literalism.

There remained, of course, the problem of epistemological authority—whether the truth of cosmological claims was to be determined by exercise of the human capacities of sense and reason, by appeal to biblical revelation, or by some combination of the two. This was the central methodological issue in the Galileo affair, and we must not allow our enthusiasm for the local and human aspects of the struggle to obscure it. That the position adopted by church officials in the Galileo affair was not an *inevitable* accompaniment of Christian belief is clear from the fact that Christians of all stripes are now heliocentrists; and many Christians, both Catholics and Protestants, were heliocentrists in the seventeenth century. But we must also recognize that the issue of epistemological authority in areas of overlap (actual or potential) between the Bible and the natural sciences remains unresolved for some Christians to this day, as we see in contemporary battles between "creationists" and "evolutionists." This lack of resolution means that tension and the potential for conflict will continue to hover over the relationship between Christianity and science.

Finally, how shall we judge the protagonists in the Galileo affair? Making such judgments is not the historian's main business. However, I believe that we can learn something important from the exercise if we proceed with due caution. In order to avoid making a complete mess of the matter, we need to be aware of whose values we are employing—those that prevail in modern democracies at the beginning of the third millennium, or those that prevailed in Catholic Europe almost four hundred years ago, in the first third of the seventeenth century. Let us try it each way.

Viewing the church's apparatus for suppression of dissent, its police power, its threat of torture (indeed threat of burning at the stake for a heretic who refused to recant), we who live in modern democracies will surely be inclined to judge harshly. There are certainly modern examples of the attempt to impose theological censorship on scientific beliefs; but they are exceptional. And the belief that such censorship should be imposed under threat of burning at the stake is no longer, to my knowledge, a common approach to the problem. There is no question that such a comparison of cultures (theirs with ours) stands to teach us interesting and useful lessons; though if we merely use this exercise as an occasion for condemning seventeenth-century Italians for not being modern Europeans or Americans, the effort has been wasted.

But the interesting and informative judgment for a historian is one that evaluates early seventeenth-century actions not by modern values but by standards of civilized behavior widely or universally subscribed to by early seventeenth-century people. Judged in these terms, the Galileo affair takes on quite a different appearance. The early seventeenth century was a time of growing absolutism in Europe, in both the religious and political realms. The freedom to express dangerous ideas was as unlikely to be defended in Protestant Geneva as in Catholic Rome. The idea that a stable society could be built on general principles of free speech was defended by nobody at the time; and police and judicial constraints were therefore inevitable realities.

Measured against such contemporary norms and values, the proceedings against Galileo were not faultless: for example, identification of cosmological claims as "matters of faith and morals," if consistent with the biblical literalism that emerged after the Council of Trent, certainly ran against the grain of older exegetical principles, articulated in the writings of Augustine and Aquinas. There were also procedural irregularities. By and large, however, the central bureaucracy of the church and the people who staffed it lived up to widely held norms, followed accepted procedure, and even on a number of occasions treated Galileo with generosity. Had these troubles befallen somebody other than himself, it is doubtful that Galileo would have found the inquisitorial procedures objectionable. He objected not to the right of the church to enter the cosmological debate, but rather to the position it adopted in that debate. In short, examined in seventeenth-century terms, the outcome of the Galileo affair was a product not of dogmatism or intolerance beyond the norm, but of a combination of more or less standard (for the seventeenth century) bureaucratic procedure, plausible (if ultimately flawed) political judgment, and a familiar array of human foibles and failings.

Acknowledgments

I am grateful for critical comments on various versions of this article by Ernan McMullin, Michael H. Shank, and Richard Olson.

3

Christianity and the Mechanistic Universe

William B. Ashworth Jr.

In the second half of the seventeenth century, a new philosophy of nature came into prominence. Although it was presented in several distinct forms by the likes of René Descartes, Pierre Gassendi, and Robert Boyle, in all forms it treated matter as lifeless and inert, without any properties of its own. It also suggested that all natural phenomena could be explained by the mechanical interactions of matter in motion. This "mechanical philosophy," as it came to be called, was in strong contrast to the picture presented by traditional philosophies, such as Aristotelianism, and by other, newer, philosophies of nature that had been constructed in the late Renaissance, such as natural magic and Paracelsianism.[1]

The acceptance of the mechanical philosophy played a major role in the events that we collectively call "the Scientific Revolution." What concerns us here is the impact of this new philosophy on religion, specifically the Christian religion, as well as the impact of Christian thought on the mechanical philosophy. It is not immediately obvious that there should have been any interaction at all, since matter theory and theology would seem to be widely separate domains of inquiry. And yet, as we will see, the proponents of a mechanical philosophy were driven by religious concerns, the debate between different forms of the mechanical philosophy was waged on religious grounds, and the success of the mechanical philosophy was hailed as a Christian triumph. Religion and the mechanical philosophy were, in fact, inextricably linked throughout the seventeenth century.[2]

One might proceed to develop this story in any of several ways, but I think it would be helpful to have a mechanical philosophy in view before we go any further. Once we understand the basic features of one mechanical philosophy, it will be easier to appreciate the nonmechanical philosophies that preceded it, as well as the rival versions of a mechanical philosophy that would eventually be proposed. Since René Descartes was chronologically the first to publish a mechanical philosophy, we will begin with his version.

The Mechanical Philosophy of René Descartes

Descartes (1596–1650) (fig. 3.1) came of age in France at a time of considerable concern over the conflicts between traditional forms of knowledge (including religion) and the new discoveries of the late Renaissance. Where previously there had been one well-established Church, one known world, and one picture of the cosmos, there were now (ca. 1600) a variety of Christian churches, a recently discovered New World, and several new cosmologies. In France this resulted in what has been called the "skeptical crisis," with writers such as Michel Montaigne (1533–92) despairing that perhaps we can never know truth with certainty.[3] Descartes reacted to this skeptical crisis by deciding to doubt everything that he had been taught. He found that through methodical doubt he could arrive at certain "clear and distinct" ideas that could not be doubted no matter how hard he tried, and on these foundations he erected a new philosophy. He presented the outlines of this new philosophy in his *Discourse on Method* (1637), and he worked out most of its details in the *Principles of Philosophy* (1644).[4]

When it came to determining the "clear and distinct" attributes of matter, Descartes found that he could easily doubt most of the features that his predecessors had assigned to matter. An object before us may feel warm, smell sweet, or appear red, but Descartes was certain that these qualities could not be found in the matter itself—the stuff of which this object is composed. Here is how Descartes made the argument, in his *Second Meditation:*

> Let us now . . . consider the objects that are commonly thought to be the most distinctly known, namely, the bodies that we touch and see. . . . Take, for example, this piece of beeswax; . . . it has not yet lost the taste of the honey it contained; it still retains something of the odor of the flowers from which it was gathered; its color, shape, and size are apparent [to sight]; it is hard, cold, easily handled, and sounds when struck with

the finger. Thus all that contributes to make a body as distinctly known as possible is found in the one before us. But while I am speaking, the wax is placed near a fire: it loses the remains of its taste, its odor evaporates, its color changes, its shape is destroyed, its size increases, it is liquified, it becomes hot, it can hardly be handled, and it emits no sound when struck. Does the same beeswax still remain after this change? It must be acknowledged that it does; nobody would deny it or

Figure 3.1. Portrait of René Descartes (1596–1650), from Charles Perrault, *Les hommes illustres qui ont paru en France pendant ce siecle: avec leurs portraits au naturel* (Paris: Chez Antoine Dezallier, 1696–1700), 1:59. (Courtesy of the Linda Hall Library of Science, Engineering and Technology)

> judge otherwise. What, then, was it in the wax that I knew so distinctly? Nothing, apparently, that came to me through my senses; since all the features that came under taste, smell, sight, touch, and hearing are now altered—and yet the same wax remains.[5]

It followed for Descartes that those characteristics perceived by the senses (he called them "secondary qualities") were illusions of the senses. They represented appearance, not reality. One by one, Descartes threw out all of the traditional attributes of matter, until he came to the one feature he could not doubt. Matter, to be matter, has to occupy space. In Descartes's terms, it has to have extension. So Descartes defined matter as extension and extension as matter. It follows that matter has no other properties than the occupancy of space.

The implications of this radical redefinition of matter were profound. If matter has no properties but extension, then all it can do is move about and collide and move some more. It cannot attract, or seek, or sympathize with other matter. It can only be pushed about, and therefore everything that happens in the universe must be reducible to matter in motion and matter colliding with other matter. Descartes went further. Since extension is identical with matter, there can be no extension without matter and, therefore, no empty space. The universe must be filled with matter, a plenum. Since we do not see a matter-filled universe, some forms of matter must be imperceptible. Descartes uses the expression "third matter" to refer to matter that we perceive with our senses, and the term "second matter" to refer to the imperceptible matter that fills all space between the chunks of third matter. The tiny spaces between particles of second matter must also be filled, which requires the postulation of a "first matter" (fig. 3.2). All of this matter is in continual motion.

Descartes found that the picture that resulted from this new view of matter has a great deal of explanatory power. For example, in a universe filled with matter, particles must move in closed circles, or whirlpools. If particles were to move in straight lines, then the motion of one particle would necessitate the displacement of an infinite number of successive particles. Descartes called whirlpools "vortexes" (or "vortices"). We can imagine then that our solar system is a vortex of swirling second matter, carrying with it the large chunks of third matter, the planets. The ultrafine first matter is forced to the center, where it forms the Sun (fig. 3.3). The light and heat of the Sun are conveyed to us by pressure transmitted through the intervening second matter. We have, as a result, a mechanical model of the solar system that is quite satisfying, since it explains why the planets move and why they all move in the same plane and the same direction, and it predicts that the central body will be different from the planets that move around it.[6]

There is more to Descartes's mechanical philosophy, but further details require the introduction of God and Christianity into the picture, and so at this point it would seem desirable to sketch out the philosophy of nature that Descartes was reacting against. When he went to the Jesuit college at La Flèche, he was taught a rather different theory of matter, deriving ultimately from the writings of Aristotle, as interpreted by the Scholastic theologian Thomas Aquinas (d. 1274).

Aristotle's Theory of Substance

Two thousand years earlier, Aristotle (384–322 B.C.) had defined "matter" quite differently. His term for things that exist of themselves—what we think of today as physical objects, such as rocks or tables or pieces of wood—was "substances." But a substance must be understood, Aristotle argued, as a composite of "form" and "matter." By "form," Aristotle meant the total collection of qualities or properties that make a thing what it is. However, properties must be the

Figure 3.2. A figurative representation of Descartes's three kinds of matter. The screw-shaped and prism-shaped particles represent "first matter," which fills all the spaces between "second matter," represented by the stacked spheres. "Third matter," the ordinary matter we see and feel, is represented by the table and the wall. From Wolferd Senguerd, *Philosophia naturalis, quatuor partibus . . .* 2d ed. (Leiden: Apud Danielem á Gaesbeeck, 1685), 192. (Courtesy of the Linda Hall Library of Science, Engineering and Technology)

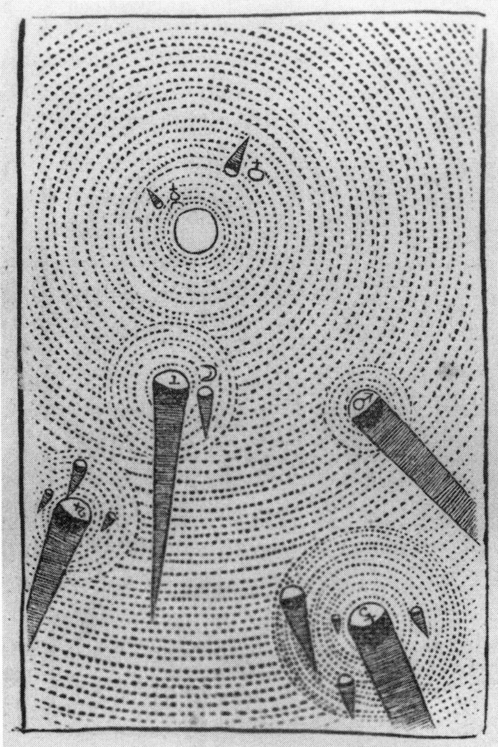

Figure 3.3. A diagram of the solar system, by a follower of Descartes's vortex theory. The Sun is the center of the principal vortex, which carries the planets around, but each planet also has its own vortex, which moves its moons about and is also responsible for gravity. From Petrus Hoffwenius, *Synopsis physica disputationibus aliquot academicis comprehensa*, 2d ed. (Stockholm, 1698), 84. (Courtesy of the Linda Hall Library of Science, Engineering and Technology)

attributes *of* something, and that something, for Aristotle, was "matter." Matter is the "stuff" of a substance; when matter has been "informed," a substance results. Aristotle also has an expression for matter that has not been informed: he calls it "first matter." But in reality, matter and form exist only as a composite; we can never encounter either one by itself.[7]

Aristotle's scheme can be nicely illustrated by his theory of the four elements: earth, water, air, and fire. Like all other really existing things, these four elements are substances. Earth results when first matter is endowed with the qualities cold and dry. Water is the cold and wet element, air is hot and wet, and fire is hot and dry. These elements may be combined in various proportions to produce all of the substances that make up the world.[8]

According to Aristotle, forms can be either "substantial" or "accidental." Qualities or properties that make a thing what it is comprise its "substantial form"; a table, for example, must have a top and legs, and be hard and have weight. These traits make up its substantial form. A table might also be brown and smooth and smell of vinegar; these properties are not essential, and they are referred to as "accidental qualities."

When it came to explaining how the world operates, or how one object interacts with another, Aristotle had recourse to what he called the "natures" of objects. Nature is a rather complicated concept in Aristotle, since it involves both form and matter, but in essence, it is simply that which determines an object's behavior. A rock, for example, is heavy "by nature," so if you drop it, it will fall, seeking its natural place at the center of the universe. A tree by nature grows up, and because that is its nature, it does not have to defy gravity to do so. A planetary sphere is inclined by its nature to move in a circle, which explains why the heavens are dominated by circular motion. For Aristotle, a great deal of the motion in the world is natural and does not require explanation by forces. Where motion is unnatural or violent, Aristotle does suggest that a motive force is necessary. But because such motion is not natural, it will cease, and the object will come to rest, as soon as the force is removed. It follows that Aristotle did not embrace a principle of inertia, or the idea that motion is conserved. But then, in the world of everyday experience, objects do slow down and stop when you leave them alone.

The picture that Aristotle painted of the material world was an eminently satisfying one, well in accord with common everyday experience, and so it is not surprising that it was fully integrated into Scholastic philosophy in the thirteenth century, after it was translated into Latin. Initially, there was some significant tension between Aristotelianism and Christianity (see chapter 1 of this volume), especially with respect to Aristotle's thoughts on the eternity of the universe and the mortality of the soul. But his matter theory was accommodated more readily, since it had considerable explanatory power, and it had no serious competition. Once it was assimilated, scholars discovered that certain aspects of this matter theory were quite useful in interpreting certain Christian doctrines. The Christian soul, for example, could be explained as the "form" of the body, the immaterial feature that gives the body its defining essence. There was even an unexpected fringe benefit in the ability of Aristotle's theory of form and matter to make sense of the sacrament of the Eucharist. Church doctrine had established that when the host is consecrated, the bread of the wafer is miraculously converted to the body of Christ, although it still appears to be bread. In Aristotelian terms, one could now say that the substantial form of the bread was changed, while the accidental form of bread remained. So Aristotle's theory of substance provided a perfectly acceptable explanation of both the observed phenomenon and the miracle.

Because of its conformity with sense experience and its compatibility with Christian doctrine, Aristotle's theory of form and matter survived intact well into the seventeenth century, where the mechanical philosophy would confront it head-on. But even before then, philosophies of nature that rivaled Aristotle's had arisen, and we need to be aware of several of these.

Renaissance Natural Magic

In the late Renaissance there arose a whole host of alternative natural philosophies, which have been variously labeled as "Hermetic" or "naturalist" or "magical"; we will use the terms "magical" and "natural magic" to refer to these philosophies. Natural magic tended to accept Aristotelian matter theory as a starting point, but it unleashed on the Aristotelian worldview a battery of new forces and influences. At the heart of natural magic was the belief that any object in the universe can potentially affect any other object, provided that a sympathy is established between them. There were many examples of natural sympathies that could be pointed to; the seven planets in the heavens seem to correspond to the seven metals in the earth, suggesting that there is a sympathy between Mars and iron and between the Sun and gold. Analogy further suggests that there might be sympathies between Mars and a warrior, the Sun and a yellow flower, or the Sun and the human heart. Sympathies ruled the magical universe, and these sympathies, it was argued, could be manipulated by the natural philosopher. If you were under the influence of the melancholy planet, Saturn, and wanted to change your mood, you might wish to establish a sympathy between yourself and a beneficent planet, say the Sun. You could do so by wearing a yellow robe, singing songs in praise of the Sun, playing your lute in a solar mode, and drinking wine from a golden goblet, and so replace your spirit of melancholy with a happier frame of mind.[9]

One of the attractions of natural magic was the fact that there are many phenomena in the world that seem sympathetic in nature. If you have a shop full of lutes and you pluck a string on one, strings on others will magically begin to vibrate, and if you check, you find that all the resonating strings have the same pitch. They seem to be vibrating in sympathy. Magnetism appears to be another example of a sympathetic force. A magnet attracts only iron, it works at a distance, and it is undeterred by a barrier. It is difficult to explain how a magnet works if you do not invoke a sympathy of some sort.

One other interesting feature of natural magic is that it placed great emphasis on experiment and observation. It did so because nature seems to be such a maze of forces that reason alone could never sort it all out. Aristotelians tended to think about how nature works; natural magicians tinkered with nature instead. One of the great personal guides to the Renaissance world of phenomena was the *Magia naturalis* (1589)[10] of Giovanni Battista della Porta (1535–1615). Della Porta described endless experiments and tricks that could be performed with magnets, mirrors, ovens, bleaches, dyes, and explosives, and he regularly used sympathies and correspondences to explain why a magnet attracts iron, how lenses magnify, why strings resonate, and how medicines cure. His book presented a view of nature in which matter can move matter over great distances by means of attraction and repulsion. Very rarely were these forces thought of as mechanical in their action.

Renaissance natural magic ran into some opposition from religious authorities, especially the Catholic church. The problem was that a magical worldview threatens the idea of miracles. In a world where anything can happen naturally, little room is left for the supernatural. In the aftermath of the Protestant Reformation, miracles were one of the defining features of Catholicism, and the church did not take kindly to attempts to treat miracles as natural phenomena, because, in effect, it undermined the whole idea of divine intervention. The church also took a dim view of those who moved beyond planetary influences and attempted to manipulate angels and demons. Such "black magic" was attacked vigorously, and most natural magicians such as della Porta carefully maintained their distance from such practices.[11]

Natural magic maintained its vigor well into the lifetime of Descartes. In 1617 Robert Fludd (1574–1637) published a very popular book, *On the Greater and Lesser Worlds,* in which he argued that, since the human body is a microcosm of the universe (the macrocosm), we can map all the correspondences that connect the two worlds, and Fludd did so with a number of attractive engravings. A common feature of all the various images of microcosm and macrocosm are the dotted lines and harmonic ties that, in effect, are the lines of force that make the universe work. And it is worth stressing again that there are no mechanical forces in this world of cosmic sympathies.[12]

The Religious Implications of Cartesian Mechanism

If we now return to Descartes and view him in the light of both Aristotelian and Renaissance natural philosophy, we see the almost shocking novelty of his mechanical philosophy. In making extension the sole attribute of matter, Descartes denied the real existence of all other qualities. Accidental forms are only names for the products of our perception; they do not really exist in nature. Substances do not have "natures" or natural tendencies of any kind. There are no such forces as sympathies or correspondences. The only allowable forces are mechanical forces. Magnetism must be explained mechanically without recourse to sympathies, and Descartes described a possible mechanical model for the magnet. Alleged sympathies that cannot be explained mechanically are denied existence in the Cartesian universe.

Matter is one ingredient of Descartes's universe. Mind is the other—the only thing besides matter that God created. Mind is immaterial and, according to Descartes, endowed only with the property of thought. The existence of mind was Descartes's initial clear and distinct idea: *Cogito, ergo sum* (I think, therefore I [a thinking being] must exist). Everything in Descartes's universe is either mind or matter, and nothing is both. Indeed, mind and matter intermingle only in humans, for mind is identical with the human soul. Since plants and animals have no soul and thus no mind, it follows that they consist solely of matter, and they should be regarded simply as mechanical automatons—nothing more than intricate pieces of machinery.

What were the religious implications of Descartes's mechanical philosophy? Since the prevailing philosophy of his time was Scholasticism, which was thoroughly imbued with Aristotle's theory of form and matter, we might expect there to have been considerable conflict. Descartes, however, saw no problems for religion in his mechanical philosophy. After all, his second clear and distinct idea was the existence of God, and it was God's perfect wisdom that guaranteed Descartes's entire reasoning process. Descartes maintained that his universe of matter was created by God, that God gave this matter a certain amount of motion at the Creation, and that he has conserved that motion ever since. Nevertheless, Descartes did envision a God who worked through natural causes and who allowed the universe to build itself by laws that he had established at the Creation. Descartes's God was like a watchmaker, who created a world that basically ran itself. This would prove to be a controversial notion, as we shall see.[13]

The proposal of other, non-Cartesian, mechanical philosophies complicated matters. Descartes had constructed his natural philosophy from the ground up, with little attention to earlier authorities. However, there were other mechanical philosophies around, one of which had been in existence long before Aristotle was born. This was the atomic philosophy of Leucippus (fl. 435 B.C.) and Democritus (fl. 410 B.C.).

Ancient Mechanical Philosophies

Ancient Greek atomism postulated that all matter is composed of atoms, which are by very definition indivisible bits of matter. Atoms are infinite in number, and they differ only in size and shape. All atoms are made of the same stuff, and they have no qualities except possibly their immutability. The differences we perceive in objects are solely the result of differences in the arrangement of their constituent atoms. Leucippus and Democritus also postulated the existence of void space. All atoms are in motion in the void, and they have been in motion forever. Atoms continually collide and react and form arrangements that we perceive, and then break up and move on. In fact, we, the perceivers, are ourselves only a fortuitous assemblage of atoms, like all other objects.[14]

The Democritean scheme allowed for material atoms and the void and nothing else. In particular, there are no really existing immaterial things. The soul, then, must be made of atoms. The atoms of the soul could be especially elegant ones, fine and proper, but they are still material. Similarly, gods were welcome in the atomic universe, but they too are composed of atoms, and are therefore no more immune from the laws of nature than any other object.

The Democritean world is truly a mechanical world. There are no Aristotelian forms, no souls, no purposes. Everything is the result of matter in motion, undergoing collisions, and assembling in chance configurations. To Aristotle and many others, such a world—without order, without purpose, and without a divine being—could not possibly be the world in which we live.

However, atomism as a natural philosophy did not immediately wither and die. In the generation after Aristotle, Epicurus (341–270 B.C.) attempted to revive it. Epicurus was primarily an ethical philosopher, who sought a prescription for human happiness, and since, in his opinion, most anxieties are the result of irrational fears, he searched for a worldview that would lay such fears to rest. He found it in atomism, with its repudiation of the supernatural, its banning of divine intervention, and its determined commitment to exclusively natural causation.[15]

But Epicurean atomism differed from that of Leucippus and Democritus. Epicurus gave the atoms an additional property, namely weight, which is what causes them to move. In effect, all atoms are continually falling through an infinite universe, as a sort of cosmic rain. Since it is difficult to see how or why a rain of falling atoms could interact (since all are falling in parallel lines at the same speed), Epicurus introduced a purely arbitrary element into the universe, which he called the "swerve." The swerve causes an atom to deviate from its straight-line descent, leading to a chain reaction of collisions and ultimately the world in which we live. Epicurus presumably introduced the swerve to forestall complete determinism and to keep a modicum of free will alive in the world. However, the swerve also keeps the Epicurean universe from being completely mechanical, since the swerve is arbitrary, an event without a cause.

As an ethical philosopher, Epicurus inspired a considerable following. As a natural philosopher, however, he was less influential. He did attract the attention of Lucretius (ca. 99–55 B.C.) whose great poem *On the Nature of Things* is an espousal of Epicurean atomism. But the apparently atheistic character of Epicureanism, with its eternal universe, infinite atoms, and material souls, attracted few adherents in the early Christian centuries or in any centuries thereafter, until the surprise of the seventeenth.

Gassendi's Revival of Ancient Atomism

Pierre Gassendi (1592–1655) (fig. 3.4) was another original thinker who emerged from the skeptical crisis of the sixteenth century with a fistful of new programs. He was a youthful participant in the campaign of Marin Mersenne (1588–1648) against both Renaissance Aristotelianism and natural magic. He was an early believer in the value of an experimental method. He turned the newly invented telescope on the heavens, tracking planets and attempting to map the Moon. In the wake of Galileo's new physics, he performed experiments on falling bodies; he constructed ingenious barometric tubes in the aftermath of Torricelli's invention of the barometer. With respect to matter theory, Gassendi was, like Descartes, dissatisfied with Aristotelian explanations and angry at magical ones. But Gas-

Figure 3.4. Portrait of Pierre Gassendi (1592–1655), from Charles Perrault, *Les hommes illustres qui ont paru en France pendant ce siecle: avec leurs portraits au naturel* (Paris: Chez Antoine Dezallier, 1696–1700), 1:63. (Courtesy of the Linda Hall Library of Science, Engineering and Technology)

sendi was much more of a humanist than Descartes, for he believed in the wisdom of ancient philosophy, and he would reject Aristotle only in favor of a wiser ancient authority. He found that wisdom, against all odds, in the writings of Epicurus.[16]

To someone opposed to Aristotelian forms and qualities and distressed about sympathies and correspondences, the attractions of atomism are obvious. Atomism dispenses with all qualities and all causes that are not reducible to matter in motion. The problem with atomism, at least in its Epicurean clothing, is that it hardly seems compatible with Christianity. Epicurus taught that the world is infinite in time and space and was uncreated. He thought that the soul is material and mortal. Epicurus believed in more than one god, and the deities of his world are material like everything else and subject to the laws of nature. There is no purpose in the Epicurean world, and certainly no divine providence. The Christian, however, believes that one immaterial God created the world out of nothing at one specific time; that there are no other worlds than this one; that purpose is everywhere; that there is an immaterial human soul capable of salvation; and that divine providence rules.

For a Catholic priest like Gassendi, seeking an alternative to Aristotle, Epicurus would thus seem an unlikely choice, but Epicurus is where Gassendi bet his intellectual life savings. In a series of works written between 1647 and his death in 1655, and culminating in his *Philosophical System,* published posthumously in 1658, Gassendi re-presented Epicurean philosophy, with modifications that would make it acceptable to Christian thinkers.[17] In his version, Gassendi agreed that the world was made of indivisible passive atoms, which differed only in size and shape. These atoms moved and congregated in a void, and all large-scale effects are reducible to differences in the sizes, shapes, and configuration of atoms. However, these atoms are not infinite in number, nor are they eternal. Rather, God created the atoms, and he made only a finite number of them. God also infused motion into the atoms when he created them, providing them with just the right amount to fulfill his providential intentions. So Gassendi was able to dispense with the arbitrary swerve.

Gassendi also reasserted the immateriality of the soul. And he maintained that the soul was capable of influencing matter, thus allowing for the reinstitution of free will.

Gassendi versus Descartes

Thus two distinctly different mechanical philosophies emerged in the 1640s. They shared the desire to remove substantial forms, sympathies, and correspondences as explanatory devices, preferring instead to reduce all explanation to the motion of tiny, insensible, and totally passive particles. However, Descartes and Gassendi disagreed as to whether these particles were ultimately indivisible (Gassendi) or divisible (Descartes), and as to whether the universe was filled with matter (Descartes) or consisted of matter moving through void space (Gassendi). They also disagreed about the role of God in shaping and maintaining a mechanical universe.

Descartes viewed God as a supremely wise Creator, who created the universe from nothing and then let it run, like a machine, by itself. According to Descartes, God created matter, put it in motion, conserved that motion, and then withdrew and allowed the universe to unfold on its own accord; God was not needed to supervise every step of the process. Descartes did maintain that God preserves the universe at every moment, implying that the conservation of matter and motion would cease instantly if God did not act to maintain it. Nevertheless, Descartes's God appears much more like a wise designer than a constant shepherd; Descartes thus left himself open to the charge that God, having created the world, seemed no longer necessary.

Gassendi had a somewhat different conception of the deity. He believed that God had complete freedom to create any sort of world he wished, and that poor mankind, with its feeble intellectual powers, was in no way equipped to deduce anything about the resulting product. We may construct laws of nature, but they are our laws, not God's, and he is not constrained by them. Since we cannot figure out the nature of the universe by rational means, we have no choice but to stick to observable phenomena and perform experiments; and to explain these phenomena, we should devise the simplest, most straightforward hypothesis available. For Gassendi, that hypothesis was atomism. We can imagine atoms, and how they move, and how they interact, and we explain virtually every natural phenomenon by these atomic interactions. But that is as far as we can go; we can never know the underlying mechanisms with certainty.[18]

Hobbes and the Early Signs of Danger

Although both Gassendi and Descartes emphasized the role of God in the creation and maintenance of the universe, while excluding the soul from mechanical explanations, it was apparent, from the example of Democritus himself, that the mechanical philosophy might be theologically dangerous in the wrong hands. The danger became manifest in the work of Thomas Hobbes (1588–1679). Hobbes became an atomist in the 1640s, perhaps as a result of encounters with Gassendi, and by the time of the publication of his great *Leviathan* (1651), he had become a thoroughgoing materialist. That is, Hobbes maintained that the universe consists of matter in motion and nothing else. There are no forms, no spirits, no nonmaterial entities in Hobbes's world. There are no forces other than impact. Souls exist, but they are material, rather than spiritual. Even God himself is a material rather than a spiritual being. Hobbes denied the possibility of free will. He ruled out all absolutes; good, evil, justice, and injustice became relative terms, which acquired meaning only through human definition. While Hobbes was not an atheist, he held religiously dangerous views, including doubts about the authenticity of Scripture. As a result, the "monster of Malmsbury" came to represent, in the eyes of many, the religious peril presented by a mechanical philosophy.[19]

There were several ways of dealing with the threat of Hobbes while still retaining a mechanical philosophy. One could argue that mechanism alone is not sufficient to explain the world—that spiritual entities of divine origin are necessary. This would be the route taken by the Cambridge Platonists. Or one could argue that a functioning mechanical universe would be possible only if God designed it and that indeed mechanism demonstrates the wisdom and providence of God. This would be the path taken by Robert Boyle.

More, Cudworth, and the Spirit of Nature

The Cambridge Platonists, as their name suggests, were a small group of scholars who taught at Cambridge University in the 1640s and 1650s and who preferred the writings of Plato (427–347 B.C.), and of Neoplatonists such as Plotinus (A.D. 205–269/70), to the more traditional Aristotle. They sought to establish a rational Christian religion, arguing that all of the tenets of Christianity could be demonstrated by reason from Platonic principles. Among the group who went by the name of Cambridge Platonists, two in particular wrestled with the religious implications of the mechanical philosophy: Henry More and Ralph Cudworth.

More (1614–87) was a fellow at Christ's College, Cambridge, who discovered Descartes in the mid-1640s.[20] He was attracted immediately by Descartes's version of the mechanical philosophy. He admired the way it explained, so rationally, the mundane properties of matter, and More was largely responsible for introducing Cartesian natural philosophy to England, where it would flourish. But by 1650 More had grown critical of Cartesianism because he thought that mechanism alone could not explain many phenomena, such as gravity or magnetism. There was no way, More argued, that the interaction of colliding particles in a vortex could produce a downward tendency in an object, as Descartes maintained. There must be something else in the universe, something nonmaterial and nonmechanical that provides direction in such instances. More proposed that such a substance exists, which he called the "Spirit of Nature." This spirit was incorporeal and devoid of will or reason; it pervaded the whole universe and directed matter to produce those effects that could not be produced by purely mechanical means. The Spirit of Nature was, in effect, God's agent in the world; it was the means by which God gave life to a mechanical universe without having to watch over it every minute.

This idea, that a mechanical universe cannot function successfully without the presence of immaterial spirits, would prove popular in the second half of the seventeenth century, especially in England. It seemed to provide a way of refuting the materialism of Hobbes by arguing that matter in motion is *not* sufficient to explain the world.[21] More's fellow Platonist Ralph Cudworth (1617–1688) wrote *The True Intellectual System of the Universe* (1678), a massive treatise in which he argued that the mechanical philosophy, properly understood, was not a threat to religion, but a new support, because it demonstrated the necessity of spiritual agents (fig. 3.5).[22] Cudworth called his principal agent "plastic nature," and he pointed to one of the advantages of such a nonintelligent agency—that it could explain mistakes. Accounting for imperfections in the world was always difficult when they were ascribed directly to God; they were much easier to accept if they were regarded as the result of a blind immaterial force like plastic nature.

Boyle and Corpuscularianism

Robert Boyle (1627–91) (fig. 3.6) was just as appalled as the Cambridge Platonists by Hobbes's materialism and by the apparent threat to revealed religion of an uncompromising mechanical philosophy. But Boyle found a different way of dealing with the problem.

Boyle is best remembered as a proponent of an experimental approach to studying nature. In his well-known experiments with air pumps, he discovered the elasticity of the air (and Boyle's law) and demonstrated the role of air in combustion and respiration.[23] He has also been hailed as one of the founders of modern chemistry and an opponent of alchemy (although recent research has revealed that Boyle had quite an interest in higher, as opposed to vulgar, alchemy). At heart, however, Boyle was a lay theologian who pursued scientific inquiry because it demonstrated to him the existence, wisdom, and attributes of God. We will say more about this after we examine the road that led Boyle to the mechanical philosophy.

Like many of his generation, Boyle was bothered by the forms and qualities that were used by Scholastic Aristotelian philosophers to explain ordinary phenomena. To say that a substance is red because it has the form of redness, or that

Figure 3.5. The Cambridge Platonists had difficulty reconciling Epicurean atomism with Christianity. Ralph Cudworth put Epicurus among the "atheists" on the frontispiece to his *The True Intellectual System of the Universe* (London: Printed for Richard Royston, 1678). (Courtesy of the Linda Hall Library of Science, Engineering and Technology)

an object falls because its substantial form includes the quality of gravity, seemed to Boyle to be less an explanation than an act of labeling. We don't really understand why acids burn or roses smell if we simply attribute their effects to forms that exist only as names.

In a treatise that he wrote specifically on this subject, *On the Origin of Forms and*

Qualities, Boyle proposed that a theory of matter, to be useful, should have explanatory power. It should help us understand how an effect can occur. It should be plausible and intelligible, and it should be as simple as possible. Notice that Boyle does not say that it should be true. To discover the true constitution of matter is probably well beyond the poor powers of human reason. But we can certainly do better than substantial forms.[24]

The philosophy of nature that Boyle settled on was the mechanical philosophy, but it was not the version proposed by either Gassendi or Descartes. Boyle did accept as a working premise that all things can be reduced to matter in motion, which is the heart of a mechanical philosophy. But he did not commit himself to unbreakable atoms, or to Descartes's infinitely divisible matter having the sole property of extension; he thought that in practice there were smallest units, which he called corpuscles, but in theory there was no reason why these could not be further and indefinitely divided. Boyle accepted the void; indeed, he was

Figure 3.6. Portrait of Robert Boyle (1627–91), from *The works of the Honourable Robert Boyle,* ed. Thomas Birch (London: Printed for J. and F. Rivington, 1772), vol. 1, frontispiece. (Courtesy of the Linda Hall Library of Science, Engineering and Technology)

one of those who demonstrated its existence. Boyle called his version of the mechanical philosophy "corpuscularianism."[25]

Boyle embraced corpuscularianism for several reasons. One reason was that it made sense of experience in a plausible manner. It is hard to understand how an acid works if you attribute its action to the form of acidity. But if you imagine an acid as a tiny sharp corpuscle that can move between two other corpuscles and force them apart, then you have a mechanism that the human mind can understand. Heat can be visualized as the collective motion of many corpuscles. Cohesion might be the result of particles that lock or stick together. The resulting world picture was clear and easy to comprehend. For Boyle, that met the test of an acceptable natural philosophy.

Boyle, however, had another, and perhaps a prior, reason for defending a mechanical philosophy. If the orderly, organized, purposeful world in which we live is nothing more than an immense system of particles in motion, then it can hardly be the result of chance. It must have been designed by a God of extraordinary wisdom and providence. As Boyle put it:

> It is intelligible to me, that God should at the beginning impress determinate motions upon the parts of matter, and guide them, as he thought requisite, for the primordial constitution of things; and that ever since he should, by his ordinary and general concourse, maintain those powers which he gave the parts of matter, to transmit their motion thus and thus to one another.[26]

Boyle rejected the notion that God had to work through a Spirit of Nature or through any other kind of intermediary:

> It more sets off the wisdom of God in the fabric of the universe, that he can make so vast a machine perform all those many things, which he designed it should, by the mere contrivance of brute matter managed by certain laws of local motion and upheld by his ordinary and general concourse, than if he employed from time to time an intelligent overseer, such as nature is fancied to be, to regulate, assist, and control the motions of the parts.[27]

Finally, it was Boyle who gave us the concept of the world machine. He compared the world to a clock, whose parts were so exquisitely made that it would run perfectly long after the Maker had put it in motion. Boyle conceived this metaphor of the clocklike universe in order to support his belief that a wise God would design a world that does not need constant adjustment. But it was soon put to use as a proof of the existence of God. After all, if you found a watch on the ground, you would naturally infer the existence of a watchmaker. Then certainly a universe that runs like the most exquisite clock mandates the existence of a Creator God. Moreover, a mechanical world is not only proof of the existence of God, but testimony to his wisdom and providence. It is the best argument going against atheism.[28]

Boyle's view that the study of nature provides evidence of God's existence and attributes would lead to a movement known as natural theology. Its aim was to show that the "Book of Nature" could supplement the "Book of Scripture" in leading us to God and demonstrating his wisdom and foresight. Boyle himself left a sum of money upon his death in 1691 to fund an annual series of lectures, known as the Boyle Lectures, whose purpose was to use the Book of Nature to prove and teach the tenets of Christianity. These lectures would be one of the principal vehicles for natural theology up through Darwin's time in the nineteenth century.[29]

Although Boyle continually argued that the design of nature evidenced the wisdom of God, it is important to note, as a final point, that Boyle's God (like Gassendi's) had total freedom to create any kind of universe he wished. He was wise and providential, but he was not constrained by anything, certainly not laws of nature. There may be a general order of nature, but Boyle had no problem with God's tinkering with that order naturally, or even intervening miraculously. Indeed, Boyle believed that there are many things about the world that we will never understand because they are beyond the limits of human reason.[30]

The other interesting feature of Boyle's mechanical philosophy, a feature that also arose from his theological concerns, is its ecumenical character. Boyle tried to formulate a natural philosophy that all Christians could adopt. He carefully avoided taking a stand on the existence of atoms, so that his corpuscularian theory could be adopted by both Cartesians and Gassendian atomists. He never insisted on the truth of any of his mechanical explanations, offering them instead as possible and plausible. He even tried to appeal to Aristotelians and Paracelsians, by pointing out that the four elements, or the three principles, could themselves be reduced to corpuscles. His goal was simply to formulate a view of nature that allowed us to understand and marvel at the wonder of the created order, so that we might better appreciate the glory of the Creator. He thought that a mechanical philosophy fulfilled that aim.

Newton and the Limitations of Mechanism

Mechanical philosophers before Newton disagreed a great deal, about whether matter was atomic or infinitely divisible and about the existence or nonexistence of empty space. But they generally agreed on one principle: that bodies interact only by contact. Any other means of passing motion from body to body was deemed magical, and thus suspect. Action at a distance was nothing but a sympathy and was not to be entertained by any serious mechanical philosopher.

Isaac Newton (1642–1727) (fig. 3.7) challenged this basic assumption by redirecting the mechanical philosophy to a new focus: that of forces.[31] He proposed to explain the operations of nature by various forces that could be mathematically described, and he did so with great precision in his *Principia mathematica* (1687). The mathematical treatment of forces does not require any inquiry into their origin, and Newton generally avoided discussing causes. But it was apparent to Newton, and to his readers, that the forces that produce such effects as fermentation, animal motion, and gravitation could not be ascribed to mechanical impacts. This would be a severe problem for more orthodox mechanical philosophers.

Newton converted to the mechanical philosophy around 1664, during his undergraduate years at Cambridge. His formal education was Scholastic, and

Figure 3.7. Portrait of Isaac Newton (1642–1727), from his *Philosophiae naturalis principia mathematica*, 3d ed. (London: Apud Guil. and Joh. Innys, 1726), frontispiece. (Courtesy of the Linda Hall Library of Science, Engineering and Technology)

thus Aristotelian, but he introduced himself to the works of Gassendi, Walter Charleton (1620–1707),[32] Hobbes, Boyle, More, and, most importantly, Descartes, and he was soon won over by the explanatory power of matter in motion. Although initially enamored of Descartes, Newton found problems in Descartes's theories of light and gravity, and he came to prefer some sort of atomic model. He would remain an atomist the rest of his life. At the same time Newton seems to have discovered, probably through More, the dangers of mechanism, with its tendency to exclude spirit from nature. This drove him even further from Descartes.[33]

Initially Newton also accepted the principle that all change of motion is a result of impact, and as he forged his way to a theory of universal gravitation in the 1670s and early 1680s, he assumed that some sort of aether was responsible for the tendency of bodies to "gravitate"—to move toward one another. He had to finally reject this notion when he discovered that the planets do not act as if they were moving through any kind of aether, because observational evidence made it clear that they are not impeded in any way. The Sun attracts the earth, but it does not do so through a material medium. Gravity is not a mechanical force.

Universal gravitation was, of course, one of the great discoveries of the century, but it caused great problems because of its nonmechanical nature. Christiaan Huygens (1629–95), for example, simply could not believe that nature would employ a force that acted at a distance. Huygens had himself discovered several of the physical laws that led to the discovery of universal gravitation, such as the law of centrifugal force, and he could have taken the additional step to the law of gravitation, but apparently his adherence to the mechanical philosophy of Descartes made that impossible.[34] What was it about Newton, equally devoted to a mechanical philosophy, that allowed him to break with one of its fundamental tenets? It has been suggested that Newton's theology might have made it possible for him to accept what others found unacceptable.

Newton's religion was a curious affair. He was fiercely Christian, but he avoided taking holy orders (a customary requirement for a Cambridge fellow) because he could not subscribe to one of the major tenets of the Anglican Church, namely, the doctrine of the Trinity. Newton was an Arian, believing that Christ was the son of God, but not himself divine. He kept his Arianism a secret during his lifetime, but his private writings leave no doubt about his views. Newton was also intensely interested in biblical prophecy, and he believed that God had revealed all of future history in the Books of Daniel and Revelation, if one knew how to read them.[35] Whether Newton's Arianism or his interest in prophecy directly affected his natural philosophy is hard to say. Most historians do not see much connection.

But one aspect of his theology might have influenced his mechanical philosophy. Newton shared Gassendi's and Boyle's view that God had total freedom in his creative work, that he was unconstrained by any laws of nature. Mere humans, with our feeble powers of reasoning, can never know the fine points of the world that God created. He could have done whatever he wished, by whatever means he wished. So it is possible that when Newton's inquiries suggested that bodies attract each other by a force whose mechanism we cannot imagine, he was more inclined to accept it.[36] Newton's God could directly cause the gravitation of bodies simply by willing it, and who are we to deny it as inconceivable? After all, most divine attributes are inconceivable.

Whether or not Newton's religious views helped shape his version of the mechanical philosophy, there is no doubt that Newton used the mechanical philosophy to advance his religious views. Newton was adamantly opposed to the watchmaker God of Descartes, who created a universe that ran by itself. Newton's God, the God of prophecy, was always active in the world, and Newton took it upon himself to show that a mechanical universe could not maintain itself without God's continual supervision. In a famous query that he appended to the Latin edition (1706) of his book on optics, Newton points out that material bodies are passive; they do not move of their own accord. Therefore all of the motions and changes that characterize our world must have some source other than the natures of those bodies. Newton called these sources "active principles" and held them responsible for gravity, magnetism, fermentation, and other apparently nonmechanical forces.

Active principles might seem out of place in a mechanical philosophy; to call something an "active principle" seems little better than to label it a "substantial form." Newton was aware of this, and he speculated during his career about how active principles might operate. At times he inclined to the idea that they were the result of God's direct intervention; at other times he pursued a material explanation, such as an aether, or an immaterial one, such as a spirit. In the end, he could not resolve the problem. But he did believe that active principles exist and that they are God's means of ordering and bringing activity to the world and of exercising divine providence. Newton believed that it was utterly impossible that our universe could have arisen by blind chance through the mere action of passive laws of nature.[37]

Newton went a step further; he argued that without divine intervention, the universe would run down. He calculated that the planetary orbits were inherently unstable, so that the solar system would eventually break down without some adjustment. Newton proposed that comets were the means by which God fine-tuned the cosmos and brought everything back into proper alignment. Gottfried Leibniz (1646–1716), the German philosopher, was appalled at the suggestion that God was such an inferior craftsman that he had to keep tinkering with his Creation to keep it running properly.[38] But Newton was unabashed. Leibniz was in the Cartesian tradition and saw God as the all-wise Creator who could make a universe that would run forever. Newton rejected that vision, because it made God essentially unnecessary once the universe was created, and this meant that it could lead to atheism. Newton preferred a God who was always with us, always reminding us of his presence.

For Newton, then, a mechanical philosophy without God was unimaginable. A universe of blindly moving atoms could never fortuitously produce an orderly world such as ours, and even if it could, that world would rapidly grind to a halt. A God of wisdom was necessary to create it, and a God of providence was necessary to maintain it. The fact that God chose, in this instance, to make a universe that ran like a clock, on mechanical principles, was interesting, because it was a system of nature that humans could understand. However, our understanding is unimportant unless it leads us to continual awareness of the universe as a product of divine handiwork.

Conclusion

Newton, Boyle, Descartes, and Gassendi all subscribed to some version of the mechanical philosophy. They also believed in an all-wise, all-powerful God who had once created and still preserved this universe of matter in motion. None of these natural philosophers saw any conflict between the two beliefs; in fact, one might go so far as to say that they found these two creeds, Christianity and the mechanical philosophy, inseparable and equally necessary.

It is important to appreciate how remarkable and unexpected this development was. The mechanical philosophy had been greeted with outrage by the early church fathers, who regarded mechanism as a path to atheism. It survived the Middle Ages only as an object of abuse. It aroused little interest even in the eclectic Renaissance. And yet in the seventeenth century, the mechanical philosophy was revived, refurbished, and embraced by Christian philosophers, who not only came to see a mechanical universe as intrinsically Christian, but who eventually put mechanism forward as a bulwark against atheism.

4

Matter, Force, and the Christian Worldview in the Enlightenment

Thomas H. Broman

In the minds of most people, historians and nonhistorians alike, the eighteenth-century Enlightenment occupies a pivotal position in the evolving relationship between science and religion. For it was during the Enlightenment that the cultural landscape of Europe was first reshaped in a way that enabled "science" and "religion" to emerge as separate and hostile camps in a long polemical struggle. That the Enlightenment did play this part in defining the relationship between science and Christianity was asserted by no less an authority than Peter Gay, whose two-volume masterpiece, *The Enlightenment: An Interpretation* (1966, 1969), represents the twentieth century's greatest synthesis on the subject. In the eighteenth century, Gay wrote,

> the evidence for a growing disenchantment, a growing component of critical rationalism in the minds of educated Christians, is overwhelming. For religious men sensitive or learned enough to participate in the currents of the century this was a time of trouble. The dangers of atheism and materialism, the threat of secularism, had been cried up for centuries. . . . But in the age of Enlightenment realities seemed to bear out the predictions of the most pessimistic Christians.[1]

The cause of this skepticism and disenchantment toward traditional religious belief, as Gay makes clear elsewhere in his study, was what he labeled "the recovery of nerve," a growing sense that humanity could acquire a scientific comprehension of the world and put that knowledge to use for its own benefit.[2] As a direct result of this renewed confidence, Gay concluded, the Enlightenment became an age of secularization, a time when "religious institutions and religious explanations of events were slowly being displaced from the center of life to its periphery."[3]

Gay's presentation of the Enlightenment as characterized essentially by a confrontation between science and religion was one that many eighteenth-century contemporaries would have shared. As early as 1733, the historian, dramatist, and philosopher Voltaire (the pen name of François-Marie Arouet) observed that a century before, the great mathematician and natural philosopher René Descartes had been persecuted and driven from his native France because of "the darkness of the School [i.e., the Aristotelian philosophy favored by the Catholic church] and the prejudices of popular superstition." By contrast, Voltaire continued, Isaac Newton was venerated in England, his homeland, because he had had the great good fortune "not only to be born in a free country, but at a time when . . . reason alone was cultivated and society could only be his pupil and not his enemy."[4]

Voltaire's dim view of the Catholic church was anything but unique; to the contrary, he was in good company and far from being the most radical member of the club. In his monumental history of the decline and fall of the Roman Empire, for example, the historian Edward Gibbon condemned the "intolerant zeal" of early Christians, while lampooning their beliefs and acts of piety in scornful and occasionally hilarious footnotes. Of Origen, the early third-century theologian who is said to have castrated himself in conformity with Jesus' injunction to amputate parts of the body in avoidance of sin (Matthew 18:8–9), Gibbon dryly remarked that, as it had been Origen's general practice to interpret the Scriptures allegorically, "it seems unfortunate that, in this instance only, he should have adopted the literal sense."[5] The mathematician Jean-Antoine-Nicolas de Condorcet, meanwhile, dispensed with even this facade of urbane wittiness when he composed his *Sketch for a Historical Picture of the Progress of the Human Mind* at the end of the century:

> Disdain for the humane sciences was one of the first characteristics of Christianity. It had to avenge itself against the outrages of philosophy, and it feared that spirit of doubt and inquiry, that confidence in one's own reason which is the bane of all religious beliefs. . . . So the triumph of Christianity was the signal for the complete decadence of philosophy and the sciences.[6]

Small wonder, then, that the Enlightenment's historical identity has been forever marked in the minds of those who have come after it by such uncompromising denunciations. For Peter Gay and other defenders of the modern secular world, there can be no question that the Enlightenment was a positive force in history. The title given by Gay to the second volume of his work, "The Science of Freedom," deliberately invoked images of the Enlightenment as a movement away from superstition and political repression and as a program of free inquiry and criticism loosened from the shackles of religious intolerance. For conservative Christians, in contrast, the Enlightenment placed Western culture on the path of "secular humanism" leading away from God and toward social chaos and spiritual bankruptcy. On one thing alone do the two sides agree: that the Enlightenment initiated the seemingly eternal "war" between science and religion, in pursuit of which so many trees have become unwitting martyrs.

As enduring as these perceptions are, however, they can be deeply misleading, about both eighteenth-century attitudes toward religion and the Enlightenment's place in Western cultural history. The extreme antipathy shown by Gibbon and Condorcet toward Christianity was one that few eighteenth-century intellectuals shared. Voltaire and many others, including the critic and playwright Gotthold Ephraim Lessing, may have railed against the clergy and satirized its pretensions and institutions, but their criticisms of religious belief itself were more muted. Just as important, the eighteenth century also witnessed the birth of a number of major evangelical movements, such as Pietism in southwestern Germany, the First Great Awakening in North America, and Methodism in Great Britain. Thus at the very moment when a group of intellectuals—and a small group at that!—was proclaiming the virtues of science over what they labeled religious "superstition," a much broader stratum of society was turning to religious belief with renewed vigor.

Thus, the very terms by which we comprehend the Enlightenment must be reassessed. If enlightenment represents, in Gay's words, "a growing disenchantment" and secularization, we might do well to consider that our sense of disenchantment did not suddenly begin in the eighteenth century. Instead, its origin might arguably be found as far back as that moment when, according to the Book of Genesis, God gave Adam dominion over the earth. A world subject to human dominion was not just something to be wondered at and venerated as God's divine creation, but also to be understood and turned to human use. Therefore enchantment and disenchantment, religious belief and enlightenment, appear to have been simultaneously built into our perceptions of the world almost from the moment when our remote ancestors began reflecting about their place in it.

If the ambivalent stance between enlightenment and enchantment is an ancient one in human experience, what happened in the eighteenth century to sharpen that ambivalence into a conflict between science and religion? This is a complicated question that bears on the aims of Voltaire and other partisans of the movement, and also touches on their conception of "humanity," in the name of which they frequently claimed to speak. We will return to this matter in the conclusion; for the present, however, as good an answer as any might be found in a comment made by Immanuel Kant in his *Critique of Pure Reason* (1781). "Our age," he wrote,

> is in especial degree, an age of criticism, and to criticism everything must submit. Religion through its sanctity, and law-giving through its majesty, may seek to exempt themselves from it. But they awaken a just suspicion, and cannot claim the sincere respect which reason accords only to that which has been able to sustain the test of free and open examination.[7]

In describing his era in this way, Kant made the systematic and most of all the *public* examination of beliefs of all kinds the cornerstone of the Enlightenment. Kant's invocation of publicity suggested that criticism was not to be merely a set of specialized debates among experts, but a broad and serious discussion by the public of its own values, desires, and knowledge. To be sure, it could not have been a matter of indifference to Christian churches that the partisans of Enlightenment were elevating public criticism to the authoritative status once reserved for religious dogma.[8] Not surprisingly, therefore, to the extent that religion was subjected to criticism, it was often because religious institutions attempted to suppress or limit criticism.

As suggested by Kant's allusion to critical practice, the partisans of the Enlightenment deployed science as a tool in aggressively combating the authority of Christian churches in social and cultural life. They did not simply discover what had previously been a latent conflict between "science" and "religion" and bring it to light; to the contrary, they created the conflict and trumpeted it as a way of supporting their own notions of truth. Thus, the conflict between science and religion that we believe to have broken out during the eighteenth-century Enlightenment does not necessarily represent an inevitable confrontation between two antagonistic ways of looking at the world. And to the extent that we continue to believe in a struggle of this kind, we remain trapped by the categories created and bequeathed to us by Voltaire, Gibbon, and the other critics of Christianity.

A very different view of the relationship between science and religion can be obtained simply by looking closely at the kind of scientific work done in the eighteenth century. When we do so, we soon perceive that the pursuit of science, far from representing an alternative to religious belief, was in fact deeply tied into the most significant metaphysical and religious questions of the time. Broadly stated, eighteenth-century natural philosophers were obsessed with the nature of forces such as gravity and electricity. The problem posed by such forces was whether physical matter was endowed with inherent forces, and therefore in some sense self-sustaining and self-moving, or whether the forces acting on matter could be thought of as external to matter itself. The ramifications of this problem found expression in subjects as diverse as chemistry, mechanics, medicine, and animal reproduction. In what follows, we will examine the problem of force in the eighteenth century from three viewpoints. First, we will explore the problem at the most general level by looking at how forces were thought to cause motion and gravitational attraction. Then we will take up the most highly charged and contentious debates on this subject, those centering on how forces work in animal movement and reproduction, and the implications of living forces for the nature of the soul. Finally, in the last section we will examine what were known as the "subtle fluids" as objects of natural-philosophical study and popular fascination, before turning again to the larger cultural consequences and resonances of these debates.

Force and Matter

It is an item of everyday experience that bodies collide with one another. Pool players use a cue stick to strike one ball, setting it in motion, so that the latter may strike another. Baseball players hit a thrown ball with a bat, reversing the movement of the ball. What exactly is happening in these collisions? Does one body, such as the cue stick, contain an inherent capacity for motion that is transferred to the ball it strikes? Or does the motion exhibited by the billiard ball at the moment of collision result from a source other than the cue stick? To us this might seem a pointless question, but to eighteenth-century students of nature (the "natural philosophers" referred to above) it touched on a crucial problem: God's role in producing changes in the world. If we suppose that God is all-knowing and all-powerful, one position we might take is that God directly controls everything that occurs, including motion. On this basis, then, the cause of motion when one body collides with another is not *transfer* of motion from the one body to the other, but the *creation* of motion by God in the second object and its annihilation, also by God, in the first. However, we might just as reasonably suppose that God does not create these new motions themselves, but instead has created a certain *capacity* for motion—which we can call "force"—that is present in a moving body and is transferred to another body at the moment of collision. Bodies would then go around knocking into one another without God's constant participation. Of course, the most disturbing possibility to Christians would the supposition that God has nothing at all to do in this business, and that all of the qualities exhibited by matter in motion are to be understood as arising from matter itself.

The problem of the causes and explanation of motion were certainly not novel in the eighteenth century, having already attracted considerable attention in antiquity and the Middle Ages. Yet for the purposes of scholars writing in the Enlightenment, the problem had been given a distinct formulation by Newton in his *Principia mathematica* (1687), where gravity is described as a force that makes bodies move toward each other in mathematically determinable ways. In Newton's view, if a planet is seen to orbit a much larger body, as the earth orbits around the Sun, this happens because there is a gravitational force (the existence of which Newton could demonstrate mathematically, although he could do no more than speculate on its physical cause) that deflects the earth at every instant from its natural tendency to move in a straight line, thereby bending it into an elliptical path around the Sun.

Newton's mathematical description of motion, which provided a unified account of celestial and terrestrial motion, was widely praised. But the concept of gravity itself was incomprehensible to many natural philosophers, leaving a gaping hole in the picture of the physical world created by Newton. In effect, Newton's treatment of motion reproduced the same old problem of the causes of motion, but now under the guise of gravity. How, asked the mathematician and philosopher Gottfried Wilhelm Leibniz, among others, could bodies possibly act on each other when separated by empty space? To Leibniz, Newton's use of gravity was little different from conjuring up a secret power in matter to account for this effect. For his part, Newton attempted at first to avoid the metaphysical uncertainties of gravity by claiming that it was only a mathematical description of planetary orbits in the heavens and falling bodies on Earth, not a theory of physical nature. This satisfied almost no one: not Leibniz, who soon entered into a long polemical exchange over the question with Samuel Clarke, one of Newton's champions, and not even Newton himself, whose later writings weighed the possibility of motion being communicated from one planetary body to another via a subtle, weightless, but nonetheless very material medium.[9]

As originally conceived by Newton, gravity appears to have functioned as a shorthand designation for the invariable tendency of bodies to accelerate toward each other. The cause, solid bodies, and the effect, acceleration of motion, were linked by the term "gravity." Newton never supposed that gravity was a real substance, a kind of fairy dust sprinkled over bodies to make them accelerate toward each other. To the extent that he sought to explain it at all—and he tried mightily to avoid the subject—Newton accounted for a force like gravity by picturing a physical world that was entirely dependent on God's volition and active intervention. According to Newton, God had created a world containing what he labeled active and passive principles. Matter alone was inert and passive, and without God's intervention through active principles such as gravity, Newton declared, "the Bodies of the Earth, Planets, Comets, Sun, and all things in them would grow cold and freeze, and become inactive Masses."[10]

Newton's emphasis on providing an accurate *description* of what happens in motion, rather than an *explanation* for the phenomenon, found an echo in John Locke, whose *Essay Concerning Human Understanding* (1690) would become one of the most influential works of philosophy during the Enlightenment. There, Locke investigated what he called the idea of a "power." Powers, he argued, are not intelligible in our minds as real things in themselves. Instead, the idea of power refers to the relations that we observe among objects whereby one object appears to produce changes in another.[11] Like Newton's use of gravity, therefore, Locke applied "power" as a term linking cause and effect.

To Leibniz, all this was nonsense, first because Newton lacked an explanation for why motion under the influence of gravity occurs, and second because Newton's theology was no less suspect than his physics. In Leibniz's view, Newton's system reduced God's majesty (not to mention God's wisdom) by supposing him to have created a world that needed constant support and maintenance. Against this model, Leibniz supposed God to have designed a perfect world system that was animated by a real force actually present in matter. He named this force *vis viva* (living force) and claimed that its total quantity in matter was a constant. It could be transferred between bodies, but never created or destroyed.

The argument over the reality of gravity and *vis viva* persisted with undiminished vigor throughout much of the eighteenth century. One group, represented by the French mathematician Jean d'Alembert, held the entire debate over the physical nature of *vis viva* and other forces to be meaningless. While the concept of *vis viva* could well be useful in certain limited contexts, d'Alembert conceded, the question of the material reality of this *vis* was irrelevant to its utility in studying the mechanics of motion. D'Alembert's own approach was to treat motions analytically as cases of equilibrium between bodies—in effect, dynamics was transformed into statics and the world rendered as a huge and elaborate balance. The use of such methods did not demand that d'Alembert invest anything in the reality of forces, because "force" as a concept virtually disappeared from view. Not surprisingly, d'Alembert derided forces as "obscure and metaphysical entities" that create murkiness in a science that ought to be "full of clarity."[12]

An opposing position was held by Leonhard Euler and Pierre-Louis de Maupertuis, who argued that forces properly understood were not only useful but essential for mechanics. They developed the principle of least action, which argued that nature (and God) always acted in the most economical manner possible. In line with such commitments, Euler developed the calculus of variations as a mathematical technique for describing the motion of a body or system of bodies under prescribed conditions, with least action being the condition of greatest interest. Thus, in distinct contrast to d'Alembert, for whom mechanics amounted to an instantaneous snapshot of physical systems, Maupertuis and Euler intended their mechanics to provide an account of motion under the causal influence of forces acting on bodies.[13]

The wide-ranging controversies over force and motion that have been described here posed fundamental problems for European scholars concerning the nature of motion and the goals of natural philosophy. The word "nature" in the preceding sentence has been deliberately chosen to be vague, for it signifies the profound uncertainties that attended attempts to produce a scientific understanding of the world. In studying the "nature" of motion, was one obligated merely to offer the most exact possible *description,* as Newton and Locke believed, without attempting to account at some more basic level for motion? Or, as Leibniz held, could there be no real science of motion unless its descriptive elements had been grounded in a set of concepts that also comprehended the *causes* of motion? Lurking just beneath the surface of these disputes were some difficult theological issues as well, especially concerning whether the world required God's direct intervention for its operation (what came to be known as the "theistic" position), or whether God had created a world that sustained itself (the "deistic" position). The appearance of this uncertainty over and over in various forms throughout the century suggests that it was one of the deepest and most vexing problems encountered by natural philosophers of the time. And, as we shall see in the next section, its consequences for the understanding of animal life provided some of the most acrimonious disputes of the period, as well as some of the most daring statements of religious heterodoxy.

Animal Generation and Active Matter

Historians may well argue over what would qualify as the greatest scientific discovery of the eighteenth century, but to people living at the time one such event would probably stand out well ahead of the others: Abraham Trembley's discovery of the hydra in 1740 and 1741. The Swiss naturalist first came across the little freshwater polyp in a sample of pond water he had collected while on a stroll with the two children for whom he was employed as a tutor. Interested mainly in the aquatic insects in his sample, Trembley at first paid scarcely any attention to what he thought were the plants he found clinging to his jar's sides. But slowly he began to notice some odd things about them: the plants had tiny arms that would gently wave in the water, even when the water was apparently still, and the appendages would contract suddenly if the jar was jostled. More remarkably, the polyps seemed to be able to migrate toward the light, for if Trembley set the jar in a window and rotated it one half turn, the little beings would begin walking head over foot toward the other side (fig. 4.1). Needless to say, these and other characteristics caused Trembley to doubt that he was observing a plant. But the most astonishing discovery of all was still to come: Trembley found that when he cut one of the little critters in two, the two parts grew back into two complete organisms. "I first thought of the feet and antennae of crayfish that grow back," he wrote, "but the difference is that the two portions . . . seem actually two complete animals; in such a manner that one could say that of one animal, two have been produced. There is much resemblance between what happens to this animal, and plants which grow from cuttings."[14]

The last line in Trembley's statement hints at the profound uncertainty created in his mind by the discovery. He had undertaken to cut the hydra in two in order to settle the question of whether he was dealing with a plant or an animal. On the assumption that plants can regenerate from cuttings while animals mostly cannot, he had fully expected the divided parts to die. But the hydra refused to cooperate. As a result, much of the scholarly community in Europe would soon join Trembley in seeking an explanation for the astonishing properties of the little polyp.

At the most basic level, the bewilderment caused by the hydra can be understood by the doubt it cast on widely held beliefs about animal reproduction and generation. By the end of the seventeenth century, many European scholars had reached agreement that sexual reproduction in animals involved the growth of individuals that exist fully formed in the mother's egg, with the male sperm serving to activate this tiny latent being. The doctrine of preformation made considerable sense because, in the first place, it fit nicely with Christian beliefs that God had laid down the framework of all future generations at the moment of Creation. Every living thing that would ever exist had been both conceived of and realized in the vast, creative act described in the Book of Genesis. All the generations of organisms that would ever exist were nested within one another, like an infinite series of Russian dolls, awaiting only the proper moment to begin growing into an adult.[15] The development of new organisms was therefore a matter of enlargement.

Thus, the first source of support for the doctrine of preformation was an impeccably orthodox one. A second kind of support for preformation would strike us at first glance as more religiously suspect: the idea, originating in the seventeenth century, that animals were machines, governed by the same laws of mechanics as inorganic bodies. One good example of this kind of thinking about bodies as machines came in the writings of German physician and university professor Friedrich Hoffmann, who defined medicine as "the art of properly utilizing physico-mechanical principles, in order to conserve the health of man or to restore it if lost." To clarify his meaning, Hoffmann added that "size, shape, motion, and rest are the entire basic states of simple bodies. From these, therefore, the reasons for all natural phenomena and effects are to be sought."[16] Hoffmann's

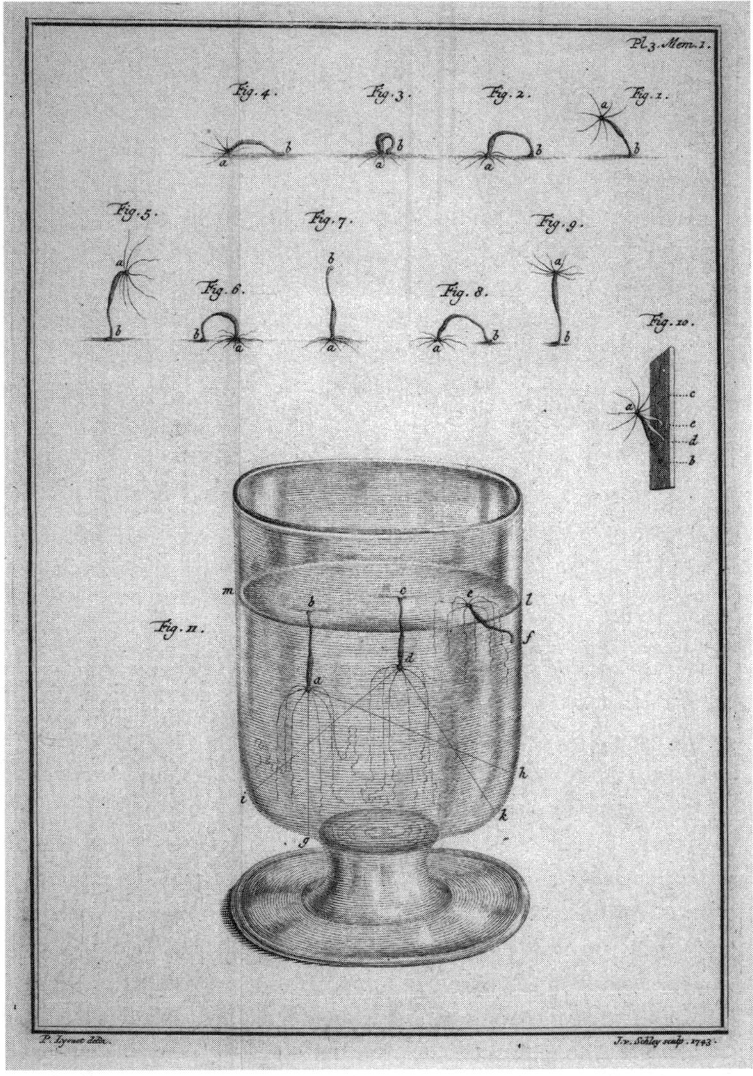

Figure 4.1. The hydra in its various configurations. From Abraham Trembley, *Mémoires pour servir à l'histoire d'un genre de polypes d'eau douce, à bras en forme de cornes* (Paris, 1744). Photo courtesy of the Department of Special Collections, Memorial Library, University of Wisconsin-Madison.

contemporary, the Dutch university professor Hermann Boerhaave, took this line of thinking even further, describing the body's solid parts as

> either membranous pipes, or vessels including the fluids; or else instruments made up of these, and more solid fibers . . .; whenever they shall be put into motion we find some of them resemble pillars, props, cross-beams, fences, coverings, some like axes, wedges, levers, and pulleys; others like cords, presses, or bellows; and others again like sieves, strainers, pipes, conduits, and receivers; and the faculty of performing various motions by these instruments, is called their function; which are all performed by mechanical laws, and by them only are intelligible.[17]

Therefore, for scholars who tended to think of the body in terms of mechanical principles and to liken bodily functions to the workings of machines, the idea that the body-machine existed preformed in each succeeding generation was much more acceptable than the alternative doctrine of epigenesis, which held that new organisms arose from undifferentiated blobs of organic matter. It also permitted them to maintain a clear distinction between the machinelike body and the soul, which was consistent with both Catholic and Protestant teachings.

This vision of human and animal bodies as machines found a considerable number of opponents. Critics such as Georg Ernst Stahl, Hoffmann's colleague at the University of Halle in Germany, pointed out that mechanical models of living organisms were utterly unable to account for what appeared to be the coordination and goal-directedness of animal functions. In Stahl's view, the agent responsible for this coordination is the *anima,* a term that, since antiquity, had been used to designate the source of vital functions in animals. The *anima* was not the soul of Christian theology—that would be the *spiritus.* The *anima* was thought to be the principle that distinguishes animals from plants by allowing them to display hunger, responses to stimuli, and so forth. Outside the Latin-drenched debates between scholars, meanwhile, the French man of letters Bernard de Fontenelle targeted the machine analogy's weaknesses in a characteristically pithy remark. "Do you say that beasts are machines just as watches are?" he asked. "Put a male dog-machine and a female dog-machine side by side, and eventually a third little machine will be the result, whereas two watches will lie side by side all their lives without ever producing a third watch."[18] Yet despite its inadequacies, the machine metaphor expressed something significant about the ways in which Europeans tended to think about animal (and human) bodies: as admirable products from the workshop of a Master Craftsman, indeed as machines considerably more perfect and wonderful than the most intricate watch.

However wonderful these samples of God's handiwork were, the hydra appeared not to have come from the same workshop. Its ability to reproduce from pieces appeared to undercut the divine wisdom encapsulated in the doctrine of preformation. Observations by other naturalists soon began to pile up further uncertainties. Shortly after Trembley's discovery, Charles Bonnet, a fellow Swiss naturalist, began a series of experiments on the regenerative capacities of snails and other invertebrates. He discovered that certain snails could grow entirely new heads after decapitation, and that certain worms, even after being chopped into several dozen pieces, could eventually form into several dozen complete organisms. This work attracted considerable notice and comment, although Bonnet himself, a devout Calvinist and a firm believer in preformation, did not interpret his own work as undermining preformation theory. Instead, Bonnet opined that the minute "germs" capable of re-creating complete adult animals were not only to be found in the sperm and ova involved with sexual reproduction, but also distributed over the entire body.[19]

Doubts planted by the hydra about reproductive preformation were serious enough, but still more troubling were questions that lurked beneath the surface of the debate. If new organisms could be created without having been preformed in the parental ovum or sperm, how in fact did they come about? Bonnet's idea that germs of new organisms were spread throughout the body was one possibility, and for Christians an attractive one, because it preserved the essentials of the preformationist theory. But other scholars allowed their thoughts to proceed in a more radical and disturbing direction. Maupertuis, whom we have already seen weighing in over the question of *vis viva,* published his thoughts regarding generation and reproduction in *Vénus physique (The earthly Venus),* a quasi-pornographic little treatise that appeared anonymously in 1745. In the midst of passages extolling the manifest pleasures of sexuality, Maupertuis proposed a theory of animal generation that made attractive forces between particles the moving force in the creation of new offspring.

> Why should not a cohesive force, if it exists in Nature, have a role in the formation of animal bodies? If there are, in each of the seminal seeds, particles predetermined to form the heart, the head, the entrails, the arms and the legs, if these particular particles had a special attraction for those which are to be their immediate neighbors in the animal body, this would lead to the formation of the fetus.[20]

There it was, the most disturbing implication arising from Trembley's discovery: living beings could arise spontaneously from organic matter, without the help of God's creative hand or the presence of an immaterial *anima.* Nor was Maupertuis the only person to present such ideas. Similar ideas were also put forth in the late 1740s by Georges-Louis Leclerc de Buffon and John Needham, who conducted several experiments demonstrating that tiny "animalcules" could arise from different kinds of organic matter that had been placed in tightly sealed flasks and thoroughly heated. Needham also found that under the microscope the black, powdery material from blighted wheat contained numerous white fibers. When water was applied to the sample, those white fibers appeared to become animated. Like the hydra's amazing reproductive powers, therefore, these experiments demonstrated what Buffon and Needham believed was the spontaneous generation of tiny animals from organic matter.[21]

What made these results so controversial was not, of course, whether blighted wheat could generate little animalcules, or whether snails could grow new heads. The more general concern was whether animal life consisted of some immaterial principle, such as Stahl's *anima,* that was joined in some unspecified way to organic matter and made animals self-moving and responsive to their surroundings. Do the movements and changes seen in living organisms result from intervention by God, either directly or through some intermediate agent such as the *anima,* or does living matter contain within itself the properties and forces that we call life?

Most troubling of all were the implications that such ideas had for the understanding of the human soul, so central to Christian theology. The belief in such an immaterial principle, which both gave us our unique mental powers and connected us with God, was very old. The ancient Hebrews, for example, described Adam's first awakening as that moment when God breathed on him (Genesis 2:7), an action that transferred a spiritual presence into Adam. The ancient Greeks too believed that humans were inhabited by a kind of "air" *(pneuma).*[22] The centrality of the immaterial soul to Christianity was also underscored by the apostle Paul, whose letters repeatedly pictured humans as a meeting point between a wicked, sinful body and an immortal, godly soul or spirit. The lesson of Jesus' death and resurrection, as Paul never tired of pointing out, was that our material existence had no final hold over us, and that as spiritual beings we too could hope for a final victory over death and sin.

Thus, for many educated Europeans in the eighteenth century, it was but a few short steps from *materialism,* the idea that organic matter possessed the power to organize itself into living beings, through a similar claim about the spontaneous origins of human life and the denial of the immortal soul in Christian belief, and finally to atheism, the complete denial of God's existence. This association between materialism, spontaneous generation, and atheism, it should be noted, did not follow by logical necessity. Instead, the link arose more out of their historical juxtaposition than because materialist doctrines such as spontaneous generation inescapably demanded the denial either of God or of the human soul. John Needham, himself an exiled English Catholic who had taken holy orders, refused to accept that spontaneous generation opened the path to atheism. Needham valiantly declared that the spontaneous generation of organisms could just as well be part of God's divine plan for the world as could doctrines of preformation, and he saw no reason to abandon the idea that humans, at least, possessed an immortal soul. But his voice was drowned out in the excited tumult caused by his and Buffon's work and by the hydra. By the 1740s, when this work began to appear, the association between materialism and atheism had already had centuries to solidify and had been made explicit in works such as *De rerum natura (On the nature of things)*, by the ancient Roman poet Lucretius, and in the unnervingly unprovidential *Leviathan* of Thomas Hobbes. Moreover, the espousal of materialist doctrines by various radical groups during the English Civil War of the 1640s made materialism appear as not just unacceptable from the standpoint of religious orthodoxy, but socially and politically threatening as well.

In light of these traditions, it is scarcely surprising that Needham's defense of the religious orthodoxy of spontaneous generation went largely unheeded. Much more typical was the position held by the Italian priest and naturalist Lazzaro Spallanzani, whose almost visceral rejection of spontaneous generation prompted him to undertake a series of carefully devised experiments to refute Buffon's and Needham's claims. Spallanzani was satisfied that his own results conclusively demonstrated that what Buffon and Needham had believed was the generation of animalcules had actually been caused by inadequate precautions against contamination of their plant and animal samples. But as subsequent events during the nineteenth century would show, the doctrine of spontaneous generation and its associated imputations of atheism and political radicalism could not so easily be driven out.

Keeping Body and Soul Together: Albrecht von Haller and Irritability

As wide-ranging and important as these debates over animal generation and reproduction were, they formed but one variant of a still larger debate over the forces of living matter. And just as the publication of Newton's *Principia* had given the debate over the forces of inorganic matter its distinctive contours for eighteenth-century natural philosophers, so too did the work of a single gifted experimenter and natural philosopher, the Swiss physician Albrecht von Haller, place the general problem of organic forces at the center of scholarly attention. Haller was a complex and unhappy man. An insomniac who became addicted to opium to combat his sleeplessness (and later wrote a medical treatise on opium addiction), Haller regarded his own life as a succession of miseries. Yet he was also possessed of deep religious piety and talents so varied and considerable that he became known to posterity for two completely different reasons. To historians of science and medicine, Haller is known as the discoverer of the forces of irritability in muscles and sensibility in nerves. But to literary scholars, he is one of the most prominent German poets of the late 1720s and 1730s, someone whose works communicated lofty ideas and moral lessons in vivid and evocative verse.[23] As disparate as they are, Haller's scientific work and his poetry were both the products of a person whose entire life consisted of a sincere and often anguished attempt to comprehend the world as God's creation.

Haller's first immersion in the problems that would later occupy his attention came during his time as a medical student at the University of Leiden. In the mid-1720s, when Haller studied there, Leiden was arguably the leading center of medical education in Europe, owing chiefly to the renown of Hermann Boerhaave, whose advocacy of mechanistic explanations of animal function we have already encountered. Like his teacher, Haller believed in the mechanical explanation of natural phenomena and in the doctrine of preformation. But Haller soon parted company with Boerhaave in his thinking about the role of forces in nature. The impetus for this development in Haller's thinking, as for everybody else, was news of Trembley's polyp. Haller abandoned belief in preformation and began considering whether organisms could arise from organic matter in the manner suggested by Buffon, Needham, and Maupertuis.

Haller's flirtation with these ideas did not last more than a few years, as evidenced by his critical comments on Buffon's theory of generation, which appeared in volume two of a German translation of the French naturalist's *Histoire naturelle (Natural history)*. Buffon had proposed that both males and females produce "semens," which mingle in creating new organisms. The offspring's specific resemblance to its parents arose from what Buffon labeled the "internal mold" *(moule intérieur)*, which provided the pattern for how the particles from parental seminal fluids would be organized in the offspring. Buffon had also proposed what he described as "penetrating forces," forces analogous to the attractive forces that cause chemical substances to react in characteristic ways, as the agents responsible for actually creating the newly organized embryo.[24]

Haller rejected Buffon's theory as inadequate for explaining how organized offspring can be created out of parental sperm and ova and how species can remain true to form through many generations. In the first place, he saw no anatomical evidence for the existence of a maternal semen. Moreover, the internal mold was no real explanation at all, because it failed to account for how just the right particles were brought to their proper position in the developing organism. Even were there such a mold, Haller continued, the penetrating forces to which Buffon had assigned the task of assembling the new organism were incapable of carrying out the work. "I do not find in all of nature," he wrote,

> the force that would be sufficiently wise to join together the single parts of the millions and millions of vessels, nerves, fibres, and bones of a body according to an eternal plan. . . . M. Buffon has here the necessity of a force that seeks, that chooses, that has a purpose, that against all the laws of blind combination always and infallibly casts the same throw.[25]

Yet although Haller rejected Buffon's theory of generation, the idea that forces analogous to Newton's gravity or the forces of chemical attraction could usefully be invoked in animal physiology clearly appealed to him. In various publications that appeared in the 1740s, he progressively articulated the idea that different forces could be found in specific anatomical structures. Finally, in 1753, Haller produced what would be his most influential and controversial statement on the subject of vital forces, in a paper titled "De partibus corporis humani sensilibus et irritabilibus" *(On the sensible and irritable parts of the human body).* In this relatively brief publication, Haller described two distinct vital forces: irritability, located in the muscles, and sensibility, located in the nerves. Using the same theory of knowledge that had been adopted by Newton and Locke, Haller identified his forces of sensibility and irritability as the regular effects observed in certain experimental manipulations. Irritability, he claimed, was present in any part of the body that contracts when touched with a needle, for example, or when alcohol or caustic chemicals are applied to it. By repeatedly poking and prodding various parts of the body, Haller found that irritability was overwhelmingly centered in muscle tissue. Sensibility, by contrast, presented a more complicated case. Haller defined sensibility as inherent in any body part "which upon being touched transmits the impression of it to the soul." In animals, where the existence of a soul was not so clear, he defined those parts as sensible "the irritation of which occasions evident signs of pain and disquiet in the animal."[26] This ability to transmit pain, he discovered, was the exclusive business of nerves.

By insisting that irritability was uniquely the property of muscles and sensibility that of nerves, Haller attempted to combat Georg Ernst Stahl's claim that vital functions such as muscular contraction depended on the direct participation of the *anima* or some other immaterial cause. Haller argued instead that muscles continued to display irritability even after the animal had died or after the muscle in question had been detached from the rest of the body, situations where there was no question of participation by an *anima.* In a manner similar to Newton's exposition of gravity, Haller sought to avoid speculating on the ultimate source of vital forces such as irritability and sensibility, but in this endeavor he was betrayed by his own experimental method. Like gravity, irritability could be discussed at an exclusively "phenomenological" level, namely as the regularly observed association between stimulus (such as poking with a needle) and response (contraction). But in contrast to irritability, sensibility could be demonstrated only by the registration of the irritation in the experimental subject as "evident signs of pain and disquiet." Sensibility therefore required for its demonstration the appeal to a consciousness or animal soul in the subject that was not directly observable by Haller the experimenter. This made the interpretation of sensibility far more complex.

This fact would ensnare Haller in numerous controversies with his contemporaries. One such dispute pitted him against the Edinburgh medical professor Robert Whytt, like Haller an erstwhile student of Boerhaave's, whose own theory of muscular motion posited a "sentient principle" (analogous to Stahl's *anima*) resident in the nerves and distributed throughout the body. According to Whytt, it was this sentient principle that perceived external stimuli, even in cases where the perceptions never came to consciousness in the brain, and prompted muscles to react.[27] For his part, Haller refused to accept that sensibility or a "sentient principle" could be distributed over the entire body. He insisted that irritability was a force innate to muscles alone and completely separate from sensibility, not subordinate to it, as Whytt believed.

Unfortunately for Haller, the claim that irritability was somehow innate to muscles made him vulnerable to the charge of materialism, because he appeared to be attributing the power of contraction not to something external to matter, such as a soul, but to matter itself. Even before Whytt and Haller began their dispute, a materialist interpretation of Haller's ideas about irritability had been aggressively placed before the educated public by Julien Offray de La Mettrie, another former student of Boerhaave's, in his short treatise *L'homme machine* (*Man a machine,* 1747). Citing a large variety of medical and anthropological evidence, La Mettrie mounted an argument to demonstrate that what was normally presumed to be a mental and spiritual life independent of our physical constitutions was nothing other than the product of the peculiar organization of matter in the human animal. He concluded in the most provocative manner possible:

> Thus the soul is merely a vain term of which we have no idea and which a good mind should use only to refer to that part of us which thinks. Given the slightest principle of movement, animate bodies will have everything they need to move, feel, think, repent and, in a word, behave in the physical sphere and in the moral sphere which depends on it.[28]

As one piece of evidence for this provocative claim, La Mettrie cited the irritability found in muscles. Worse still, from Haller's point of view, La Mettrie dedicated his scandalous book to him, claiming to have been inspired by Haller's early comments on irritability. The deeply pious Haller angrily lashed back, initiating a polemical exchange between the two that was cut short by La Mettrie's early death in 1751. It was lucky for Haller that La Mettrie departed the scene when he did: we can only imagine what he would have done with Haller's paper on irritability and sensibility.[29]

Despite all the caution and painstaking experimentation that went into his work on vital forces, Haller found himself stumbling into the same metaphysical and religious trap that had ensnared Newton. Whether one was speaking of "gravity" or "irritability," it seemed to be nearly impossible to attribute forces *to* matter without simultaneously making those forces an inherent property *in* matter. Haller apparently had hoped to avoid this problem by distinguishing between sensibility, as the soul's portal into the body through the nerves, and irritability, conceived as a force of vital action entirely distinct from the action or participation of the soul. In this way, Haller hoped to have the soul present in the body but not be coextensive with living matter. But the idea of an immaterial soul somehow residing in the brain and in contact with the rest of the body via nervous sensibility was no more easily swallowed by most of Haller's contemporaries than was Newton's idea of gravity as action at a distance.

The Subtle Fluids

As the preceding discussion of the debates over vital forces has suggested, for many natural philosophers the investigation of nature was inseparable from their most basic religious convictions. Not only were "science" and "religion" not in conflict in the thinking of Whytt and Haller (and even La Mettrie); one suspects that those scholars could not have even conceived of conducting their work in the absence of its broader ramifications for religious belief. Of course, this is not to say that Haller, Whytt, and their contemporaries cared only about the religious issues; their disagreements over sensibility and irritability also focused on medical and pathological issues such as the causes of disease.

The complexities of the relationship between force and matter and their implications for religious belief extended well beyond the circles of physicians and natural philosophers whom we have been discussing. The sensation created by Trembley's polyp pervaded not just university lecture halls, anatomy theaters, and academies of science, but also coffee houses, salons, and reading societies. Nor were tiny beasts like the hydra the only source of intriguing questions about force and matter. There was a further class of phenomena that raised the same questions, and did so in vivid and highly accessible ways. These phenomena were provided by the subtle fluids, substances that were believed in the eighteenth century to be lacking the mass and extension through space of normal matter, but whose effects could be studied and manipulated. Any number of such fluids, including light and heat, were believed to exist. But for most eighteenth-century intellectuals, two such fluids stood out as objects of particular fascination: magnetism and electricity. Above all, it was electricity that came to represent for many Europeans the pervasive power of unseen forces at work in the world. As the English natural philosopher and Unitarian minister Joseph Priestley observed in 1775, the study of electricity enjoyed "one considerable advantage over most other branches of science, as it both furnishes matter of speculation for philosophers, and of entertainment for all persons promiscuously."[30]

The capacity of amber and certain other substances to attract feathers, hair, and other bits of stuff after being rubbed with a cloth had been known since antiquity, but it had long remained a curiosity and largely marginal to scholarly interests. A similar curiosity was presented by the lodestone, a peculiar mineral that attracted iron powerfully and, when made into the shape of a floating needle, invariably oriented itself in a certain direction. In the fifteenth and sixteenth centuries, however, the expansion of trade and maritime travel by various European states, along with the consequent need for more sophisticated methods of navigation, focused new interest on the lodestone. The most comprehensive treatment given to magnetism came with the publication of *De magnete* (*On the magnet,* 1600), by the English physician William Gilbert. Gilbert's interest lay primarily with magnetism, as suggested by the title of his treatise, but the similarity between electrical and magnetic phenomena led him to discuss electricity too. In fact, Gilbert devoted a good bit of effort to discriminating between the two. He pointed out that whereas there was an intense, sympathetic bond between lodestone and iron, so that they attracted only each other, amber was far more promiscuous in what it attracted after rubbing. On that basis, Gilbert defined electricity as an invisible fluid that flows from one body to another. He also found that some other substances besides amber displayed the power of attraction.

For eighteenth-century scholars habituated to thinking about forces in nature, electricity resembled gravity in its capacity for acting on bodies at a distance, but with one huge difference. Unlike gravity, which stubbornly resisted all attempts at experimental manipulation, electricity could be produced, transferred, stored and released at will, often with powerful and visually spectacular consequences. This made it an immensely popular plaything and a stock-in-trade for countless scientific lecturers and entertainers in eighteenth-century Europe, and all sorts of parlor tricks and games using electricity were devised by people with ample supplies of both time and money on their hands. One trick involved insulating a boy from contact with the ground, for example by suspending him by silken ropes, electrifying him, and then observing how he could attract little bits of paper into his hand (fig. 4.2). Another favorite variant was to have an unsuspecting gentleman approach a lady who had been insulated from contact with the ground (by having her stand on a small disc of wax, for example) and then charged with static electricity. When their lips came close enough almost to touch, a spark would leap between the two—symbolizing either the power of sexual attraction or the dangers of pressing one's case too ardently. Perhaps most impressive of all was the "beatification" performed on a young boy by Georg Mathias Bose, a German theologian. Bose was said to have created an electrical demonstration where he placed a small boy on resin cakes (to insulate him from the ground) and then began using an electrical generator to begin building up a charge inside the boy. Before long, reported a witness,

> the electrical matter accumulated in such immense quantities inside him that first his shoes, then his legs and knees appeared to be covered in fire. Finally his entire body was bathed in light and surrounded in the manner sometimes used to depict the glory of a saint by encircling him in light.[31]

Electricity seized the imaginations of Europeans—and of Americans such as Benjamin Franklin—in ways both serious and frivolous. The diversity of phenomena it presented, including the way in which electrical charges could be stored in enormous and sometimes dangerous quantities, its transmission through metal wires to locations quite distant from its point of generation, and the visible sparks it made when discharging, all suggested that electricity might indeed be a universal source of action in the cosmos. This possibility did not escape the notice of contemporaries. John Wesley, the founder of Methodism, produced the following speculation in 1759, which identified electrical phenomena with an all-pervasive universal fire:

> It is highly probable this is the general instrument of all the motion in the universe; from this pure fire (which is properly so called) the vulgar culinary fire is kindled. For in truth there is but one kind of fire in nature, which exists in all places and in all bodies. And this is subtle and active enough, not only to be, under the Great Cause [i.e., God], the secondary cause of motion, but to produce and sustain life throughout all nature, as well in animals as in vegetables.[32]

Far from betraying Wesley's ignorance of natural philosophy, his remark was virtually a commonplace among many of the writers who attempted to comprehend the role of electricity in nature. Furthermore, the belief that life itself

Figure 4.2. A demonstration of electricity. Note the bits of paper being drawn to the boy's left hand. From Jean Antoine Nollet, *Essai sur l'électricité des corps,* 5th ed. (Paris, 1771), frontispiece. Photo courtesy of the Department of Special Collections, Memorial Library, University of Wisconsin-Madison.

depended on the actions of an electrical or magnetic fluid received powerful experimental support late in the century from the Italian anatomist Luigi Galvani, who discovered in 1791 that a frog's leg could be made to contract spontaneously when hung from a brass hook, and the hook touched to an iron beam or railing. Because a similar contraction could be observed when the leg muscle was stimulated with an electrical discharge, Galvani concluded, not without reason, that he had detected a kind of "animal electricity" that was the basis of vital actions.

Galvani's work with animal electricity became a topic of intense interest throughout Europe and North America, and scores of other experimenters rushed to explore the new phenomenon. Nor was electricity alone among the subtle fluids in attracting attention to its role in life. The German physician Franz Anton Mesmer developed a fantastically successful medical practice, first in Vienna and then in Paris, by advocating the use of magnets in the treatment of a variety of medical ailments. Mesmer catered to a largely female clientele who sought him out for his amazing ability to channel the currents of magnetic-nervous fluid in their bodies. At the same time, he developed his ideas about what he called "animal magnetism" into a grand doctrine of a universal fluid that was responsible for sustaining all life.[33]

There can be little question that the popularity enjoyed by electrical and magnetic phenomena as objects of scientific investigation stemmed in part from their entertainment value. But we would be mistaken in seeing this appeal as lacking in larger cultural resonances. The popular and scholarly engagement with subtle fluids hints at the same deep-rooted uncertainty over the nature of matter and force that we have seen present elsewhere in the culture of the Enlightenment. In some cases, such as in the writings of Joseph Priestley, the religious implications of the fundamental relationship between force and matter could be made astonishingly explicit. In natural philosophy, Priestley declared in his *Disquisitions Relating to Matter and Spirit* (1777), there could be no significant distinction between matter and force, because every particular "solid" body was in fact the product of the attractive forces that bound its constituent particles together. Since those particles too were bound together by attractive forces, there was an infinite regress toward a vanishing point at which there could be neither "forces" nor "matter," but only a continuity of existence. On this basis, Priestley rejected the traditional Christian belief in the radical disjunction between matter and spirit, the image of humans as both sinful flesh and immortal soul, and the idea of God having three distinct "persons"—as Father, Son, and Holy Spirit. In contrast to this Trinitarian doctrine, a centerpiece of Protestant and Catholic theology, Priestley held to the Unitarian belief of God as having only a single essence. The Unitarian vision of a world supported by an undifferentiated spirit-matter continuum offered an unsettling challenge to the more comfortable duality taught by Christian writers from the apostle Paul down to Martin Luther. But in an era when people came to see themselves as surrounded by insubstantial and energetic fluids, when life itself gave every appearance of being the product of some more fundamental electrical or magnetic power, the old pieties of Christian doctrine began to look ever more vulnerable.

Conclusion: Science, Religion, and Criticism in the Enlightenment

As I hope the preceding discussion has made clear, there is little in the scientific controversies of the eighteenth century to suggest that science and religious belief emerged at that time as inimical ways of looking at the world. Far from rejecting the relevance of religious belief in making inquiry into nature, scholars as diverse as Newton, Haller, and Priestley saw religious belief and scientific inquiry as mutually supportive. Haller could no more set aside his religious beliefs when he studied vital forces than he could his dissecting implements; Priestley's *History and Present State of Electricity* was but one side of his more fundamental concern with matter and spirit. Yet we would be doing an injustice to the Enlightenment if we failed to note that some of its proponents did mount a strong attack on religious beliefs, especially certain Christian ones. La Mettrie may not have been declaring the fundamental impossibility of reconciling science and religion in *L'homme machine,* but he evidently sought to attack the specifically Christian belief in the immortal soul by drawing conclusions from recent scientific work. David Hume, the Scottish philosopher, published a controversial criticism of the reasonableness of miracles in his *Inquiry Concerning Human Understanding* (1748). After showing that no evidence could possibly establish any event as miraculous, Hume concluded that "the Christian religion not only was at first attended with miracles, but even at this day cannot be believed by any reasonable person without one [a miracle]." In short, Hume was saying, no rational person could believe in Christianity.[34]

What then are we to make of these more radical attacks on Christianity in assessing the Enlightenment's place in the longer history of science and religion? One way to understand such attacks as La Mettrie's or Hume's is not so much as an attempt to eliminate religious belief itself, but instead as an attempt to limit the role of the Protestant and Catholic churches in public life. In this effort, critics of Christianity were aided by a dramatic expansion of the space allowed to secular culture during the eighteenth century. Whereas in the sixteenth and seventeenth centuries much of the output of European writers and artists had been concerned with religious themes, that was demonstrably less true after 1700. During the Enlightenment, it became possible to write about the state of the economy, to advocate programs of social improvement, or to describe the latest experiments in electricity without having to situate those discussions in a larger religious context. Religion did not cease to be a matter of supreme importance for educated Europeans; it simply started becoming less a matter of supremely *public* concern.

It may seem paradoxical to claim that an age in which La Mettrie and Hume, along with many other writers, could launch vitriolic attacks on Christianity was one in which religion was becoming less a matter of public concern. But their very ability to mount those attacks and not suffer retribution for them reveals how much religious belief had ceased structuring public life. The same point can be made by looking at the widespread public interest in science. The attention generated by Trembley's polyp and the spectacular powers of electricity indicates both the public's ability to be informed of new scientific developments and its receptivity to that knowledge. The public became consumers of new developments in science and other branches of cultural life by buying experimental apparatus, subscribing to new magazines that followed the latest trends and fads, and enrolling in lecture courses that brought to them the newest ideas about nature.

Yet if religion was becoming less central to *public* life in this way, and if, as we saw at the outset of this article, writers such as Immanuel Kant could declare the Enlightenment above all else as an age of public criticism, it was another matter entirely with religious belief within the private sphere. At the outset of this article, I indicated that during the eighteenth century there arose a number of religious movements, such as Methodism in England, that stressed the cultivation of an intensely personal inner religiosity. For followers of these movements, salvation was not to be found in acts of collective piety and observance, but in the one-to-one bond between an individual believer and God. The effect of such movements was to transfer much of the weight that religion had previously exerted in public life (for example in education, in elaborate religious celebrations, observance of feast days, and so on) onto each person's conscience, as beans can be transferred on a balance, a shifting of weight that underscored the importance of personal conversion.

Thus, instead of hailing or condemning the Enlightenment as the creator of the conflict between "science" and "religion," might we not more usefully understand it as the origin of another fundamental boundary, that between "public" and "private"? Science became a matter of broad public interest precisely because it required none of the intensive self-searching and prayer that the new religiosity demanded. Science was something accessible to enlightened and universal human reason, and as such could be debated, criticized, and propagated in the name of the collective good of humanity. At the same time, matters of faith and salvation began to withdraw into the confines of a single person's relationship with God, a relationship sustained by other beliefs and values that might not find their echo in the wider public sphere. If this line of thinking is correct, it might suggest that what appears to many people to be a basic antipathy between "science" and "religion" conceals within itself a deeper and fundamental tension between public culture and private life.

5

Noah's Flood, the Ark, and the Shaping of Early Modern Natural History

Janet Browne

The story of Noah's Ark and the Flood is deeply embedded in Western minds. As the Old Testament narrates (Genesis, chapters 6 through 9), God was angry at the sinful behavior of humans on Earth and wished to punish mankind. He sent a Deluge for forty days, which covered the earth and killed every being that moved. He looked kindly on Noah, however, and told him to build a ship and to take with him his family and two of "every living thing of all flesh." Noah built his Ark under God's instructions, closed its doors, and was ordered to keep his cargo alive until the ship came to rest and the waters receded. Then he was to release the animals so that they could be fruitful and multiply. Noah, his three adult sons, and their wives, the only humans left alive, were blessed by God. God made a covenant with humans (a promise to all mankind) that he would not send another flood to destroy the world. From the three sons, Shem, Japheth, and Ham, descended all the peoples of the postdiluvial world.

Science, religion, literalism, and allegory have always been closely intertwined in understanding this narrative. Above all, the story presents a powerful parallel with the biblical Creation story—as a Christian message, Noah's Ark is second in theological importance only to Adam and Eve and their expulsion from the Garden of Eden.[1] It is concerned with punishment and deliverance and God's promise to mankind. Just as significant, it foreshadows the story of Christ, presaging the foundation of the Christian church and the rites of baptism.[2] Over the centuries, Noah's story has helped Christian thinkers of every denomination to examine and explain the basic tenets of their faith, especially the central doctrine of salvation.

As a more practical message, moreover, it also provides an account of the early history of mankind and other living beings and their point of origin on Mount Ararat. It tells of the natural processes of death, reproduction, population growth, and migration. Furthermore, the story carries lasting cultural content in its depiction of total global destruction, accompanied by the preservation of a small group of living beings who go on to restock the earth. This motif of destruction and salvation is widely understood, far beyond the walls of a seminary or church: few science fiction stories, for example, could operate without this scenario. And the long artistic tradition, in paintings and sculpture, of depicting Noah riding out the Deluge should not be ignored. From the simple red and black wall paintings in medieval English churches to early printed books with woodcut illustrations and the work of the world's greatest artists, these pictures of Noah made—and make—the Christian story accessible and full of meaning. Even the toy Arks crafted for children carry something of this educational symbolic meaning.

Such a deep-seated theme in Western culture has naturally generated a wide variety of scholarly comment. Modern biblical scholars note that the original Deluge story is based on two narratives, one composed rather later than the other, giving rise to important discrepancies in interpretation. Anthropologists consider the Flood as a highly significant foundational legend, or primal myth, shared in many ways with different cultural groupings: creation stories based on water are very common, appearing in ancient North American, Hindu, Polynesian, and Babylonian tales.[3] Archeologists have made several attempts to locate possible sites for the final resting place of the Ark or evidence for a period when the ground was extensively covered with water, and newspapers occasionally report significant wooden finds.[4] And twentieth-century flood geology or Creation science, which attributes most of the fossil record to the action of Noah's Flood, has gained a substantive public profile in recent years.[5] In short, history, legend, and symbolism are inextricably linked. Noah's Ark and the Flood evidently pull on many different threads in past and present culture.

Yet it is commonplace to read of a clash between belief in the Bible and the facts revealed by the developing world of science; it is often said that belief in the real former existence of the Ark is "unscientific" and has been exploded by a properly "scientific" investigation of the evidence. It is the purpose of this case study to look again at early accounts of Noah's Ark and the Flood to show that such a view is oversimplified. Thinking about Noah's Ark provides an excellent opportunity to consider the various ways in which science and religion have overlapped, merged, and moved apart through the centuries. From about 1500 to 1800 the idea of an Ark, and Noah's story itself, played key roles both in European religious doctrine and in the emerging body of thought that came to be known as science.

This case history highlights the changing forms of explanation that scholars have accepted over the centuries—especially the different criteria that were developed for judging those explanations. Was the Bible, literally interpreted, to be the final, most authoritative source of explanation for the natural objects and other living beings that people saw around them? Or should a more elastic exegesis be adopted? In what manner did the accumulation of stores of practical information about the world impinge on such views? This essay shows how far scholars moved toward a greater dependence on nonbiblical answers to their questions about nature during the seventeenth century—momentum that continued steadily through the eighteenth century. At the same time the field of natural philosophy was itself being defined as the study of the natural world, encompassing all the subjects we now divide into physics, biology, chemistry, and geology. The best way to handle such a significant theme in the history of Western culture is to take many currents of thought into account.

The Historical Ark

The text of the Bible has always been given a privileged status in Western culture because of its central place in the Hebraic-Christian tradition. Put briefly, the words recorded in the Book of Genesis were believed to be the words of God. These words were often interpreted allegorically and analogically by the early fathers as metaphors for divine action and the divine plan.

But it is also important to recognize that from very early times many people considered that the Book of Genesis also deserved special attention because of its value as a real historical record. The Bible described the Creation of the earth, the origin of animals and plants, the origin of men and women, and their subsequent history. Much of this was very direct and specific. Names, weights, measures, length of time, and the emotional motives for the actions of the first created humans were given. All those chapters filled with lists of people begetting people were perceived as having a literal purpose: they provided a complete chronology for the history of the world, running from Adam and Eve, through their sons Cain and Abel, to Noah and his three sons, generation by generation, accompanied by dates and names. For an erudite scholar equipped with various translations of Genesis and canonical Greek or Hebrew texts, it was possible to count backwards from the modern era and calculate a chronology for the earth. This was how the celebrated Protestant scholar and founder of biblical chronology, Joseph Justus Scaliger (1540–1609) calculated that the Deluge took place 1467 years after the Creation.[6] This was how James Ussher (sometimes spelled "Usher") (1580–1656), an Irish Protestant bishop working nearly fifty years later, reputedly calculated the exact date of the beginning of the world, which "fell upon the entrance of the night preceding the twenty third day of" October, 4004 B.C. Ussher's dates were for many decades printed in the margins of the Authorized Version (King James translation) of the Bible and came, therefore, to be considered authoritative by generations of Anglican Christians. Giving such an exact date for the Creation is frequently mentioned nowadays as a sign of laughable, or overstrict, religious authority, although to be fair to Ussher it is not entirely clear how far he was personally responsible for the dating with which he was credited.[7]

Even so, the point is evident. From medieval times to the seventeenth century, and sometimes beyond, a great number of biblical scholars in Europe defended a literal, historical interpretation of the Scriptures. Implicit in this view was the assumption that by carefully studying the natural world and matching it to the written words of God, human beings could reach a greater understanding of the Bible and even perhaps begin to perceive the ways in which a supreme being worked. Human observation, it was thought, could help provide a judicious account of the world's origins based on God's own narrative. Naturally enough, the exact relationship among faith, human logic, reason, and the evidence drawn from natural phenomena was much discussed by churchmen through the centuries: can one legitimately come to belief in the existence of God through logic alone, for example, or is faith a different kind of experience altogether? Should evidence drawn from nature be subservient to textual authority? These well-known and age-old debates played a leading role in the divergence of the Catholic and Protestant churches in Europe during the sixteenth century, although such discussions can rarely be characterized as simple black and white opposites. The Catholic Scholastic tradition, for instance, took a far more favorable view of reason and philosophy than did the first generation of Protestant reformers; indeed, Martin Luther (1483–1546), the German ecclesiastical reformer and translator of the Bible, had a particular antipathy toward the highly logical, rational systems of Aristotle. A long list of early theologians from Augustine to Aquinas and beyond insisted on the complementary status of reason and revelation. Yet Protestantism in a more general way dispensed with a great deal of church tradition and encouraged scholars to investigate biblical history using the practical knowledge of their day.

In all of this, the story of Noah's Ark and the Deluge was central. Through studying Noah's Ark, the best theological minds of the seventeenth century found it possible to discuss fundamental questions about the Creation and the early history of the globe.

Among those people most intrigued by the Ark during the seventeenth century was the German Jesuit Athanasius Kircher (1602–80).[8] In 1634 Kircher became professor of mathematics at the College of the Society of Jesus in Rome, the greatest center of Jesuit scholarship in Europe at the time. Kircher was staunchly committed to the idea that the study of nature and the understanding of natural phenomena would reinforce Christian doctrines. Natural philosophy, he claimed, did not seduce the faithful from the church; it actually strengthened religion by demonstrating how God's work could be explained in terms that human beings could understand. To Kircher, the Bible was in essence a record of real events, although he acknowledged that there were passages in the Genesis narrative that needed to be interpreted more liberally. In 1675, in a large, handsomely illustrated volume called *Arca Noë,* he took it upon himself to explain the details of Noah's story in the light of then-contemporary natural philosophy.

Making full use of the historical, literary, philological, and scientific studies of his day, Kircher first sought to answer the old question about the size and shape of the Ark. Was it big enough to accommodate all of the animals? He minutely scrutinized the dimensions given in Genesis. Like other biblical commentators of his day, he enjoyed arguing over the length of an ancient cubit. In the end he concluded that Noah must have used the common cubit (eighteen inches or so) for his calculations; and that the boat consequently possessed three floors or decks. If Noah allowed for adequate aisles and passageways, then three hundred animal stalls could be built on the lowest deck, each big enough to house an elephant (fig. 5.1). The second floor provided a huge storage hold, containing food enough for a city of animals, and the top deck, according to Kircher, must have held around two hundred bird cages and accommodations for human passengers.

Since Kircher listed only 130 different "beasts" (by which he meant four-footed animals), 30 pairs of snakes, and 150 kinds of birds, there was no overcrowding. Noah did not need to take fish on the Ark, he said, nor those reptiles and insects that spawned from putrefaction. It was not necessary to take hybrid animals like mules, in Kircher's view, or those species known to exist in faraway countries that had been altered by climate (he was thinking here of North America) or, according to the theories of the day, those that had degenerated from their original created state. Trees and plants were not included in his account of Noah's cargo either, because the Bible expressly said that the Ark carried only "living things that move upon the earth." When Noah sent the raven, and then the dove, out to test the subsidence of the waters, trees and shrubs were only then being revealed after their drenching. With all of these exclusions, the number of animals was small enough for Kircher's moderate-sized Ark.

Figure 5.1. Athanasius Kircher took scholarly interest in explaining the literal details of Noah's Ark. From Athanasius Kircher, *Arca Noë* (Amsterdam, 1675). Courtesy of the Wellcome Library, London.

Kircher's book was merely one in a continuing series of biblical commentaries, and long before his time scholars had begun to query the issues he covered so extensively. The problem of the size of the Ark could not be dismissed out of hand, for example. Many people asserted that Moses must have meant geometrical cubits when describing the Ark's construction rather than the smaller common cubits, which were one-sixth their size. Since these measurements may have referred to only one deck out of three, they ought perhaps to be tripled for the final dimensions of the vessel. One erudite scholar, the Dutch mathematician Willebrord Snellius (1581–1626), calculated that the floor space should be more than half an acre in size. Noah's daily life was endlessly examined, and the logistics of stabling, feeding, and cleaning the animals were worked out in exhaustive detail.

Other questions of a literal nature arose over provisions for the livestock. What did they eat, and where was the fodder stored? Did Noah bring additional sheep and goats for feeding the carnivores? What of the disposal of dung? Extra decks were added to accommodate proper food supplies, to make room for gangways and workrooms, and to house all the domesticated beasts not mentioned by Moses. Additional partitions and stables were introduced to keep unclean animals away from the rest and to provide ventilated living quarters for Noah and his family. Not surprisingly, the proliferation of such ad hoc alterations ultimately transformed the traditional silhouette of the Ark. Where earlier writers had thought of it as a houseboat, like the toy Arks made today for children, Kircher and others believed it to be more like a rectangular block perched on an invisible keel: Kircher's *Arca Noë,* for example, has the appearance of a modern hotel. The iconography of this transformation is striking and bears no relation to alterations in contemporary ship design.

The Weight of Numbers

Kircher and his generation had good reason to puzzle over the size of the Ark. For just at this point travelers began to bring to Europe an unprecedented wealth of unusual creatures from other parts of the world. Exotic animals and plants had already been brought to Europe for centuries. The spice trade with the Dutch East Indies (Indonesia and Malaysia), the old silk route across Nepal, and other well-established sea and overland links established this possibility from the earliest times. Ever since the days of the Roman Empire, improved means of communication had made the transport of foreign animals and plants across Europe feasible. The lions and other beasts exhibited in the Roman circuses, or Hannibal's elephants crossing the Alps, are no doubt a cliché, but they existed. Pliny wrote of the arrival of the first elephant in Rome in 275 B.C. and himself saw a rhinoceros in the circus in 55 B.C. Adventurers and sailors brought back tales of the foreign lands they visited and, as often as not, a parrot or monkey as well. European physicians and apothecaries sought exotic herbs and roots in faraway countries for their nostrums, especially ginger, nutmeg, rhubarb, and ipecacuanha. From the New World came the Jesuits' bark (cinchona, which was then known as quinquina), the giant sunflower, tobacco plant, dahlia, cocoa, nasturtium, and many other species, either as dried herbarium specimens, as sketches or pictures, or as seeds to grow in the physic gardens springing up in European cities and university towns. Trade in tea, coffee, furs, sugar, and spices underpinned growing European economies. In 1588, for instance, a few potato tubers were given to the French botanist Charles de L'Écluse (Carolus Clusius) by the papal legate, who obtained them from Spanish America; and L'Écluse cultivated them for distribution in the Netherlands. This took place several years before the traveler Sir Walter Raleigh, as popular anecdote has it, presented specimens to Queen Elizabeth I after his adventurous voyage to Virginia.

As for animals, it is evident from his *Historia Animalium* of 1551 that the Swiss naturalist Conrad Gesner (1516–65) knew about unusual species like the guinea pig and opossum from South America and the bird of paradise from Indonesia. The opossum had not been mentioned in any of the ancient Greek or Roman texts that Gesner used as his reference sources, nor did it have any moral or educational fables built around it like other animals known only by reputation.[9] Even though the opossum was new to him, he gamely included it in his printed menagerie (fig. 5.2). When pictures of a sloth and a toucan appeared in André Thevet's *Les Singularitez de la France Antarctique* (1557), Gesner plundered them to add to the next edition of his volume. Adding individual species in ones and twos to the total needing to be carried on the Ark evidently caused him no great intellectual problem.

Enormous catalogues by Ulisse Aldrovandi, Johannes Jonston, and Edward Topsell were produced early in the 1600s, in which the authors similarly listed all the species known to them. Strange animals like the kangaroo, rattlesnake, and anteater, which had been sighted in distant countries, were reported to European naturalists. Often the reports were fragmentary and difficult for the authors and artists back home to understand. Although the artists did their best with the information available, many of these early descriptions and illustrations have an eccentric appearance to modern eyes. The fables and magical properties customarily associated with the organisms were usually included, too. Earlier, when Pope Leo X received a gift of a rhinoceros from King Manuel of Portugal in 1517, he commissioned Albrecht Dürer to draw it; this well-known engraving shows

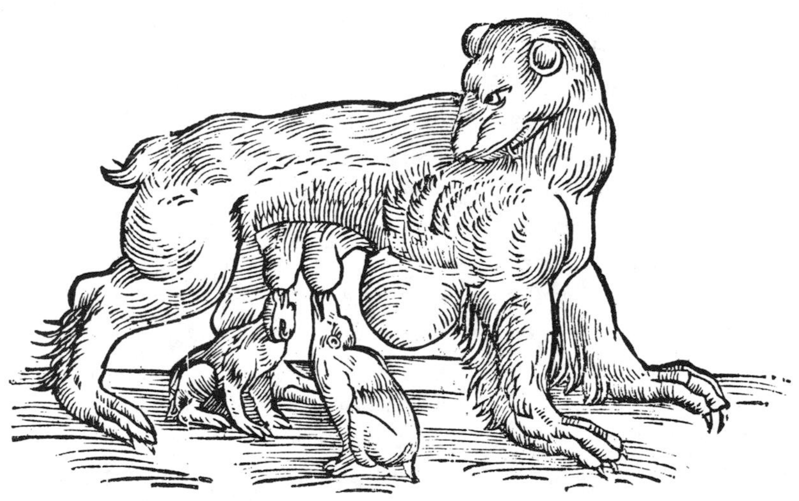

Figure 5.2. Knowledge of unknown species increased dramatically during the sixteenth and seventeenth centuries. Conrad Gesner based this illustration of an American opossum on verbal reports of an "apish foxe" that had a "bagge" under its belly "wherein she keepeth and carrieth her young ones." From Edward Topsell, *The Historie of Foure Footed Beastes* (London, 1607). Courtesy of the Wellcome Library, London.

the rhinoceros's leathery, folded skin almost as if it were ceremonial armor, an image that reinforced its perceived invincibility. The specimen itself is perhaps the one on display in the Paris Natural History Museum. Inevitably, these drawings reflect early stylistic devices and concerns. When the first chimpanzee known to science arrived in Europe in 1640 for the menagerie of Prince Henry of Nassau, it was described by the Dutch anatomist Nicolaas Tulp as an "Indian satyr." Another chimpanzee, arriving a few years later, was drawn as if it were an elderly human being, leaning on a stick, and was after its death anatomized by Edward Tyson (1650–1708). Tyson thought the animal was an orangutan, at that time hardly known to Western science at all, and called it a "pygmie."[10] Rare specimens like these were often skinned and stuffed for display in the museums that were being established in Florence, Rome, Prague, and elsewhere. Museums and botanical gardens were, at that time, regarded as places where people could bring together all the scattered species of God's Creation, almost as if they were re-creating the Garden of Eden or Noah's Ark. The English gardener, John Tradescant the younger (1608–62), who collected in North America and imported many rarities, called his museum in Lambeth, near London, the "Ark."[11]

In explaining the existence of such species previously unknown to the West, scholars displayed both the weaknesses and strengths of seventeenth-century natural history. The Bible did not mention these animals, yet the authors' attempts to defend the story of Noah's Ark showed consummate skill in assessing the means of migration, population statistics, and reproductive rates and demonstrated a fine appreciation of the distinctive features of the natural world. The Spanish Jesuit historian and traveler, José de Acosta (1540–1600), for instance, listed many species that were to be found only in the Americas. One hundred years later, biblical commentators pondered how they had gotten there. Could they have been carried by boat from Mount Ararat with the first human inhabitants? If one accepted that human beings and animals disembarked from the Ark at the same time, and that human and earthly history were broadly synchronous, the suggestion is perfectly reasonable. But some scholars were quick to disagree. Early on, the Flemish philologist and critic Justus Lipsius (1547–1606) took the view that men would hardly travel willingly with rattlesnakes and bears. He proposed that Africa was once connected to the New World by the former continent of Atlantis, now sunk, and that animals had crossed from Ararat on dry land all the way.

Most experts in the following generation contented themselves with less extravagant hypotheses about the arrival of animals in America. Kircher, for one, was reluctant to admit that the surface of the earth had undergone such dramatic changes after the Flood and thought it more likely that animals swam from island to island, or were brought by human beings in boats. Some species, he suggested, may have escaped from ships moving around the world long after biblical times. The Dutchman Georgius Hornius (d. 1670), who favored an overland migration theory, believed that animals crossed from Europe to North America in winter, when the northern seas were frozen. Sir Matthew Hale (1609–76), a celebrated English jurist, attempted to explain the unusual features of American animals through their migration from Mount Ararat and subsequent degeneration, the latter being brought on by the effects of hybridization, mutation, domestication, and local environmental factors. If all of these fail as explanations, Hale added, there is always the possibility that American species are spread worldwide and that when Africa and Asia are more thoroughly explored, the New and Old Worlds will be found to have carried roughly the same populations.

At the same time, naturalists were busy in Europe, traveling through unexplored regions, collecting plants and animals as they went. They added the names of a multitude of local plants and animals to existing lists and confirmed the presence of other species that were misidentified. Toward the end of the seventeenth century Britain's premier natural history cataloguer, John Ray (1628–1705), recorded the existence, in the known world, of some 500 species of birds, 150 of quadrupeds, and approximately 10,000 kinds of fish and invertebrates of one kind or another. These totals represented a considerable increase over Kircher's figures of only twenty-five years before. Plainly, there was an abundance of previously unknown forms of life. Unless alternative suggestions could be made, all these had to be enrolled on the Ark's list of passengers.

Equally plainly, the Ark was not infinite in size. Literal interpretations of the Bible, therefore, introduced a dilemma of a magnitude unheard of in medieval times: all the animals had to get into the ship, yet numbers were becoming too great for this to be feasible. Possibly even more important was the difficulty of explaining how the various animals had gotten from Ararat to the places where they are today. Because animals were so diverse, so widely distributed over the globe, and usually separated from their presumed geographical starting point by vast mountain ranges, deserts, and oceans, migration from the Ark's final resting place on Mount Ararat was becoming increasingly difficult to explain.

Encountering the Waters

Although many seventeenth-century naturalists, particularly those in Britain, recognized that the Ark was becoming overloaded, they did not confront the dilemma directly. In fact, they avoided it. It is hard to say, in retrospect, whether this was a deliberate policy, designed to skirt the necessity of making alternative, nonbiblical suggestions, or whether, like so many other issues in the history of science, the overloaded Ark was obscured by more obvious, more pressing, or more answerable questions in natural philosophy and religious doctrine.

Either way, from the 1670s or so natural philosophers moved away from thinking directly about the Ark and turned to investigate the Flood. In religious terms, such a shift in focus fitted the changing times. From 1670 to 1700, the Flood became a major explanatory device for European theologians, especially Protestant theologians, who wished to stress how the surface of the earth had changed materially from its original, supposedly pristine state into a corrupted, degenerate condition, occupied by humans after the Fall. This reshaping of the earth was achieved by the Noachian Deluge, which in turn provided church doctrine with a dramatic dividing line between original perfection and debased human times. The purpose of the Flood in this way of thinking was to show humans the error of their ways. How much the change in perspective actually had to do with the size of the Ark or the question of a literal interpretation of the Bible is debatable. Most likely, the priorities of religious teachers were altering in tune with contemporary demands, and the problem of squeezing increasing numbers of animals onto the Ark had little direct effect on ecclesiastical developments.

There was empirical evidence as well. There now seemed to be tangible evidence that a great mass of water had once spread across Europe. A growing number of seventeenth-century investigations into the topography and structure of the land and a wish to produce explanations of what was seen stimulated remarkable new studies of mountains, rivers, rocks, strata, and possible organic remains, the majority of which, at the start, interlocked smoothly with the notion of a great Deluge. Naturalists, mineral prospectors, farmers, local historians, and antiquaries surveyed the topography of their home regions, in Britain and on the Continent, and came to believe that much of the present dry land was composed of silt, sands, and gravel beds, indicating a watery origin. Old sea shells could be found on mountain slopes and sometimes appeared embedded in rocks far removed from the existing sea. Animal or plant remains, dug up from the ground, hinted at the possibility of organisms that had lived long before, perhaps even before the waters came. To the majority of natural philosophers, it seemed as if an extensive Deluge was an incontrovertibly real event in the recent past (fig. 5.3).

But there were difficulties. The first and most obvious of these was the possibility that the Flood had not covered the entire world. At least one natural philosopher early on decided that the biblical account must be interpreted much more loosely. The French Calvinist intellectual and diplomat (subsequently a Catholic, after forced conversion) Isaac de La Peyrère (1594–1676) doubted that the Flood was universal or that Mount Ararat was the geographical source of all

Figure 5.3. The physical extent and origin of the waters of the Deluge were highly problematic issues. From Athanasius Kircher, *Arca Noë* (Amsterdam, 1675). Courtesy of the Wellcome Library, London.

postdiluvial species. He argued in 1655 that organisms could not have migrated as widely as a literal reading of Genesis demanded. What if the Deluge had been a local event, confined to Europe and the Middle East? The Noah story would then refer only to the Bible lands. Elsewhere the various peoples of the world could have lived on, accompanied by a full complement of birds, beasts, and plants, untroubled by the catastrophe.

Such arguments were heretical in Catholic France, and La Peyrère's *Prae-Adamitae (Men before Adam)* was roundly condemned by the Paris parliament of

1655. The essence of the preadamitic theory—which actually dated from very early medieval times—was that this world is merely the latest in a sequence of divine creations. Each of the earlier worlds had been destroyed owing to God's disapproval. One of the founders of Latin Christianity, Saint Augustine (354–430), criticized such views as heresy in *De civitate Dei* (*The City of God,* 413–26). La Peyrère's project was fundamentally a theological campaign that would separate Jewish history from that of the rest of the world. Yet La Peyrère was indirectly drawing on irrefutable evidence of savage tribes living in previously unexplored countries and on contemporary talk about the purported existence of the antipodes. Both of these implied the probable existence of human races not descended from Adam. Those scholars who, like La Peyrère, held to the preadamite view were consequently branded skeptics or even atheists. His writings seriously called into question belief in the Bible as the sole authoritative account of human origins.[12]

Ostracized by the church, La Peyrère fled to Brussels, where he was arrested in 1656 on the order of the archbishop of Malines. On his release, he was escorted to Rome to sign a public retraction in the presence of Pope Alexander VII, ending his days in a seminary near Paris. His early iconoclasm made him a notorious figure, the source of theological uproar. Yet La Peyrère's attitude toward Genesis is clearly important in the context of seventeenth-century thought. In his view there was no need to fit large numbers of animals into the Ark. There was no need to suppose that animals migrated from Ararat or that every human being was descended from Noah and his sons. Using the same means of inquiry as his contemporaries, La Peyrère managed to support a thesis fundamentally opposed to conventional beliefs. And although his arguments were exclusively applied to the origins of humankind, it seems that the way was now marked for more secular interpretations of the early history of the world itself. One perceptive scholar, the Dutchman Abraham van der Myl (1563–1637), promptly asserted that the animals and humans found in the New World had lived there from the beginning and had not arrived by migration from Ararat. Van der Myl pointed out that native American animals were unlikely to trek halfway across the globe to board the Ark in the Bible lands and afterward trek all the way back again. For van der Myl, it was more likely that there had been many distinct creations.

Deeply disturbed by La Peyrère's writings, Matthew Hale (the Matthew Hale who discussed animal migration) suggested in 1677 that the Frenchman's theory revealed a series of theological problems more perplexing even than those faced by commentators on the Ark. What, for example, would be the theological status of preadamite humans: had they too been divinely created; would they also be saved? La Peyrère's argument, moreover, stipulated that the earth had no beginning and showed no prospect of coming to an end. Without the twin ideas of a Creation and an eventual Day of Judgment, the Christian faith would lose two of its most vital doctrines. La Peyrère's view made nonsense of the central tenets of both the Catholic and the reformed Protestant churches.

So, in *The Primitive Origination of Mankind* (1677), Hale attempted to refute these "immoral and irreligious principles" by proving that all humans must have originated with Adam. There, he introduced a line of reasoning that not only counteracted La Peyrère's conjectures about preadamite humans, but also remained central to theoretical natural history for sixty years, when it was codified in Linnaeus's species concept. Hale's point was simple and effective. Through the normal processes of reproduction, human beings increase in number with every generation. A family of two, he calculated, would grow to eight over a period of thirty-four years. Moving backward, it was obvious that every generation was smaller in numbers than the one that followed. Logically, this process of diminution must end in a single couple—Adam and Eve—for whom there could be no natural origin. Here, said Hale, one must assume a Creator. The same argument applied to the animal kingdom: all natural groups of animals tend to increase in number, indicating that the species began with only two representatives. After God, only Noah could have provided the modern world with two of everything. The Mosaic account, therefore, was thoroughly vindicated.

Hale's argument temporarily eclipsed La Peyrère's theologically subversive theories about preadamite humans and the eternity of the world. But there was still room for doubt about the second half of La Peyrère's unconventional thesis, the limited physical extent of the Flood. Edward Stillingfleet (1635–99), a brilliant young English theologian and historian, later bishop of Worcester, agreed with La Peyrère that the Flood may well have been a local event. This controversial view was given wider circulation by the Anglican theologian Matthew Poole in a compendium of biblical criticism published in 1669. The same suggestion, reiterated by the French Protestant theologian Jean Le Clerc in his influential series of biblical commentaries, provoked a frenzy of opposition in the form of Flood-based (diluvialist) writings in Britain. By such means, British natural philosophers at the end of the seventeenth century were pushed into defending the truth of the Bible, literally interpreted, by trying to prove the case for a universal Deluge.

Relics of the Flood

Fossil evidence played an ambiguous role in these developments. For a long time these objects were considered to be inorganic—that is, not of animal or plant origin. They were mostly described as formed stones or curiosities or viewed as antiquities that would provide information about local archaeology, a country's mineral wealth, or recent history.[13] In 1565, for instance, Gesner pictured what he called a crab made of stone. Aldrovandi similarly wrote about sea urchins chipped out of rocks lying a great distance from the sea. Each author placed these "stony" objects in the same classification group as other stony items such as arrowheads or medical curiosities such as gallstones expelled from the bladder. But there was diversity of opinion, as seen in Leonardo da Vinci (1452–1519), who argued (half a century before Gesner) that the shells found in the rocks of northern Italy were the remains of species common today in the Mediterranean. To these men the word "fossil" meant anything that was dug up. A wide variety of mineral ores, gemstones, natural crystals, useful rocks (such as coal, slates, and flints), and suchlike, as well as the objects that we now call fossils, were included in this single category. The term "fossil" took on its modern meaning only in the late eighteenth century.

Moreover, early naturalists found it genuinely hard even to propose the idea that such objects might be the remains of living beings. There was the problem of definition, for one thing. The objects were composed of solid rock or minerals, which immediately called into question the possibility of their being organic: a shell-shaped stone could not be categorized as a mollusk when it did not have any of the internal parts or defining characteristics of a living being; nor could the impression of a stony leaf reveal the features necessary for botanical classification. Gesner consequently paid no attention to this problematic organic-inorganic divide, and instead based his classification scheme on the external shape and resemblances of objects, placing stones that looked like geometrical figures in one category, and those that resembled living structures such as snails, leaves, or rams' horns in another.

A second difficulty lay in the fact that resemblances to living animals and plants were generally distorted or difficult to spot. Although some fossil objects displayed significant similarities to living organisms so that the analogy could not be overlooked, many others were more puzzling. In Britain, the seventeenth-century doctor and naturalist Martin Lister (1638–1712), a friend of John Ray in the newly established Royal Society of London, took the view (though not without misgivings) that most fossil objects were sports of nature produced by some "plastic virtue" in the earth. Large coiled structures such as the ammonite, often two feet in diameter, were so different from any snail shell known to him that he doubted whether they had ever formed part of any animal. He called them shell-stones. The townsfolk of Whitby in Yorkshire were accustomed to call these stones (which occurred abundantly in the limestone nearby) snake-stones, because of their coiled appearance (fig. 5.4)

Robert Hooke (1635–1703), a contemporary of Lister and Ray in the Royal Society, accepted that some fossil objects were the remains of plants and animals. He thought shell-stones must be the remnants of antediluvian organisms that had been killed by the universal Flood. Cautiously, Lister reminded him that this view meant that shell-stones should be found indiscriminately all over the place, flung here and there by the surging waters, rather than appearing only in particular places, and embedded inside rocks, as they did. John Ray also criticized Hooke's proposal, for he was particularly worried by the larger philosophical question of extinction. If these stony organic beings were indeed different from those of today, and had been destroyed by the Flood, it meant that they must now be absent from God's creation. Why would God allow the disappearance of animals that had once existed? Ray was too sophisticated a scholar to think that these species had simply missed their chance of getting onto the Ark. The folksy picture of (say) a pair of unicorns arriving too late to climb aboard was not what worried Ray. Instead, the age-old concept of plenitude was threatened. If nature was complete (the world containing every being that could exist), then extinction was a logical impossibility. To him, as for many others, there could be no gaps in

De figuris lapidum, &c.

culo, eiuſdémq; ferè latitudine & craſſitie, è fuſco pallidus. Pars vna, quæ inferior eſſe videtur, zonis trãſuerſis modicè incuruis, quinis diſtinguitur, quæ p latera etiã inflexæ aſcẽdũt. Superior autẽ pars læuis eſt. In extremo tuberculũ rotundum prominet: cui in parte oppoſita reſpondet cauitas. Satis durus & grauis eſt.

DE LAPIDIBVS QVI SERpentes & inſecta referũt. Cap. XV.

Lapis huiuſmodi, nõ multò maior q̃ hîc apparet, inuẽtus eſt in mõte quodã Heluetiæ, ſerpẽtis in ſpirã reuoluti eſ-

Conradus Geſnerus. 168

ti effigie, ita vt caput in circumferentia promineat, extrema verò cauda p centro ſit. Erat & alia pars lapidis, foris aliqua ex parte ambiens tanq̃ alterius maiuſculi ſerpentis minorẽ hunc ambientis: quam in hac figura nõ addidimus, quòd fracta eſſet. Hic lapis apud me eſt, ſatis durus & fuſcus, inferiore parte planus & informis, niſi q vno in loco ſerpentini corporis breue indicium apparet.

Lapis hic admirabilis inuentus eſt in agri Tigurini torrente, quẽ Toſam nominant, pugno maior paulò, vnciarum xxj. pondere, è fuſco albicans, intus verò, ſi ſcalpatur, candidus: mirè

Figure 5.4. The organic origin of fossils was much discussed by seventeenth century philosophers. There were great theological difficulties to resolve if these were really the remains of antediluvian animals and plants. From Conrad Gesner, *De omni rerum fossilium* (Tiguri, 1565). Courtesy of the Wellcome Library, London.

nature.[14] There were other worries for Ray as well. Extinction unavoidably suggested to him some kind of imperfection or incompleteness in the original Creation. But Christian theology dictated that the world had been created perfect. What kind of supreme divinity would design animals with imperfections? Furthermore, if the objects were once animals and plants, how was it possible for them to have gotten inside rocks? These questions were set out most fully in Ray's final work, *Three Physico-Theological Discourses* (1693).

For a few years around 1670 it was therefore decidedly problematic to suggest that fossilized shells and bones were the debris of animals killed in the Deluge. By encouraging the careful observation of nature in order to understand the Bible, men like Lister and Ray in fact found themselves facing problems about extinction, and queries about the definition of fossils, that religious doctrine could not immediately solve.

In 1669 Hooke's arguments received independent validation from the Danish anatomist and supporter of Descartes, Nicolaus Stensen, usually known as Steno (1638–87). Steno's *De Solido intra Solidum naturaliter Contento Dissertationes (Dissertations Concerning a Solid Contained in a Solid)* was based on Descartes's *Principles of Philosophy* (1644). Steno explained the successive transformations undergone by the surface of the earth in Tuscany by using Descartes's idea that all matter was composed of particles in motion. It was clear to Steno that most of the geological beds he was dealing with had been deposited by sedimentation—that the particles had been laid down, in sequence, underwater. This cosmogony, which supposed that the primitive earth had once been a star, like the Sun, made up of particles of matter in motion that separated into different layers, earth and stones on the outside (the crust), water and metals to the inside, agreed broadly with the first verses of Genesis. He then suggested that seashells must have become entombed in sediments depositing out in layers while the waters of the Flood lay over the land. But Steno left the question of the origin and dispersal of the waters unanswered. His readers wondered exactly how he intended finding enough floodwater to cover the Tuscan hills, which rose up to the Apennine peaks standing at a height of around three thousand feet.

Hooke tackled the problem in a different way, arguing that the waters could have been brought on by a major rearrangement of the earth's surface through a series of devastating earthquakes. He suggested that earthquakes had been more potent in the past and cited many ancient records, including the legend of a drowned Atlantis, to support his view that they could cause drastic changes in geography in a short period of time. It seems that he imagined the Flood waters surging up out of the sea as the continents rent and moved under the pressure of earthquakes—in itself an appropriately naturalistic interpretation of Genesis 7:11: "the same day were all the fountains of the deep broken up." The animals and plants that we encounter as fossils, he claimed, were subsequently caught among the shifting layers of the crust. He also wrote of the earth deteriorating from a pristine golden age, undergoing a continuous process of aging and decay. All this was very bold, and mostly treated by his contemporaries as such.

However, it is important to note that religion was here providing Hooke and others with a significant explanation for the phenomena that appeared in the natural world around them. Religion provided the framework within which to fit phenomena or objects that we would now consider part of science—fossils, earthquakes, sedimentary deposits. The story of the Flood supplied meaningful explanations for things that were otherwise difficult to understand. Putting it another way, belief in the Bible as a historical record at this point encouraged the development of interpretive categories and investigative techniques that proved constructive in the long-term development of natural history. These categories included the idea of a flood, the collection of fossils, classification of animals and plants, and the calculation of human population statistics. That such categories of thought were influential throughout European natural science can be inferred from the manner in which modern historians refer to the seventeenth century with labels like "the classificatory era" and "the age of observation." The twentieth-century French philosopher Michel Foucault notably claimed that the intellectual ethos (or episteme) of the period was essentially observational, descriptive, and classificatory, concerned predominantly with the surface appearance of things. Biblical literalism was a way of thinking about the natural world that encouraged careful observations, innovative investigations, and analytic accounts of natural processes based on accepted logical principles.

Natural Theology at the Royal Society of London

Such a marked combination of scientific and religious views was most fully expressed in the writings of natural philosophers associated with the Royal Society of London during the closing decades of the seventeenth century. The fellows of this new society played a leading role in the intellectual life of Restoration England and in the ferment of knowledge generating and generated by the Scientific Revolution, thereby having a lasting impact on Western thought. Much ink has been spilled in analyzing their Protestant and Anglican commitments: for this was a strongly royalist and churchly group, composed of gentleman scholars, the nobility, virtuosi, ordained parsons, medical men, and deans of cathedral towns, who came together at a particularly striking moment in British theological and political history.

This was a time of high emotion, religious schism, financial crisis, and great political unrest. The Civil War between Royalist and Parliamentary troops (called Cavaliers and Roundheads, respectively) ended with the beheading of Charles I in 1649. The period of rule by Oliver Cromwell, Lord Protector of the Commonwealth, was followed in 1660 by the Restoration of the monarchy—the coronation of Charles II. After all of this turmoil, it was important both politically and socially for ambitious natural philosophers to be seen as firm adherents of the Church of England, the established state religion. The Royal Society—founded in 1660 by Charles II with a royal charter, as its name suggests—was an expressly new kind of scholarly venture. Its fellows fully intended to demonstrate that the natural world was a stable, orderly set of phenomena, ruled by God in his divine wisdom; this, in turn, would justify the monarch's rule over the country through the divine right of kings. The fellows thus turned to nature to demonstrate that everything was governed by law.

Particularly, they emphasized "natural theology," already a well-established branch of religious thought in Europe, which became widespread in British culture during the late seventeenth century and reached its height in the writings of William Paley in the early years of the nineteenth century.[15] In this system, the apparent design exhibited by animals and plants, and by the human body, and the natural laws that apparently governed human society and the environment, were regarded as revealing the goodness and wisdom of God. At a deeper level, the same argument could be used to prove the actual existence of God—a God who designed it all, as the Bible says. Of course, as in all major shifts of theological opinion, many different layers and varieties of meaning were attached to this form of natural theology, and there were important distinctions to be made between arguing from design and arguing to design. Exactly what kind of evidence was thought to be acceptable and exactly what kind of natural law was being proposed were also highly debatable questions.[16] Nevertheless, this new breed of "natural philosophers" developed specific principles of scientific reasoning and emphasized a policy of investigating nature through observation and experiment, all with the intention of understanding and praising the ways of God.

Robert Hooke was a central figure in the Royal Society, famous for a remarkable book detailing his investigations in microscopy, the *Micrographia* (1665), in which he displayed the wonders of previously unseen worlds of the small. John Ray, Robert Plot, John Woodward, Thomas Burnet, and William Whiston also featured largely in the Royal Society. Their desire to understand the Book of Genesis through the new principles of inquiry drew them into studies of the Flood.[17]

For example, Thomas Burnet (c. 1635–1715), a lively and slightly unorthodox cleric, stated in 1680 that the main events of Earth history were the Creation, the Deluge, and the final conflagration on the Day of Judgment, illustrating these stages (including the Ark bobbing on a watery sphere) on the frontispiece of his

Sacred Theory of the Earth. The Deluge was a necessary event in Burnet's scheme, because it destroyed the pristine, paradisaical world that had proceeded directly from the Almighty's hand (fig. 5.5). By simple operation of the laws of nature, said Burnet, the original creation was destroyed and the postdiluvial earth formed out of the ruins. As for the origin of the floodwaters, Burnet ingeniously proposed that the earth contained a watery core that rushed out of great cracks in the crust and covered the surface, which was then much flatter. Modern mountain ranges were the crumpled ruins of the original crust, and today's oceans hid the cracks.

John Woodward (1665–1728), a physician and talented natural historian, objected to Burnet's proposal that all changes on the surface of the earth represented a decline or decay from original perfection, and in his book *An Essay toward a Natural History of the Earth* (1695) he denied that these geological alterations were haphazard and meaningless. Woodward thought the Deluge was a complete

Figure 5.5. The engraved frontispiece to Thomas Burnet's *Sacred Theory of the Earth* (London, 1684), showing the stages of Christian history in a clockwise direction, with Jesus straddling the first and last. Noah's Ark and the Deluge are represented on the third globe, before the waters recede. Courtesy of the Wellcome Library, London.

refitting of the globe to adapt it to the needs of fallen humanity. God had provided a world suitable for his creations, whatever their state of grace. Woodward, who recognized the depth at which fossils lay, developed a theory in which the Deluge dissolved all the former land. Bit by bit, the sediments settled out in layers, with animals and plants caught inside them. This scheme became known as "diluvialism."

Woodward's views on the religious significance of a worldwide Flood had the support of John Ray and William Whiston. They too believed that the Deluge had been a real event, although they differed radically on the source of the diluvial water. Indeed, Woodward's diluvial views remained influential until about the 1830s, especially his belief that fossils could be understood only as remains from the Flood. Diluvialism played a leading role in British and American geology right through until it was replaced by ideas of an Ice Age. Ray, like Woodward and Burnet, thought that vast reservoirs of subterranean water rose to the surface, although Ray set great store on a heavily increased rainfall, which upset the natural cycle of drainage. Whiston (1667–1752), a mathematician and theologian, followed a more cosmological argument based on Newton's discoveries and suggested that a giant comet had spun across the earth's orbital path. Floods resulted from the earth passing through the tail of the comet. Each in his own way attempted to explain the increase of water by natural, not supernatural, means.

These proposals were firmly based on observational evidence: naturalists across Europe carried out meticulous studies of sedimentary rocks, of rivers and the saltiness of oceans, meteorological patterns, rain and evaporation cycles, earthquakes and mountain building, and a host of additional topics that bore directly on the theme of a universal Deluge. With such intriguing material to investigate, it is not surprising that scholars were preoccupied with the idea of the Flood to the exclusion of other components of the Noachian story.

By 1700 a steadily increasing focus on the real existence of the Deluge tilted opinion toward accepting fossils as the remains of antediluvian animals and plants. About this time, the Swiss physician and natural history collector Johann Jacob Scheuchzer (1672–1733) had no difficulty in assuming that Noah's Flood had taken place more or less exactly as the biblical text suggested. He was impressed by Woodward's arguments in particular. In 1709 he produced a magnificently illustrated volume depicting specimens of fossilized plants from his own museum collection (his "cabinet of curiosities"), calling it *Herbarium Diluvianum (Plants of the Deluge)*. In Scheuchzer's opinion these fossils were clear testimonials to the existence of the Flood. In another book, *Physique Sacrée* (*Sacred Physics*, 1731–33), he set out to illustrate scenes from biblical history, starting with the Creation and Flood. He shows the Ark with its doors closed, ready to sail, and the heavens opening, the hills crowded with terrified people left behind to die; in the surrounding border he includes pictures of fossil shells, plants, and insects, a sign that he considered these to be the remains of antediluvian animals (fig. 5.6). In a subsequent illustration he showed the waters creeping up the shore to cover shells on the sand.

To Scheuchzer, the biblical story also implied that the remains of drowned human beings would be found. Perplexed by the complete absence of human bones in sedimentary deposits, he seized on a newly discovered fossilized skeleton and claimed it to be "a man who was witness to the Deluge," illustrating it in his *Homo diluvii testis* of 1726. Much later (and partly because of his very accurate drawing) it was identified as a large fossil amphibian.

On these grounds, the popular view that religious orthodoxy stifled early geological and natural historical researches cannot be sustained. On the contrary, efforts to combine the Genesis story with what was known about the natural world opened up important new perspectives in the world of learning.

But it is necessary also to say that science and religion did not enjoy a completely happy marriage. What emerges is that the increasing emphasis on empirical inquiry and the widening understanding of the many difficulties raised by the attempt to integrate science and religion led to some areas of conflict. Continued reliance on the Bible as a source of geological knowledge pointed up difficulties that could not be resolved with ever larger doses of biblical interpretation. As Don Cameron Allen suggests in *The Legend of Noah,* it is likely that natural philosophical analyses ultimately produced so much internal disagreement that the story of the Ark and the Flood could not remain a coherent historical account. Allen argues that the story either had to be believed as a miracle or be reduced in status to one of many ancient legends. While this opinion is overly dramatic and surely applies only to some natural philosophers, it seems that the integrity of the Noah story did begin to fragment in the hands of a growing number of scholars.

The History of Living Beings

By the time the outstanding eighteenth-century naturalist Carl Linnaeus (1707–78) turned his attention to the world of nature, both natural history and the intellectual environment had changed to a marked degree. It is difficult in this respect to exaggerate the impact of growing numbers of species and natural history specimens known to European naturalists in the early eighteenth century. Increased geographical exploration and colonization on a worldwide scale, coupled with a continuous updating of existing animal and plant catalogues, saw an avalanche of species new to Western science. Where Ray described 150

Figure 5.6. Johann Jacob Scheuchzer published magnificent illustrations of the Noachian story, including this scene of the people left behind when the Ark's doors closed, ready for departure. Fossils in the decorative border were believed by Scheuchzer to be the remains of antediluvian species. From Johann Jacob Scheuchzer, *Physique Sacrée* (Amsterdam, 1732–37). Courtesy of the Wellcome Library, London.

quadrupeds, less than half a century later Linnaeus counted nearly twice this number of mammals alone. Similar increases were recorded in every major class. During his lifetime Linnaeus listed the staggering total of 14,000 species, of which roughly 5,600 were animals. Linnaeus sent his own well-trained botanists out as collectors and voyagers, each traveling on a government sailing ship; these men sent back ever larger collections of plants and animals.[18]

It was not just Linnaeus. The period was patently one of collections, of natural history cabinets, museums, and private herbaria, and of the foundation of great national or institutional showcases such as the Musée d'Histoire Naturelle in Paris and the Museum in Saint Petersburg.[19] Every European head of state wanted to explore, and perhaps claim, new countries. Every nation wished to establish profitable trading routes and economic monopolies—or trading posts and colonial settlements, as took place in the Canadian interior and on the east coast of North America. These aims were always interlinked, stimulating many monarchs and many early governments to finance expeditions of exploration. Such activities were repeated by enterprising sailing-ship captains, financial entrepreneurs, and commercial organizations too. In the middle years of the eighteenth century travelers were seemingly everywhere. Pehr Kalm traveled across North America from 1748 to 1751. Pehr Osbeck sailed to China. Fredrik Hasselquist went to the Middle East, and Peter Loefling to Spain. Johann Gronovius collected plants in Virginia. Carl Peter Thunberg managed to enter imperial Japan. Johann Gmelin traveled through Siberia. John Bartram and his son William pushed through the Carolinas to marvelous effect; the rich variety of natural species that they identified in North America astounded Europeans. A little later, the French monarch Louis XV initiated the great sequence of French naval voyages that took Louis Antoine de Bougainville (1729–1814) to the Pacific Islands in 1766 and Jean François de La Pérouse (1741–88) to Asia (La Pérouse died in a shipwreck off the coast of Vanikoro). In Britain, King George III sent James Cook (1728–79) to claim the great southern continent, later called Australia, for the British crown. There were plenty of adventurers eager to collect exotic plants for the stately homes and palaces of Europe or to sell unusual animals to museums or menageries.

Natural history also benefited from the contemporary enthusiasm for an unspoiled, primitive countryside, and collecting items from faraway places became all the rage. The strangeness of the new material contributed to a growing recognition of the unbounded variety and richness of nature in other parts of the world. Besides such noted collectors as Georg Cliffort (1685–1760) in Holland, Sir Hans Sloan (1660–1753) and Sir Joseph Banks (1744–1820) in England, and René Antoine Ferchault de Reaumur (1683–1757) in France, there were also many who collected on a smaller scale in their homelands or on their travels through Continental Europe. Naturalists in Britain and western Europe were bombarded with new species from all sides.

It is easy, then, to understand how the rapid escalation in the number of species affected Linnaeus's views on the early history of life and the biblical story. Toward the beginning of the eighteenth century, the idea that all animals could have been passengers on the Ark was becoming highly implausible. New forms of life were being discovered every day. Even if there had been a universal Flood, as Ray and others argued, the Ark would have been too small to hold all the species now known to science. Linnaeus had little choice but to confront the dilemma that the previous generation had managed to avoid.

An orthodox Protestant, Linnaeus amalgamated the story of the Garden of Eden with some elements of the story of Noah's Ark. He proposed that all living beings, including humanity, were created by God on a high mountain (probably Mount Ararat) at the time when primeval waters were beginning to subside. Drawing on his knowledge of the recent elevation of the coasts of Sweden, backed up by reference to beds of shells now far removed from the sea, Linnaeus claimed in 1744 that dry land was ever increasing at the expense of the waters. Extrapolating backward, he concluded that in the beginning only one small island had been raised above the surface of a worldwide sea. This must have been the site of Paradise and the first home of living beings. He claimed that every species was created as an original pair, so making it probable that parts of the story of the Garden of Eden were true and that all animals had, for a time, shared the same locality. These aboriginal pairs were the ancestors of every living being in existence today, unchanged since the divine hand of the Creator had made them. At the end of his life, however, Linnaeus noticed the uncomfortable fact that some plants could hybridize; and he was obliged to modify his scheme to allow some limited hybridization between species after the first act of creation.

The Ark itself played no part in Linnaeus's story. "Is it credible," he asked in 1744, "that the Deity should have replenished the whole earth with animals to destroy them all in a little time by a Flood, except a pair of each species preserved in the Ark?" To him, this was as absurd as believing that the Deluge came almost without warning and went as suddenly as it came. Most naturalists of his day, he stressed, would reject these suggestions as being contrary to practical experience.

In fact, Linnaeus could not conceive how all the animals of the world—including the fifty-six hundred species he had personally named and catalogued—were to be preserved in a wooden barque and then to repopulate the earth for a second time. He rejected this proposition as a true account of early history while acknowledging the allegorical meaning of the story. Yet it should be mentioned that Linnaeus was not prepared to relinquish other key aspects of the Bible. He saw no reason to contest the divine origins of Creation, and he retained much of the Flood story in the form of a once universal ocean. The significant point is that Noah and the Ark were the first part of the story to go. Still, Linnaeus recognized the Ark's value as an explanatory model and incorporated its essence into his own semisecularized account; he too had a mountain, a vast expanse of water, and pairs of animals migrating from a single geographical center. Linnaeus had the best of the story without the intellectual impediment of a real Ark.

Within a few decades, even this piety seemed out of touch with changing philosophical and religious values. As the Enlightenment took hold in skeptical minds across Europe, it became commonplace to suggest that the biblical record—while still held by many to be a true historical account—referred only to one race of mankind. The skeptical French philosopher Voltaire (1694–1778) summed up Genesis as a collection of folktales and stated that a universal Flood was a scientific impossibility. His contemporary J. B. Mirabaud of the Académie Française echoed these sentiments, saying that "in those rude days, men . . . knew only those parts of the world surrounding them." Mirabaud believed that Noah was just a simple man—a carpenter—who assumed that everyone who was not accompanying him on board the Ark must have perished in the waters.[20]

Simultaneously, but a little differently, the French naturalist Georges-Louis Leclerc de Buffon (1707–88) turned his mind to the origin of the earth. Buffon was director of the Jardin du Roi, a prestigious establishment of museums, gardens, and menageries in Paris. Buffon, who was interested in mathematics, geology, chemistry, and natural history, published his *Histoire naturelle* in thirty-six volumes from 1749 to 1804.[21] This work matched in grandeur of scale the other vast undertaking of eighteenth-century France, the *Encyclopédie* of Diderot and D'Alembert. In this *Histoire naturelle* Buffon described the whole of the natural world from the beginnings of the earth to the start of human civilization, breaking off where the Encyclopedists began. At all points Buffon took an enlarged, entirely secular view of the subject. Though parts of it were condemned by the faculty of the Sorbonne (the leading French university), the book's general popularity and wide distribution were clear signs of the changing times.

In his volumes on geology (1749, 1778), Buffon wrote of the earth's long historical development and suggested that animals and plants had changed over time according to the environment. He proposed that the earth had solidified as it cooled from an original molten mass detached from the Sun by a passing comet (basing this conclusion in part on experiments with metal spheres), and was therefore much older than previously thought, probably some seventy-five thousand years old, a marked contrast to the fixed biblical time scale then broadly believed. Buffon later extended this time span even further, to three million years, arguing that the earth went through successive stages, which he called "époques," roughly corresponding with the biblical "days" of Creation, if these were interpreted sufficiently liberally. Animals appeared (fifth epoch) just before the continents emerged from the primeval ocean (sixth epoch). He attributed their origin to spontaneous generation rather than divine creation. While rejecting the stories of the Garden of Eden and Noah's Flood, Buffon nevertheless left the door open for those who wished to read a Christianized version into his proposals. Fossils, for him, represented the remains of the first animals on earth, which became extinct as conditions changed.

These ideas represented the epitome of French Enlightenment culture, expanding the time scale of the past and showing how this immense period had been filled with dramatic changes in geology, topography, living beings, and climate, all without any direct intervention by the Creator. Taking care not to step into atheism or scientific materialism, Buffon made sure that God was no longer responsible for day-to-day explanations of nature.

The notion of a literal Ark had sunk so far from the view of leading intellectuals by the 1770s that it was easy for the German zoologist Eberhardt Zimmermann (1743–1815) to ridicule the whole idea in his *Specimen Zoologiae Geographicae Quadrupedum* (*Essay on the Geography of Animals*, 1777).[22] Zimmermann poked fun at the proposal that only one pair of each kind of animal was created. On this reasoning, coupled with Linnaeus's original mountain, the first pair of lions would soon eat the first pair of sheep, then the goats, cows, llamas, buffaloes, zebras, and so on, in quick succession. Finally, even the lions would die from hunger after they had eaten everything they could catch. Zimmermann wrote that it was far better to believe that every animal was created in the area where it was destined to live, under the same climate that it now enjoys, with the same food rations already in abundant supply.

Once Zimmerman's idea became established, as it quickly was, there were no further pressing reasons for scholars to continue believing in Noah's Ark as the sole explanation for the origin and distribution of natural beings. Once scholars had established beyond any reasonable doubt the impossibility of every living thing coming together in the Ark, or indeed having a common source on Mount Ararat, it became possible to ask other kinds of questions about the origin and geographical distribution of species. Similarly, but for different reasons, the idea of a worldwide Flood in the full biblical sense was looking more and more improbable. From around 1790 or so, many people regarded the Flood as simply one of a series of upheavals or revolutions that had taken place in the history of the earth. This too turned naturalists' minds toward other questions such as the age of the earth and the many changes it must have experienced. Natural philosophers began to think of the Flood as primarily a geological event, one that separated the earliest history of the earth from the period in which humans existed: an event that integrated empirical evidence and religious belief under the revitalized general label "diluvialism." It was not Noah's Ark and the source of the waters of the Deluge that came to perplex and fascinate late eighteenth-century naturalists; it was the history of the earth before the diluvial period and the origin of species.

Conclusion

The main features of this case history have been the interplay between reliance on empirical data gathered in the field and the status of authoritative religious sources that addressed the same issues. The first important natural philosophical debate about Noah's Ark concerned the adequacy of its size, and yet at almost the very moment when this debate appeared to be resolved through a combination of biblical interpretation and practical knowledge of animals, seafarers and naturalists arrived in Europe bearing large burdens of strange beasts, birds, and plants that also had to be squeezed on board. At broadly the same time, natural philosophers attempted to understand fossil remains and their place in the story of the Deluge. They puzzled over the definition of fossils as organic remains. They tried to assemble enough water to cover the whole earth; and it is not surprising that they found it necessary to reduce the size of their floods, first to the inhabited world and then to the Bible lands alone. When this process of ad hoc reinterpretation of the biblical text was added to uncertainties arising from close textual examination of the Scriptures, widespread doctrinal difficulties about the theological status of preadamite man, and growing schism between Catholic and Protestant, the historical accuracy of the traditional account was seriously threatened. The Ark as a real historical event was for the most part discarded by the natural philosophers of the eighteenth century, although the revelatory, allegorical, and metaphorical meanings of Noah's story remained secure.

But as this case history reveals, empirical investigations into the Bible story also produced many significant results. The controversies arising from speculations about the meaning of fossils and the geographical extent of the Flood brought new ideas about the age of the earth and the prehistory of mankind into the open. These were set within a devout religious context but were flexible in their interpretation of ancient writings and chronologies. The attempt to explain how animals arrived in the New World and how Noah's children repopulated the earth stimulated the study of indigenous societies and early human migrations. The search for appropriate sources of water led to distinctive observations about comets, earthquakes, rainfall, and sedimentary strata.

As time passed, however, these careful investigations and explanations created an increasing number of other problems. By pursuing their research with such energy, natural philosophers found themselves disproving what the sacred books said was true. In a way, the biblical Ark was a victim of its own success. By the middle of the nineteenth century men like Charles Darwin and Charles Lyell ridiculed the whole idea. In retrospect, European naturalists had come to accept that the story of the Ark could be understood on two levels, the historical (which they mostly rejected), and the allegorical or religious (which many continued to value). Foundational legends such as this one concerning the Ark can obviously survive even the destruction of the evidence that formerly established their existence.

6

Genesis and Geology Revisited: The Order of Nature and the Nature of Order in Nineteenth-Century Britain

Mott T. Greene

Imagine this. It is the year 1821 and you are thirty-six years old. You are a graduate of Oxford University, and you hold the first faculty position in geology (ever) in that university—a position created especially for you. You are also in holy orders, having taken ordination as an Anglican priest at the age of twenty-four, soon after your graduation. (This was not a surprising step since you come from a long line and a large family of Anglican clergymen.) You are a brilliant lecturer and an excellent writer, widely traveled in Britain and on the Continent, and you are acquainted with every great geologist in Europe. There is more: you have just made a discovery, a fantastic discovery, in a cave in Yorkshire. This discovery is going to make you the leading scientist of your generation. It will make students and faculty jostle for seats in your lecture hall. It will make you a best-selling author and the subject of not only newspaper and magazine profiles but long articles in learned journals. It will lead to prizes, offices, money, honors, and fame.

You are already well known (though not quite famous) for a little book called *Vindiciae Geologicae,* which was also the title of your inaugural address, given just two years ago when you assumed your faculty position at Oxford.[1] This little book contains your plan and pledge to "vindicate" the study of geology at Oxford by dedicating geology to the greater glory of God and by showing that a wide variety of geological phenomena visible on the surface of the English countryside are the remains of the great biblical Deluge, Noah's Flood. You are thus the author of the "diluvial theory" and the founder of modern "Flood geology."

However, you have a problem. Your fantastic discovery in the Yorkshire cave, if you pursue it, is going to undo your diluvial theory, at least in part. It will show that the biblical Deluge is *not* responsible for a great many phenomena attributed to it, and it will show that diluvial geology is *not,* in spite of what you have just written and promised, a sufficiently broad focus for geological theory, nor a very useful means for organizing geological observations. Your discovery is certain to appall and enrage biblical realists (the correct designation for what are often called biblical literalists), including many faculty colleagues who praised your recent book. Finally, you will not easily escape the charge that you are contradicting the account of the Creation given in the Book of Genesis, and therefore some will perhaps say that you are impious, or ungodly, or even an apostate. What are you going to do?

The appeal to principles is difficult. You are absolutely committed to religious truth and absolutely committed to scientific truth. In your previous experience these truths have always been mutually reinforcing, but now they stand opposed. Both are real and both worthy; how can you reconcile them? Your choice is not a struggle of right against wrong, but a struggle of right against right, in which no course of action will easily satisfy all your commitments. The question is not one of advantage and opportunity, but of conscience, and here lies a serious difficulty: you (the young professor of geology) discover almost at once that you have not one, but *two* consciences. You have a moral and religious conscience, and you also have a scientific and intellectual conscience, and you must try to satisfy them both. Will you be able to do it?

This was the real-life predicament of the Reverend William Buckland, D.D. (1784–1856), in the year 1821 (fig. 6.1). His dilemma, and the means by which he resolved it, are an important part of the story of the impact of scientific discoveries on religious beliefs in the decades before Darwin. The Reverend Buckland's scientific and spiritual struggles, and those of his contemporaries, have been told a number of times, and are the basis of *Genesis and Geology* by Charles Coulston Gillispie. They are also the basis of *The Great Chain of History: William Buckland and the English School of Geology (1814–1849)* by Nicolaas Rupke.[2] Gillispie and Rupke explored the encounter between revealed religion and geological discovery in the nineteenth century because they saw that it had some worthwhile lessons to impart about the relationship of science and belief. Among the issues it raises are the role of expert testimony and authority in religion and in science, the means by which these are established and preserved, the function of common sense in scientific work and religious practice, and how and to what purpose the Bible and the world are to be read.

Figure 6.1. The Reverend William Buckland, professor of geology at Oxford, in 1823. Courtesy of Roderick Gordon and Diana Harman.

The Bones from Kirkdale

In July 1821 a group of quarrymen discovered a small limestone cave near the town of Kirkdale in Yorkshire (a large county in the northeast of England) (fig. 6.2). The cave was littered with bones that they had discarded by spreading them on nearby roads, since their aim was not paleontology but quarrying limestone. A country doctor, passing along the road, noticed the strange sizes and shapes of some of these bones and returned later with several friends to examine them. Together they collected a large number of specimens and sent them off to museums in London and elsewhere to be identified.

Buckland received a large box of these bones in November 1821. Using the method of comparative anatomy so recently and so brilliantly developed by Georges Cuvier in France, Buckland was able to determine that the bones were a mixture of the remains of many creatures, including not only deer, bears, and foxes but elephants, hippopotamuses, rhinoceroses, and hyenas. Elated, Buckland wrote to a friend that it was hard to believe that the bones of so many large animals—animals that could never have lived peaceably together—could have gotten into the cave other than by some common calamity, by which he meant, of course, the great global Deluge at the time of Noah.

When Buckland examined these bones carefully, however, they told him a different story. He began to suspect that what the quarrymen had discovered inside the cave were not flood remains, but the accumulated debris of a long-resident den of hyenas. For one thing, the upper surfaces of the bones lodged in

Figure 6.2. The cave and bones of Kirkdale, Yorkshire. From William Buckland, "Account of an Assemblage of Fossil Teeth and Bones of Elephant, Rhinoceros, Hippopotamus, Bear, Tiger, and Hyaena, and Sixteen other Animals, Discovered in a Cave at Kirkdale, Yorkshire, in the Year 1821," *Philosophical Transactions of the Royal Society of London,* 112 (1822), 171–236, p. 236.

the mineral matter of the cave floor had been polished smooth, a phenomenon known from other animal dens where generations of predators had reclined and walked repeatedly on the bones of their prey animals. For another, the bones of the elephants, rhinoceroses, and hippos (though not of the hyenas) were gnawed and broken. Soon after Buckland made this observation, a traveling menagerie came through Oxford with a captive hyena, and Buckland took the opportunity to offer some large bones to the animal to chew. After getting the bones away (he did not say how) he was able to determine that the puncture marks and grooves in the bones he had offered the live hyena, and the way the bones had been broken by the hyena's jaws, were identical to the punctures, grooves, and fractures on the bones from the cave.[3]

One might think it astonishing that the bones of animals today found only in tropical regions should turn up in the muddy floor of an English cave. Yet the discovery, however exciting, did not astonish. A number of published books in the previous twenty years had already documented beyond question that the fossilized remains of many tropical forms of plants and animals were widely distributed across the landscape of northern Europe. Cuvier had shown that distinct assemblages of such remains could be seen in successive layers of rock in the region around Paris; he had interpreted these rocks and their contents as a series of successive creations and destructions of the living world, long before the appearance of the first humans.[4]

The absence of any human fossils among the remains kept the notion of "antediluvian worlds," even those containing recognizable and existing species, out of direct conflict with Scripture. Geologists, therefore, had experienced little difficulty convincing a broad spectrum of the scientific and religious establishments in Great Britain to accept such worlds as a distinct possibility, though not all were convinced or ready to be convinced. Yet some enthusiasts looking for geological evidences of the Deluge were happy to suppose that, in these remains, they had evidence of a cataclysm so powerful that it had swept the bodies of tropical animals as far north as England. Others, bent on the same quest by a different path, asserted that the animals had been drowned in place by the Flood, and that England's former tropical climate had been permanently altered by this same giant event. In no case was the Deluge account of the Bible placed in doubt.

Even with the supporting body of evidence from cave deposits and fossil-bearing rocks in Europe, and the increasingly popular idea of antediluvian worlds, the bones represented a real problem for Buckland's own diluvial theory. They were quite clearly not those of animals that had died in some sort of cataclysm, nor had they been swept into the cave. The skeletons (except those of hyenas) were not complete; they appeared to be parts of carcasses only, typically the long bones of the legs. Moreover, since the roof of the Kirkdale cave was intact, one could not argue that the bones had fallen or been swept in through a hole or fissure—something observed at other cave sites in England and on the European continent. These bones from Kirkdale were clearly specimens from an antediluvian world, yet they were not in any sense "flood deposits." They cast doubt on the idea that the contents of many other caves (with intact roofs) were the result of flooding, as had previously been supposed. All this had the makings of a serious setback for the diluvial theory as an organizing principle for geology.

If the bones in the hyena den at Kirkdale were a challenge to the diluvial theory, it is not immediately obvious why this should have created great difficulties for Buckland. He had known of and accepted evidence for the existence of antediluvian worlds for quite some time, as had other geologists before him. James Hutton (1726–97), one of the founders of modern geology, had written thirty years before of "revolutions on the surface of the world" with the replacement of one ordered sequence by another, usually interrupted by some convulsion in between.[5] Cuvier had, only a few years before Buckland's inaugural address at Oxford, published a widely read short work, *Discours sur les revolutions de la globe* (*Discourse on the revolutions of the world*, 1812), in which he had come to the same conclusion as Hutton. Cuvier, however, based his interpretation on detailed paleontological and geological evidence from the region around Paris; even more significant, he offered a spirited defense of the chronology of Genesis for determining the beginning of human time, though not the beginning of the world.[6]

Buckland's discoveries and conclusions did not force him to change his opinions about the world or about the Bible. He continued to believe in the reality of the Deluge described in the Scriptures and still hoped to find universal geological evidence for it. If the evidence from Kirkdale cave could not be applied to this objective, Buckland had no scientific or theological reservation concerning the notion of a succession of creations before the event told of in Genesis, because these revolutions interfered neither with his sense of divine agency nor with his reading of the text of Genesis.

Buckland's Dilemma

Buckland's dilemma—his crisis if you will—arose instead in deciding how he would now vindicate geology as a legitimate subject for study at Oxford, a university that was principally a theological seminary.[7] If the subject matter of geology turned out to be concerned almost entirely with evidence bearing on matters that had transpired before those described in the Bible, and to consist mostly of a chronology of former worlds, it was not at all clear of what use this material would be to prospective clergymen.

The original promise of Buckland's geology at Oxford had been very attractive: If one could ascribe the broad distribution of superficial gravel and the fossil remains contained in it to the action of the universal Deluge described in the Book of Genesis, Anglican clergy would be provided with powerful support from science for an unending source of homilies concerning God's providential intervention in nature. These homilies could be substantiated by such activities as looking out the window, or plowing one's field, or walking on the shore. The point of diluvial geology was not to provide general support for belief in the Bible by pointing to evidences of divine action in the natural world. No one publicly, and few privately, doubted divine action. To demonstrate God's work was the task of all of natural theology, which had been a robust enterprise for almost two thousand years. Rather, the specific point of discovering evidences for a great flood was the providing of empirical proof, outside Scripture, for God's occasional and providential intervention in the course of Nature, in this instance to punish human wickedness.[8]

The issue at hand was not the Creation of the world, but evidence of God's providential intervention to punish evil—long after the Creation. No man of science writing in England in Buckland's time disbelieved in divine providence as it related to the design and population of the world. Not every one spoke of it, and some remained indifferent to it, but not one repudiated it; indeed, a clear majority expressed enthusiasm about it.[9] Because avowed atheism and public blasphemy were crimes actively prosecuted in England until about 1830, the mere mention of providence cannot be taken as an index of belief. Yet for most writers, belief in divine providence was not only sincerely held but thought to be scientifically well supported by a vast literature containing examples of how the world had been prepared as a fitting abode for life. Even naturalists such as Hutton, who had spoken less of the action of God than of the works of Nature, and Erasmus Darwin, the grandfather of Charles and a speculative thinker often mistakenly called an atheist today, both devoted themselves unceasingly to the documentation of the providential ordering of nature.[10] Buckland's generation, however, worried most about the judicial intervention of God in history, and they sought evidence of it in the natural world.

In spite of a conscientious dilemma concerning his scientific duty on the one hand and his duty to bolster support for scriptural authority on the other, Buckland did not hesitate to publish his geological discoveries about the Kirkdale cave. He accepted without reservation the evidence that this collection of fossils in Kirkdale cave had *not* been produced by some startling catastrophe. The layered accumulation, the broken bones, the teeth marks, the complete skeletons of hyenas amid the partial skeletons of large herbivores all pointed to the hyena-den theory. Working quickly and enthusiastically, he put his scientific results together and presented them before the Royal Society of London during three successive weekly meetings in February 1822.[11] The context in which he placed this work before the national scientific elite apparently had nothing to do with theology, the Bible, or the Deluge. He put his hyena den in the purely empirical context of other caves with other kinds of remains in England and elsewhere, speaking as a scientist to other scientists about distribution of rock, soil, and bone.

The response gratified and relieved Buckland. The lectures to the Royal Society created a scientific sensation of a very enthusiastic and positive sort. Magazines in England, Scotland, Germany, and France reported their contents and conclusions for more than two years.[12] In recognition of the significance of his work the Royal Society in 1822 awarded Buckland the Copley Medal, a prize given annually for the greatest achievement in science that year. Before awarding the medal to Buckland, the society had never given it to a geologist; thus the bestowal of the Copley Medal on him represented a vindication of geology.[13]

From this point, Buckland's geological career took off with astonishing speed. Within a year he was acknowledged as the world's leading expert on cave deposits and on the geology of caves. Contemporaries immediately recognized his research as a milestone in geology and paleontology, which marked the beginning of a new field of science now called paleobiogeography, which reconstructs the character and geographical distribution of communities of plants and animals in past geological periods.

Buckland's report to the Royal Society of his change of mind regarding the geological meaning of Kirkdale, and of similar caves with intact roofs, satisfied his scientific conscience and allowed him to make great professional strides as well. However, the hard part—addressing the impact of his cave research on his theory of the Deluge and on his plans to vindicate geology at Oxford—remained. He still needed to find some way to reaffirm the truth of Scripture with particular reference to the Flood of Noah and to provide proof, by association, for the historical accuracy of the early chapters of Genesis.

To carry out his religious mandate, he decided to use the same evidence to produce an equally important publication that would add to the evidence for the Deluge, or at least preserve what he had already claimed for it. At the time of his addresses to the Royal Society, Buckland had already begun work on this book, to be titled *Reliquae Diluvianae (Relics of the Deluge).* He had collected a considerable number of observations concerning the superficial deposits and the topography of England that he took to be confirming evidence of a universal flood covering England in recent geological time, a flood he equated with that mentioned in Genesis. He had intended to expand and supplement these researches for his book. But how could his cave researches and hyena dens fit into this plan? If the dens had not resulted from the Deluge, and even pointed away from the Deluge as an explanation for the cave deposits (till then one of the best geological evidences of the Flood), what good could come from presenting his findings in a book on the "relics of the Deluge"? But how could he leave them out when he was world famous for these researches, and any omission of them in a book on the Deluge would be interpreted as an inability successfully to unite his scientific results with his commitment to scriptural authority?

Buckland had to find some way to bend and shape these results in the service of his former plans. Rupke has plausibly suggested that Buckland had an additional impetus to caution: keeping his job. His salaried position as reader was only four years old, and it would be years more before his finances would become secure. (Only after being made canon of Christ Church in 1825 did he draw a clerical salary.)[14] In any case, Buckland set to work on his book with a new approach that he hoped would allow him to keep his commitments in balance and please both scientists and theologians.

Relics of the Deluge?

Buckland, faced with a pressing need to "vindicate his vindication" of geology and to bolster support for the diluvial remnants of his theory, did not dither. In the year following his address to the Royal Society he brought out a book under the planned title *Reliquae Diluvianae; or, Observations on the Organic Remains Contained in Caves, Fissures, and Diluvial Gravel, and on Other Geological Phenomena, Attesting to the Action of an Universal Deluge.* The book he produced, however, differed somewhat from his original idea for it, largely because of his attempts to solve the problem of vindicating geology as a subject appropriate for a theological seminary while changing the character and content of that vindication.[15]

Buckland's strategy for preserving his role as a scriptural geologist appeared first in the dedication of *Reliquae Diluvianae* to Shute Barrington, the Anglican bishop of Durham. In his dedication Buckland defended both the idea that geology supplies proofs of the universal Deluge and the idea that such proofs are essential to the defense of the veracity of the Book of Genesis. The latter point, the theological point, continued to put him on the side of biblical realists. The former point, that geology provides evidence of the Deluge, was clearly intended to defend the inclusion of this scientific work in the curriculum at Oxford University.

The text of the book reveals a pattern of argument on which Buckland would increasingly rely in his subsequent writings on the subjects of geology and natural theology. He began the book with an affirmation of the truth of Scripture, but then proceeded to a detailed and purely scientific outline of his researches on specific caves and the results of other scientists' work on similar caves. He distinguished sharply between the geological evidence and the historico-theological conclusions to be drawn from it. He concluded each presentation of geology with the claim that there is, in the Kirkdale cave and in similar venues, a thin top layer of mud that must be the residue of the Flood of Genesis. Although this Flood was no longer a global cataclysm, it served as a worldwide geological marker separating the prehuman history of the earth from the scripturally recorded history of the earth and its repopulation by divine intervention.

Buckland's defense of Genesis gave neither pleasure nor comfort to biblical realists accustomed to reading Scripture as the record of the entire history of the earth and the cosmos. As Rupke has pointed out, in spite of its bold front of defending the Deluge, Buckland's work seriously undermined traditional diluvialism in three ways. First, it drastically reduced the portion of the geological record—whether surface gravel or stratified deposits—that could be ascribed to the action of the Deluge. Second, it reduced the power of the floodwaters to transport such material, or fossil bones, by arguing that the tropical fossils found in the caves had not been swept north from tropical regions thousands of miles away by a cataclysmic inundation and its aftermath. Finally, it undermined the widely held theory that the Deluge had occurred in an alternation of land and sea: the sinking of the continents, and the rising of the sea floor, spilling the ocean waters onto the continents and drowning their resident fauna. This theory had been popular not only among biblical geologists but with Cuvier himself. Buckland's theory, to the contrary, asserted that the land surface of the hyena dens had been dry both before and after the Deluge.[16]

Yet Buckland's science continued to serve the cause of religion. The Oxford geologist had maintained the complete veracity of the Genesis account to his own satisfaction, while slipping as quietly as possible past the question of what Genesis was an account *of.* On this point there were still difficulties to resolve, and through them we are brought face to face with the salient issue in any attempt to reconcile science and Scripture: not *whether* the two should be reconciled but *how.*

The Reading of Scripture and the History of the World

It would distort the story of Genesis and geology if we assumed that the Bible had always been read in a fixed fashion unless challenged to accommodate some scientific fact. The reading of the Bible has a recorded history as complex and varied as the reading of the world by science, with changes typically occurring independently of scientific research. The Bible has long been read simultaneously in a *literal* and a *figurative* sense. Until about the seventeenth century there had been a strong alliance between a realistic reading of the Bible—at once literal and historical—and a figurative reading that provided a basis for doctrine and a means to unify the book as an instrument of moral instruction. After the Protestant Reformation, however, these two ways of reading the Bible began to diverge and even to stand in opposition to each other. Increasingly, the literal reading of the Bible (what the words say) was separated from the figurative reading of the Bible (how the words form a connected and prophetic history).[17] By the early decades of the nineteenth century, when Buckland was working on his cave studies at Oxford, the question of how to read the Bible had become a lively topic of debate without any reference to science at all.

As biblical scholars wrestled to determine which inspired passages were historical and which were allegorical, apparent differences between the Genesis narrative of the Creation and the geological narrative of Earth history took on heightened significance. It was not just that in Genesis the writing is realistic and compelling, but that God is also close to humans, present, visible, directly and evidently concerned. Buckland and his contemporaries, devoted to maintaining the idea of the providential intervention of God in the present, eagerly sought for evidences to support God's interventions in human affairs at the beginning of things, and they responded critically to evidences that turned attention away from such interventions.[18]

An allied issue in the study and reading of Scripture in the early 1800s, also independent of any scientific influence, was the vexing question of authority over the text of the Bible as a whole. Martin Luther's attempt at reform in the early sixteenth century had quickly accelerated into a revolt against the entire Roman Catholic hierarchy of authority—priests, bishops, cardinals, and the pope—in favor of the "priesthood of all believers." It matters little that Luther himself and the German princes who supported him rapidly reformulated an orthodoxy and commenced a vigorous persecution of all who would not submit to it. The fundamental principle remained alive in spite of attempts to modify it: namely, that the word of God is plain and accessible when read by a believer with an open heart. Thus understanding God's word does not require exegesis or interpretation by an authoritative and trained specialist in order to be made clear; it requires only individual initiative and "common sense," if we may be allowed by this term to indicate merely the ability of the common run of humanity to understand God's will through Scripture without theological guidance or sacramental warrant, which was certainly Luther's ideal.

Long before there was a science of geology, the Scientific Revolution of the seventeenth century had already created a contest between scientific and religious authority in interpreting the world, since it embodied intellectual trends that made it run *exactly counter* to the main movement of the Reformation. Foremost among these was a revolt *against* common sense. It is common sense that we are standing still on an immobile Earth while the Sun, the Moon, and the stars revolve around us. It is common sense that if you shoot an arrow straight up into the air, it must exhaust the force with which you shot it before gravity can take hold of it. Further, if the earth is rotating, it is common sense that the arrow must fall behind you, as you and the earth will have rotated while the arrow was aloft. These commonsense ideas are all wrong, but they are not obviously wrong, and demonstrating that they are wrong took an immense amount of effort, industry, and persuasive power over almost two hundred years. In the end the Scientific Revolution produced a new physics of motion in which empirical experiments, not clear ideas, would carry the day.

By the eighteenth century the court of experience had replaced the cult of common sense within the scientific community. Men of science had increasingly come to regard as true only theories supported by experimental and observational evidence. They felt themselves *morally* bound to relinquish belief in any theory that had been contradicted by empirical evidence—no matter how much one was attached to that theory. Finally, they eschewed appeals to hypothetical arguments that the evidence might be wrong. If one had no evidence, he might still have a right to speak but no longer a right to be heard.

Genesis and Geology

Given the challenge to any pretension of authority over the reading of the Bible and the Protestant impetus toward a literal and commonsense reading of Scripture without the aid of specialist expertise, and given the opposed and equally emphatic rejection of commonsense certainty by science, it would have been remarkable if some sort of controversy about "Genesis and geology" had not broken out soon after the science of geology appeared in the nineteenth century. The debate about Genesis and geology that emerged in Great Britain with renewed strength in the 1820s after the publication of Buckland's book was, however, neither a theological nor a scientific free-for-all. It was, as I suggested, not about supplanting religion with science, or about ignoring science in favor of religion; it was a debate about reconciliation of the two. "Genesis and geology" is not code for religion versus science; it refers instead to constant adjustments in the reading of the Bible and nature in order to keep their relationship harmonious. Although there were exceptions, the tone of debate was generally moderate, in part because of the diffuseness and diversity of opinions on the theological side. Religious authority was further rapidly eroding in the 1810s and 1820s with the emancipation of Catholicism, the repeal of the tests of religious orthodoxy as a condition of government employment, and the explosive growth of dissenting and nonconformist Protestant sects, especially evangelicals. On the scientific side, the debate remained moderate because geology was still somewhat insecure and anxious to be accepted as a real science; it had only a tentative foothold in the universities, and its professional structure of associations and publications was just beginning to develop. It courted the good opinion of the world and wished to present itself in the best light.

In any case, in the early 1820s there rapidly emerged a wide variety of opinions on Buckland's work, which, because it was an outstanding combination of world-class science and devout theological interest, became a sort of test case to see how far the claims of natural theology could be pushed and how much detail it could encompass. It also tested the ability of geologists to harmonize their work with Scripture.

The broad sweep of opinion notwithstanding, the overwhelming response from men of science and clergy alike was positive. Buckland's public change of mind about geology had forced him to revise his reconciliation of geology with Genesis, but it had not brought about a change of heart with regard to religion. He had claimed to find some evidence for the Deluge but not as much as he had hoped. His book was seen to advance geological science and to maintain the alliance of science and belief, not least because it seemed to inform the reading of Scripture with scientific evidence.

Buckland, while reporting his cave geology and finding some evidence for a universal Deluge, managed to make an even bolder claim with regard to the geological sequence of the antediluvian world: he interpreted each creation and subsequent destruction as a sign of God's providential intervention. "We see at once," he wrote, "the proof of an overruling Intelligence continuing to superintend, direct, modify, and control operations of the agents which he originally ordained."[19] In this way Buckland argued that the violent (as he supposed) action of geological agencies—floods, earthquakes, eruptions, and rapid uplift of mountains—were the instruments of God's providence. Geology not only failed to challenge Scripture; it revealed the specific agency of God's continual and providential superintending of the world. This was indeed a "vindication of geology" and a presumptive reason to continue to teach it to prospective clergy. Buckland appeared to have carried the day for natural theology.

This lively interest concerning the relationship between geological and scriptural matters persisted in England throughout the rest of the nineteenth century, with a good deal of jostling and debate. The avid pursuit of natural theology in Britain often astonished continental Europeans. The German geologist C. F. Schoenbein, listening to an address by Buckland, noted that "the English have a peculiar love of regarding nature from a theological point of view, and the celebrated Oxford geologist, as he proved by his last geological work, is no exception to the rule."[20] As Gillispie has observed,

> Ever since Newton, natural theology, if not quite a distinctively British approach to God, had at least been elaborated in far greater detail and with much more enthusiasm in Britain than in other countries. British theologians and scientists, so many of whom rested the proof of the existence and activity of God on physical evidence were, therefore, more distressed than leaders of religious opinion elsewhere when empirical argument began to move, if not away from Him, at least in an irrelevant direction.[21]

In the 1820s geology in Britain had begun to move in a direction largely irrelevant to the biblical narrative, as geologists abandoned theoretical debates about the origin of the earth's surface in favor of detailed fieldwork that reconstructed the history of life on Earth. Amateur natural theologians quickly moved into the intellectual vacuum created by the departure of professional geologists. If the prelates of official science would turn geology away from natural theology, the amateurs would steer it back again and create a scriptural geology of their own, guided by common sense and religious zeal. The controversies that sometimes arose created a rather difficult situation for Buckland and others like him, who tried to stay out of the line of fire. In this Buckland was largely but not completely successful. As Rupke has pointed out, Buckland's fiercest opponents were not fellow geologists but biblical realists, critical of his backing away from a single identifiable geological stratum associated with Noah's Deluge.[22]

William Buckland and Charles Lyell

Buckland, a popular and inspiring lecturer, taught hundreds of students. Most of the great geologists of the next generation, including those forcing changes in his own work, had been his pupils at Oxford. In addition to geologists in training (a small minority of any class to be sure), his audience included prospective clergymen, lawyers, and doctors, as well as young men simply passing through to an Oxford degree. Among them in the years 1817 and 1818 was a young Scot named Charles Lyell (1797–1875), who sat in on Buckland's lectures in mineralogy and geology while preparing for a career in law. Lyell seems to have been a very good student. While doing fieldwork in the summer and fall of 1817 he solved a problem on the geology of caves that had been posed by Buckland and sent the solution to his professor.[23] Perhaps on the strength of this contribution he was elected to the Geological Society of London at the age of twenty-two. Lyell eventually passed the bar exam, but spent little time in the practice of law compared with the time he spent pursuing his real passion: geology. By 1823 he had been elected secretary of the Geological Society of London and by 1826 he had become a fellow of the Royal Society of London. Such a rapid rise suggests strong patronage from Buckland and others as much as great gifts, but there is no doubt that Lyell was an excellent observer and an even more excellent student and companion of observers greater than himself (fig. 6.3). Like Buckland before him, he rapidly developed a wide international acquaintance, facilitated by travels in Germany, France, Italy, and the United States. Drawn to large philosophical issues concerning geology, he devoted much of his effort to writing review articles and broad surveys, frequently based on the researches of others.

In 1828, returning from one of his many trips abroad, he conceived a plan to write a general introduction to the science of geology, including the principles by which geological structures and processes should be evaluated and understood. Because of the tendency of British geologists to collect individual field reports without fitting them into a larger whole, no English-language geologist had attempted a synthesis for more than a decade. Lyell's conversations with European geologists and his own observations had convinced him that there was no reason to postulate great catastrophes in Earth history, even to explain the most extraordinary dislocations that one saw in mountain ranges and volcanic regions.

Figure 6.3. Sir Charles Lyell, who was knighted in 1848.

Lyell believed that, given enough time, very small incremental changes could have accomplished everything in the geological past. Rather than imagining that the earth had been more violent in the past that at present, he assumed that geological and meteorological causes had proceeded through all time at the same rate and level of intensity as they were occurring at in the England of his own day. As he later put it, "the present is the key to the past"—a lesson he had learned from Buckland, who had given a bone to a hyena to see if the teeth marks matched those on the Kirkdale bones.

In Lyell's geological scheme, as opposed to Buckland's, the mighty forces at play were not titanic and intermittent global convulsions but the coral polyp building up the reef, the raindrop hollowing out stone, a slight elevation or depression in an earthquake, an occasional and modest outpouring of lava. Over time these processes had built the greatest structures on Earth. As Buckland had suggested in his lectures, constant erosion under flowing surface water had caused constant deposition somewhere else—and explained the building up of thick sequences of strata over limitless spans of time. Because Lyell saw the whole history of nature as completely uniform, he accepted the name "uniformitarianism" for his approach.

Lyell's three-volume *Principles of Geology* (1830–33) went through many revisions and became arguably the most influential and widely read book on geology ever written in any language.[24] Although rambling and anecdotal, the book offered a great travelogue through time and space, a kind of geological grand tour written by a superb stylist. It entertained and informed not only fellow naturalists but general readers as well. Most important, it aimed to revise common notions about Genesis. Lyell's travels and conversations and reading had convinced him that many theologians and geologists—including his old teacher Buckland—were misleading the public by suggesting that a literal interpretation (by which he meant a realist interpretation) of the Creation narrative in Genesis remained tenable. Without ever mentioning Genesis, he set about to show that an accurate reading of the geological record disallowed the traditional view.[25] Lyell wanted to establish his uniformitarian scheme as a general framework for geology, providing a narrative structure and plot for the thousands of accumulated but uncoordinated observations of geologists over the previous half century or more.

In scriptural terms we could say that Lyell gave the literal facts and words of geology their own figurative meaning, separate from that of the Bible. He took "literal" geological facts and wove them into a great novel of Earth history, showing an equable and majestically slow course of nature. He melded the idea of providence into the idea of design, treating creation and destruction not as alternating epochs but as parallel and simultaneous incremental processes in a figurative history united by the identity of the kind and rate of change. Lyell headed off direct confrontation with Scripture by specifying that geology itself says nothing of the creation or the end of the world. Both are concealed from us, though with patience we can decipher the history in between. He aimed to free geology from any references to what was then called the "Mosaic" record (a term derived from the belief that Moses had written the first five books of the Bible). Lyell demonstrated that one could, in fact, write the story of the earth without reference to the Bible. Indeed, he treated the whole matter of the Flood in two and a half pages of his third volume, in a brief section titled "Supposed Effects of the Flood." There Lyell allowed wearily that he had "been led with great reluctance into this digression"[26]

Lyell concluded his work with the following testimony:

> In whatever direction we pursue our researches, whether in time or space, we discover everywhere the clear proofs of a Creative Intelligence and of His foresight, wisdom, and power.
>
> As geologists, we learn that it is not only the present condition of the globe that has been suited to the accommodation of myriads of living creatures, but that many former states also have been equally adapted to

> the organization and habits of prior races of beings. The disposition of the seas, continents, and islands, and the climates have varied; so it appears that the species have been changed, and yet they have all been so modelled, on types analogous to those of existing plants and animals, as to indicate throughout a perfect harmony of design and unity of purpose. To assume that the evidence of the beginning or end of so vast a scheme lies within the reach of our philosophical inquiries, or even of our speculations, appears to us inconsistent with a just estimate of the relations which subsist between the finite powers of man and the attributes of an Infinite and Eternal Being.[27]

Lyell may not have uttered the word "God," but his import is nonetheless clear: geology and theology can be reconciled, though perhaps not within the bounds of geology.

The Bridgewater Treatises

After a number of years of publishing nothing on natural theology, Buckland in 1830 stirred to action. He had been extremely busy in the seven years since the appearance of *Reliquae Diluvianae.* In 1825 he had married Mary Moreland, who became his partner in fieldwork and writing. He joined the governing council of the Royal Society in 1827, remaining active there until 1849. He traveled abroad widely in the later 1820s. He often lectured on religion and science, helped establish a geological museum, and worked with Henry de la Beche (1796–1855), the first director of the Geological Survey of Great Britain.[28] Throughout the 1820s Buckland had participated in spirited debates at the Geological Society of London, of which he had been a founding member.[29]

Lyell provided Buckland with something that every public figure desires: "an enemy to the left." The phrase, borrowed from the politics of the French Revolution, suggests that one's own appearance of holding an extreme view is diminished by the presence of someone ever further out of the mainstream. Buckland had come under severe criticism from theological conservatives for his willingness to modify his views of the extent to which geology confirmed Genesis. Now, in contrast to Lyell, Buckland appeared as a shining beacon of theological orthodoxy.

Buckland employed this opportunity with his habitual resolution and skill, using the space created by Lyell to advance his new thinking about the relation of geology and theology. This all came into public view with the publication of his *Geology and Mineralogy Considered with Reference to Natural Theology* (1836), the sixth volume in a series of treatises on natural theology commissioned by the Reverend Francis Henry, the earl of Bridgewater.[30] A few years before his death in 1829 the reverend earl had set aside £8,000 (about $650,000 in early twenty-first-century American currency) to be paid to authors who would undertake to show in detail, and with abundant evidence, how the various complexities of the natural world gave evidence of design by an all-knowing and beneficent deity. One whole volume focused on the human hand as a triumph of divine design, another on the adaptation of external nature to the physical condition of man, and so on. They were all written by scholars of the first rank. Peter Roget (1779–1869), the secretary of the Royal Society of London, wrote a volume on animal and vegetable physiology. William Whewell (1794–1866), the Cambridge scholar who coined the term "scientist," contributed a treatise on astronomy and physics. William Prout (1785–1850), the celebrated chemist and physician who had shown to everyone's amazement that the gastric juices of animals contain hydrochloric acid, penned an essay on the natural theology of chemistry and the weather.[31]

Buckland was delighted to be asked, in 1830, to write one of these Bridgewater Treatises. It put him in the company of other opinion leaders in the sciences, paid a generous honorarium, and best of all allowed him to present twenty years of his work in a context that allowed him to assert his fidelity to Christian belief while providing details of the history of the earth that either had no counterpart in Scripture or seemed to contradict the account in Genesis. He set to work in 1832, the same year that he became president of the newly founded British Association for the Advancement of Science, which positioned him perfectly to make an influential statement about the harmonies of science and religion.

Buckland's Bridgewater Treatise, especially the chapter on "The Consistency of Geological Discoveries with Sacred History," focused more on the Creation story in the first chapter of Genesis than on Noah's Deluge. Although he did not use the terms, Buckland had transformed himself from a "destructionist" into a "creationist." No longer did he expect to find evidence for the Deluge in the geological remains of the world, nor did he ask for an ongoing correspondence between Genesis and geology, either in literal fact or figurative continuity. Instead, he offered a new defense of the character of geology and of Scripture. "It is argued unfairly against geology," he wrote, "that because its followers are as yet agreed on no complete and incontrovertible theory of the earth; and because early opinions advanced on imperfect evidence have yielded, in succession, to more extensive discoveries; therefore nothing certain is known upon the whole subject; and that all geological deductions must be crude, unauthentic, and conjectural."[32] Like Isaac Newton, he believed that the changing, self-correcting nature of science made it more, rather than less, trustworthy. "It was assuredly prudent," he continued,

> during the infancy of Geology, in the immature state of those physical sciences which formed its only sure foundation, not to enter upon any comparison of the Mosaic account of creation with the structure of the earth, then almost totally unknown; . . . but the discoveries of the last half century have been so extensive in this department of natural knowledge, that . . . it may therefore be proper, in this part of our inquiry, to consider how far the brief account of creation, contained in the Mosaic narrative, can be shown to accord with those in natural phenomena which will come under consideration in the course of the present essay. . . . I trust it may be shown not only that there is no inconsistency between our interpretation of the phenomena of nature and of the Mosaic narrative, but that the results of geological inquiry throw important light on parts of this history, which are otherwise involved in much obscurity. . . . If in this respect, geology should seem to require some little concession from the literal interpreter of Scripture, it may fairly be held to afford ample compensation for this demand, by the large additions it has made to the evidences of natural religion, in a department where revelation was not designed to give information.
>
> The disappointment of those who look for a detailed account of geological phenomena in the Bible, rests on a gratuitous expectation of finding therein historical information, respecting all the operations of the Creator in times and places with which the human race has no concern; as reasonably might we object that the Mosaic history is imperfect, because it makes no specific mention of the satellites of Jupiter, or the rings of Saturn, as feel disappointment at not finding in it the history of geological phenomena, the details of which may be fit matter for an encyclopedia of science, but are foreign to the objects of a volume intended only to be a guide of religious belief and moral conduct.[33]

Several elements of this passage are worthy of mention. The first is the idea that there are periods and places "with which the human race has no concern." That may sound odd coming from a geologist who studied precisely these times and places, but Buckland was merely stating that the providential and moral order that governs human existence began with the six-day Creation in Genesis, not with the original creation "in the beginning." Second, Buckland maintained that the Bible is not an "encyclopedia" of science or culture but a "guide of religious belief and moral conduct." Thus Lyell and other scientific colleagues who argued that the Mosaic narrative could not be historically true because it is so incomplete were missing the point.

How, in Buckland's opinion, could Genesis and geology be reconciled? His answer was refreshingly clear and straightforward. Buckland had no patience for attempts at plausible reconciliations based on untested scientific hypotheses. Thus he quickly disposed of two "just so" stories suggested by ill-informed amateur naturalists. The first of these ascribed the formation of all stratified rock to the effects of Noah's Flood. This opinion, Buckland declared, "is irreconcilable with the enormous thickness and almost infinite subdivisions of these strata, and with the numerous and regular successions which they contain of the remains of animals and vegetables, differing more and more widely from existing species—as the strata in which we find them are placed at greater depths."[34] He also dismissed the notion that the strata had somehow been formed at the bottom of the sea during the interval between the creation of humans and Noah's Deluge, and that later, at the time of the Flood, the land and water parts of the earth had changed places. "To this hypothesis also," wrote Buckland, "the facts I shall subsequently advance offer insuperable objections."[35]

Buckland urged a new reading of the Creation story in Genesis in terms both literal and figurative but not realistic—precisely the sort of division then current in purely biblical scholarship. Buckland considered two ways of reconciling the Creation story with the facts of geology, both of which have remained popular among evangelical Christians into the twenty-first century. The first of these, which came to be called the "day-age" hypothesis, held that the six "days" of Creation should be read figuratively as indicating indefinitely long time periods rather than actual twenty-four-hour days. This allowed the inspired narrative to be reconciled with the results of science.

Buckland also considered a second hypothesis, based on the curious gap between Genesis 1:1 and Genesis 1:2. Genesis 1:1 reads, "In the beginning God created the heaven and the earth." The next verse describes the earth as being "without form, and void; and darkness was upon the face of the deep. And the spirit of God moved upon the face of the waters." The origin of the world, described in the first verse, might have occurred long before "God moved upon the face of the waters" and began the six-day creative week associated with Adam and Eve. Sufficient time could have elapsed between these two events to accommodate all of the paleontological evidence geologists had uncovered. This so-called gap view required no altering, stretching, figuring, or massaging of the text. Although Buckland saw no sound objection to interpreting the word "day" as a long period of time, he preferred a literal to a figurative solution: "There will be no necessity for such extension [of a day to an age], in order to reconcile the text of Genesis with physical appearances, if it can be shown that the time indicated by the phenomena of geology may be found in the undefined interval, following the announcement of the first verse."[36]

Buckland made a shrewd choice. The great advantage of the gap theory over the day-age view as a reconciliation of Genesis and geology is that virtually the whole geological history of the earth took place in a space and time about which Scripture is silent and can therefore never be contradicted. Thus, from Buckland's standpoint, science and Scripture were brought into permanent harmony. Working within the framework of a gap between the first two verses of Genesis, Buckland proceeded to develop his geological history of the world as a series of events transpiring on a time scale even Lyell could have accepted, nestled between Genesis 1:1 and Genesis 1:2.

Conclusion

The scientists who debated the relationship between Genesis and geology in nineteenth-century Britain tried in different ways to accommodate their religion and their science. William Buckland, arguably the leading scientist of his day, advocated a close relationship between the exploration of nature and the worship of God. While his version of this relationship changed through time, we should not so qualify our understanding as to miss the essential point that neither he nor most of his fellow geologists saw much in their science that threatened their faith, their morals, or their confidence in Scripture as the revealed word of God. They may have experimented with and altered the means of providing such accommodations, moving from geology as evidence of providential intervention to geology as evidence of providential design, but they remained faithful to both their religion and their science. Their experience showed that science and religion, Genesis and geology, were not inevitably antagonistic and were often mutually reinforcing.

7

"Men before Adam!": American Debates over the Unity and Antiquity of Humanity

G. Blair Nelson

In 1850, the Harvard zoologist Louis Agassiz startled readers of the *Christian Examiner,* a popular religious magazine, with the claim that God had created each human race separately in its own native region and thus that most nations were "not related to Adam and Eve." Thirteen years later, Charles Lyell, the celebrated British geologist, brought out a new book, *Geological Evidences of the Antiquity of Man,* that introduced the reading public to compelling new scientific evidence that humankind had originated far earlier than the commonly accepted date for the creation of Adam and Eve, about six thousand years ago. In different ways Agassiz's and Lyell's announcements contradicted the deeply held Christian belief that Adam and Eve were the parents of the entire human race.[1] Neither scientific development involved notions of human evolution, but both reflected vigorous debates in the years before the publication of Charles Darwin's *Descent of Man* about the unity and antiquity of humanity—debates frequently obscured by the furor created by Darwin's later contention that humans had descended from apelike ancestors.

When human evolution emerged as a controversial issue for Christians about the mid-1870s, well-read believers were thus already accustomed to thinking about theories of human origins in apparent conflict with traditional readings of Genesis.[2] This essay explores the strategies that they developed to deal with the challenges raised by pre-Darwinian attempts to modify human history in hopes of shedding additional light on the relationship between science and Christianity during the often tumultuous years of the nineteenth century. But first we must begin with what Christians typically believed about the origin of humanity and the practice of science before these challenges arose.

Views on Human Origins before the 1830s

Most American Christians in the early nineteenth century read the biblical Creation stories as literal history. God, they believed, created Adam and Eve about six thousand years ago, shortly after getting the rest of the cosmos started. Descendants of the first couple multiplied rapidly, perhaps because of their extraordinary longevity, and quickly occupied the earth. In under two thousand years the first civilization had become so corrupt that God sent a flood to drown all but the family of the one righteous man, Noah. After the flood, Noah's children began to repopulate the earth. Once again human sinfulness led to judgment, and this time God chose confusion rather than extermination to chastise the human race. In an incident associated with the building of the Tower of Babel, God miraculously replaced the unified language and culture of Noah's family with the diversity of tongues that has characterized the world ever since. Relying on the lists of ages and "begats" found in the early chapters of Genesis, biblical scholars attempted to establish the date of the Creation. The most famous estimate came from the seventeenth-century Anglican archbishop James Ussher, who calculated that the earth's history had begun on "the entrance of the night preceding the twenty third day of" October 4004 B.C.[3]

This information drawn from the Bible did not stand alone in Christian thinking. Science supported and embellished it. The most important example was the concept of nature as a fixed order established by God. The intricate design of plants and animals and their perfect adaptation to their environments testified to God's design in nature and revealed God's goodness and wisdom in providing for creation. This perfect order had reigned since the beginning. No species of plant or animal had ever changed, been added, or lost. Within a species new varieties had emerged or become extinct, but the essential order of creation continued undisturbed. In the same way the human family had grown from the first couple into today's profusion of nations, cultures, and races. To understand another human race was to understand its place in the biblical narrative.

This picture of the world did not go without challenges from science. In the course of the eighteenth century scholars dramatically increased their estimate of the age of the earth. In 1775 French naturalist Georges-Louis Leclerc de Buffon, suggested that the earth was about seventy-five thousand years old, far off the mark by modern standards but a huge departure from the traditional figure of about six thousand years. The earth no longer appeared relatively unchanged through time, but had passed through a succession of ages divided by violent upheavals that had refashioned its surface. Scholars began to understand fossils as the remains of living creatures rather than minerals that merely resembled flora and fauna, and the growing catalog of strange fossilized creatures testified to a time when the world was occupied by a different and now extinct set of tenants. By the turn of the nineteenth century, Western intellectuals understood that the earth and its nonhuman populations had had a very long history before Adam and Eve.[4]

Informed Christians typically weathered this challenge by limiting human existence to about six thousand years and by theorizing that God took the extraordinarily long time of geological history to prepare the earth for its human inhabitants. Mountains rose and eroded; seas swelled and drained; species died out to be replaced by new species, ever increasing in complexity and perfection until God finally put humans on the earth. The events of the earth's long prehuman history were God's way to prepare a home for humankind.

Many Christians experienced little trouble accommodating this long Earth history to the Bible. In fact, two interpretations of chapter 1 of Genesis opened up the text to these enormous spans of Earth history, both of which became popular in the nineteenth century. One approach placed geological time in a supposed "gap" between the creation of the earth in Genesis 1:1–2 and the six days of Creation beginning in verse 3. In the other, the "day-age" theory, the six days of Creation were understood metaphorically as six great geological epochs rather than literal twenty-four-hour days. Countless books and articles appeared throughout the nineteenth century explaining how modern geology showed exactly the same order of development of the features of the earth and its inhabitants as did the Bible in the narrative of the six days of Creation.[5]

Starting in the late fourteenth century, and especially in the seventeenth and eighteenth centuries, challenges to the traditional Christian worldview came faster than ever before. As exploration increased, Europeans learned volumes about the world that lay just beyond their horizon. Knowledge of China and sub-Saharan Africa, regions that had been little more than mythological lands, grew as exploration and trade developed. The European discovery of the New World opened up whole continents populated by unknown flora and fauna and exotic peoples with unfamiliar customs and tongues. How had humans come to be *so* different if they are all part of the same family? How had they come to live in virtually every part of the globe, separated by vast oceans, virtually unaware of each other until the last few centuries?

Not all the challenges to the accepted reading of the Bible and its account of

Creation came from science. Isaac de La Peyrère, a French Protestant of Jewish descent, irritated seventeenth-century scholars with his reinterpretation of the story of Adam and Eve. In his *Men before Adam* (1655), he argued that the Bible really taught that Adam and Eve were not the first human beings, but merely parents of the Jewish people. All other races had been created earlier than Adam and Eve and belonged outside the Old Testament stories, which concerned only the Jewish people. The existence of preadamites, people before Adam, explained a number of awkward problems with certain biblical stories, such as how Cain had found a wife when he and his brother were Adam and Eve's only offspring, and who had peopled the biblical land of Nod. La Peyrère's theological work was roundly condemned as heretical throughout the seventeenth and eighteenth centuries. La Peyrère himself was forced to recant and convert to Catholicism, but his ideas were not forgotten.[6]

Refutations of preadamism flowed from European pens for more than a century after La Peyrère as Europeans wrestled with the problem of connecting the inhabitants of the New World with those of the biblical world. If La Peyrère were right, then their location in distant lands and their physical and cultural differences would need no explanation other than God's will. Few writers adopted preadamism as a solution to the problem, but the discomfort Europeans felt about their traditional understanding of human origins kept La Peyrère's ideas alive.

Yet European men of science had more to do than just ponder the place of the New World in the Christian story; they had a legion of new plants, animals, peoples, and cultures to investigate. Before the nineteenth century, disciplines such as biology, geology, and anthropology were not independent disciplines but came under the general rubric of natural history. The self-appointed task of eighteenth-century natural historians was taxonomy, the orderly classification of natural objects—plants and animals, rocks and minerals, and so on—in order to understand the natural relations among things. The binomial nomenclature used today, under which a plant or animal is given two Latin names, that of its genus and that of its species, was introduced in this period by the Swedish botanist Carl Linnaeus (1707–78). Naturalists included humans in their classification schemes, and the continuing attempt to classify the various races of humankind spawned the new science of ethnology.

Ethnologists in the eighteenth and nineteenth centuries were far more impressed with the differences among various peoples than their similarities. Sensational stories of Hottentots mating with gorillas or orangutans persisted in Europe and America, testifying to the "otherness" that such distant people represented to them.[7] They also suggested a gradation of human types from those closest to animals to those most removed. Nevertheless, few questioned the essential unity of the human family because interfertility was an accepted criterion for determining common species, and evidence abounded that mixed-race couples could produce fertile offspring. The diversity of physical appearances among peoples from different parts of the world was usually credited to the effects of environment; climate, diet, even civilization could, it was thought, alter skin color, skeletal structure, and physiological processes. Over time a population would adapt to a new environment. For example, the intensity of the Sun in the tropics darkened the skin of Africans, which made them better able to endure the torrid climate. Thus the biblical story of humankind's descent from Adam and Eve was largely supported by scientific thinking. Unity of species implied a common origin, and the human ability to adapt to environments accounted for racial diversity.

As knowledge increased, however, this happy union of science and Christianity grew strained. By the eighteenth century, Europeans had learned a great many things about the rest of the world that they were hard pressed to fit into the Bible's account of things, and some thinkers began to entertain the idea that possibly humans are not all part of the same family. The Scottish aristocrat and man of letters, Henry Home, Lord Kames, popularized this idea. In his *Sketches of the History of Man* (1774), he dismissed the environmentalist explanation as incredible. It was far more likely that humans were made for the climates of their various lands, he declared, than that they adapted to those climates after their arrival from elsewhere. He toyed with the idea that each race began as a separate creation in its own native land.[8] This theory is called *polygenism* because it involves a *polygenesis* (*poly,* many + *genesis,* creation or origin), the idea that the various races of humanity had separate and independent beginnings, that, for example, the first American Indians did not migrate across the Bering land bridge but were created in America.

Not surprisingly, most Christians rejected polygenism, arguing that the Creation stories of Genesis and the Christian doctrines of the universality of sin and the offer of redemption through Christ's sacrifice presumed a *monogenesis (mono,* single + *genesis),* a single origin for the human race. In the end, Kames choose to maintain orthodoxy, at least in print, by rejecting polygenesis in favor of the miraculous origination of the human races at the Tower of Babel. There physically distinct races, adapted to diverse environments, had come about when God confused the tower builders' tongues.

Kames found a determined opponent in Samuel Stanhope Smith (1751–1819), a staunch Presbyterian professor at the College of New Jersey (later Princeton University). Smith, convinced on both scientific and religious grounds that the environment had modified human anatomy, cited the whitening of African slaves as evidence in support of his view.[9] Ironically, Smith's defense of the doctrine of the unity of humankind was more naturalistic than Kames's refutation. Kames ascribed human racial diversity to a miracle that had not been recorded in the Bible, while Smith provided an entirely naturalistic explanation of the same phenomenon: the effects of the environment. Part of Smith's reluctance to invoke miracles arose out of his Calvinist convictions that the age of miracles had ended, but politics played a role as well. The fledgling American republic could justify its democratic ideals only if all humans were equal, potentially if not actually. If distinctions between various human races could not be explained without invoking an act of God, then the egalitarianism of American political rhetoric was unnatural.

Smith's defense of the environmentalist theory became required reading for many Americans in the early nineteenth century, but the idea that God had created the different human races separately, placing each into its own particular region, did not die at Smith's hand. Polygenism gained strong scientific support in the mid–nineteenth century, particularly in America. In fact, the plurality of human origins became so identified with American scientific circles that Europeans began referring to its advocates as the "American school" of ethnology. The polygenists promoted their views in popular as well as scholarly literature, igniting a controversy—the "unity of mankind" debate—that raged in the years before the Civil War. By 1840, most Christian leaders had accepted the old Earth of the modern geologists, and within a decade or so many had embraced Pierre-Simon de Laplace's novel nebular hypothesis about the natural development of the solar system from a rotating gaseous mass.[10] The debate over the origin of the human races, through now largely forgotten, posed a far greater scientific challenge to traditional Christianity than the age of the earth or the nebular hypothesis. Before most Americans became familiar with the name of Charles Darwin, the most contentious issue involving science and religion focused on the early history of humanity.

The Work of the American School

Both Britain and France contributed significantly to the emergence of polygenistic thinking. But it was in the United States, a nation ruled by people of primarily European descent, sharing the land with its indigenous inhabitants and millions of imported Africans, that ethnologists brought the polygenistic race theory to its highest point. A core group of four men produced the key ideas and texts that put America at the forefront of mid-nineteenth-century ethnology: Samuel G. Morton (1799–1851), George R. Gliddon (1809–57), Josiah C. Nott (1804–73), and Louis Agassiz (1807–73). Though linked by nothing more than friendship and a few cooperative publishing ventures, these four men—two physicians, an itinerant Englishman, and one of America's leading men of science—came to be associated most closely with the American school of ethnology.

Morton, the patriarch of the group, provided the first empirical support for the theory of the plural origins of humans, evidence that served as the centerpiece of the polygenistic argument. A practicing physician and respected professor of anatomy in Philadelphia, Morton in the 1830s began measuring the interior volumes of the skulls from his personal collection of hundreds of specimens. Taking several series of skulls, one for each race, he determined their volumes by stopping up the openings, filling them with tiny pellets of shot, and then measuring the volume of that shot with a graduated cylinder. The resulting data indicated that Caucasians possessed the largest mean cranial capacity at 87 cubic inches, followed by Mongolians, Malayans, and Americans (that is, Indians), with Ethiopians (black Africans) at the bottom with 78 cubic inches. Morton's racial prejudices, reflective of Euro-American views in general, unconsciously but clearly influenced his results. For example, he favored small-brained individuals in calculating the size of American Indians but eliminated small skulls from his Caucasian sample. Nevertheless, Morton believed in his own objectivity, and his detailed descriptions of his meticulous procedures and the quantitative character of his data deeply impressed other men of science (fig. 7.1).[11]

In 1839 Morton published his findings in an ethnological study of the native people of the Americas (figs. 7.2 and 7.3). He contended that, with the exception of the Eskimos, all Native Americans belonged to the same race or type, distinct

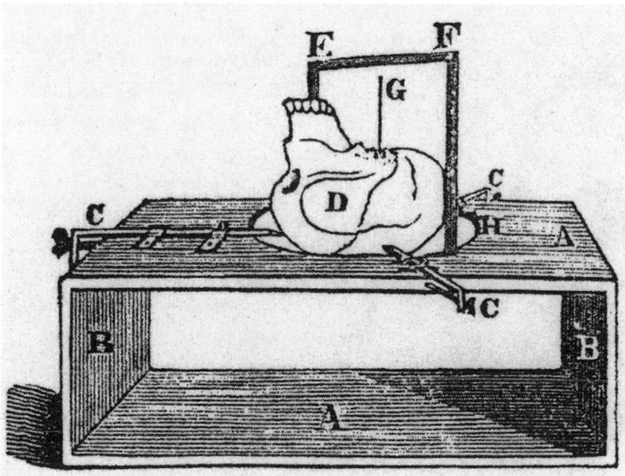

Figure 7.1. The device Samuel Morton used to support each skull while measuring its interior volume. From Morton, *Crania Americana; or, A Comparative View of the Skulls of Various Aboriginal Nations of North and South America, to Which Is Prefixed an Essay on the Varieties of the Human Species* (Philadelphia, 1839), 254.

DACOTA.

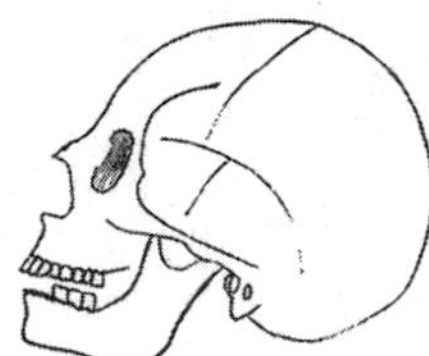 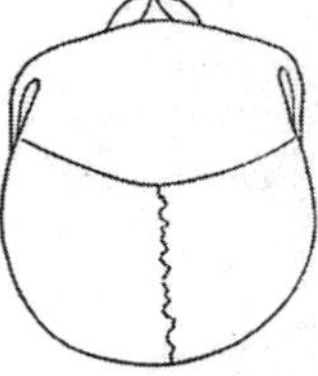 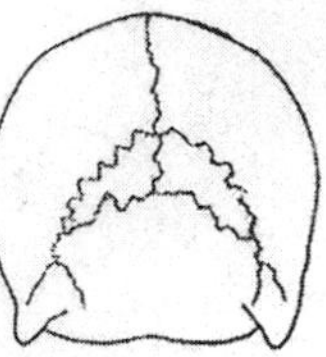

I received this skull from the late Dr. Poole, of this city, but could obtain no particulars, excepting the fact of its having belonged to a Sioux warrior of bad character, and who was killed by some act of violence on the northwestern frontier. The small squared head, the great comparative breadth between the parietal bones, and indifferent frontal development, correspond precisely with those features as observed in the individuals of the Sioux delegation already mentioned.

MEASUREMENTS.

Longitudinal diameter, ...	6.7 inches.
Parietal diameter, ...	5.7 inches.
Frontal diameter, ...	4.2 inches.
Vertical diameter, ...	5.4 inches.
Inter-mastoid arch, ...	14.7 inches.
Inter-mastoid line, ...	4.4 inches.
Occipito-frontal arch, ...	13.5 inches.
Horizontal periphery, ...	19.8 inches.
Internal capacity, ...	85. cubic inches.
Capacity of the anterior chamber, ...	36. cubic inches.
Capacity of the posterior chamber, ...	49. cubic inches.
Capacity of the coronal region, . . .	16.6 cubic inches.
Facial angle, ...	77 degrees.

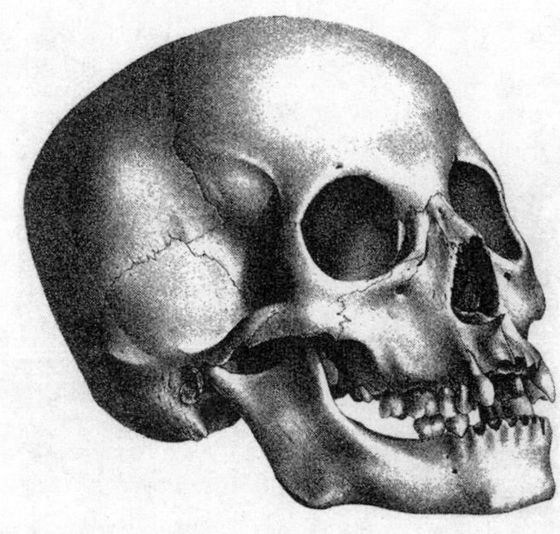

Figs. 7.2 and 7.3. Morton published his quantitative data along with high-quality illustrations of the skulls he studied, such as that of the Dakota Indian. In contrast to the precision of these measurements is the fact that he depended on his suppliers for information on each skull's former inhabitant, such as Dr. Poole's claim that this was "a Sioux warrior of bad character." From Samuel Morton, *Crania Americana; or, A Comparative View of the Skulls of Various Aboriginal Nations of North and South America, to Which Is Prefixed an Essay on the Varieties of the Human Species* (Philadelphia, 1839), 198 and plate 39.

from all other types. Morton denied that Native Americans had come to the New World from Asia, an important plank in the environmentalist theory of racial origins.[12] Five years later he published a similar study, of 137 Egyptian skulls and mummified heads, in which he concluded that the ancient Egyptians had been Caucasian—not black African, as some believed—and that racial distinctions were as old as recorded history. He also concluded, both from his personal collection of Egyptian heads and from depictions of people on Egyptian monuments, that "negroes were numerous in Egypt, but their social position in ancient times was the same that it now is, that of servants and slaves."[13] Although Morton stopped short of advocating polygenism in either of these works, his findings left little room for any other view.

Morton had acquired his Egyptian artifacts from Gliddon, an Englishman who had lived in Cairo from boyhood. The early part of the nineteenth century saw a dramatic rise in research on ancient Egypt, and Gliddon's position in Cairo allowed him to hobnob with the best scholars, collect artifacts, and turn himself into an Egyptologist of no mean ability. He began touring the United States in the 1840s, giving lavishly illustrated public lectures on Egyptian antiquities, sometimes concluding dramatically with the unwrapping of an authentic mummy.

Gliddon contributed to the polygenists' project in several ways. First, he supplied the material for what may be called the Egyptian-monument argument. American-school publications overflowed with images from ancient Egypt, as well as depictions of actual human remains such as skulls and mummified heads. The polygenists detected no differences between these remains, which dated back to over three thousand years B.C., and the skulls of modern races; and they believed too little time had elapsed between the Edenic Creation and the construction of the Egyptian monuments for the distinct racial types revealed in paintings and skulls to have developed by natural means. They reasoned that if all human development had to fit within the accepted time scale of six thousand years, the environmentalist explanation of racial diversity would need to be discarded. Although Gliddon loved to ridicule the popular biblical chronology associated with Archbishop Ussher, his own was only centuries longer.

Gliddon not only introduced his polygenist colleagues to the best scholarly literature of the day on Egyptology and biblical criticism, but edited and wrote for the American school's major publications. His contributions contained little of the rigor and restraint that typified scientific literature. He often lectured his readers on such topics as errors in the King James Version of the Bible, and by the 1850s he was doing little to mask his feelings of contempt toward the clergy and traditional Christianity. His inflated showmanship occasionally tarnished his reputation. For example, advertisements for his series of lectures in Boston promised the unwrapping of a mummy of an Egyptian priestess, to be followed by scientific inspection on stage. The local press altererd the mummy's status to princess, and Boston's scientific and social elite turned out in force. As notables including Oliver Wendell Holmes and Louis Agassiz looked on, the dramatic moment turned comic as the last wrapping fell away to reveal a male body.[14]

Gliddon's negative attitude toward organized religion was shared by Nott, a prominent physician and friend of Morton's who made his contribution from the comfort of a successful medical practice in Mobile, Alabama. Like Gliddon, Nott loved to indulge in parson bashing, privately calling ministers "skunks" who perverted truth to support their authority and social position.[15] He entered the unity debate in 1843 with an article in the prestigious *American Journal of the Medical Sciences* entitled "The Mulatto a Hybrid—Probable Extermination of the Two Races If the Whites and Blacks Are Allowed to Intermarry." The title, referring to the offspring of black-white unions, revealed both his thesis and the social concern that drove his interest in anthropology. The article commented on census statistics that indicated mulattoes were less healthy and long-lived than full-blooded Africans.[16] Nott discussed the physiological, anatomical, and intellectual differences between Caucasians and Africans, concluding that whites and blacks belonged to different species. His work caught the attention of Morton, who began a close friendship and professional collaboration with Nott in 1844. Much of their work focused on hybridity, because most naturalists believed that if two creatures could produce fertile offspring they belonged, by definition, to the same species. America held copious evidence of the fertility of mulattoes. The two physicians labored through the late 1840s and into the 1850s to prove that the production of fertile offspring could not be a criterion for identifying species. Morton and Nott regarded mulattoes as hybrids, claiming (on the basis of largely anecdotal evidence) that they were weaker and less fertile than full-blooded Africans.[17]

Nott, who preceded Morton in explicitly adopting polygenism, became the American school's leading spokesman, mediating between Gliddon's rashness and Morton's scholarly reserve in the campaign to persuade America of the truth of their common message. For Nott, as perhaps for his colleagues, ethnology possessed a pressing political function. As an ardent racist (fig. 7.4) and passionate supporter of slavery, Nott crusaded for polygenism in defense of his southern way of life. Most southerners defended slavery on theological grounds; Nott, who had rejected Christianity as a college student, chose science.[18]

Agassiz gave the polygenist campaign a great boost when he announced in the latter 1840s that Negroes and Caucasians represented separate creations. Before coming to America in 1846, the Swiss naturalist had already established an international reputation as one of the greatest authorities in the world on glaciers and fossil fish. A man of great personal charm, deep devotion to science,

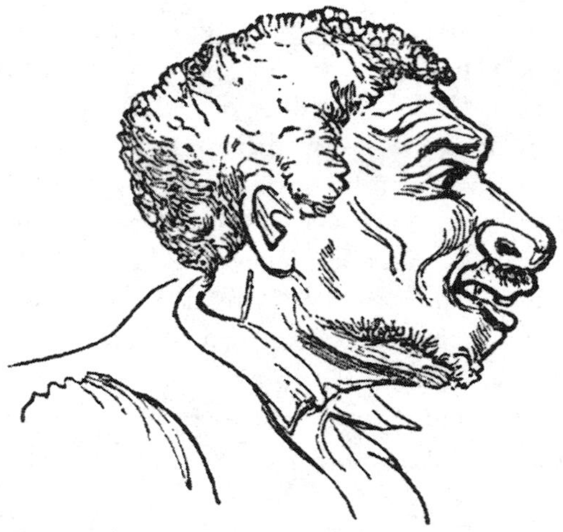

Figure 7.4. A comical sketch of a person Nott knew from the "streets of Mobile." Such caricatures fueled hostile criticism of his work. From Josiah Nott and George Gliddon, *Types of Mankind* (Philadelphia, 1854), 259.

and talent for popularizing his ideas, Agassiz quickly became the darling of both the scientific community and the American press. America loved Agassiz, and Agassiz so loved America that he made it his home, settling comfortably into a professorship at Harvard.[19] His endorsement of the American school's program was a huge stroke of good luck for them. "With Agassiz in the war the battle is ours," gushed Nott in a letter to Morton in 1850. "This was an immense accession for we shall not only have his name, but the timid will come out from their hiding places."[20]

Agassiz had come to America already advocating polygenesis for plants and animals, based on his observation that the earth could be divided into a number of areas or "zoological provinces," each characterized by indigenous flora and fauna that seldom ranged beyond their original boundaries. The evidence of distinct biological provinces made the idea of a common origin of animals from differing regions seem implausible to him, but before settling in the United States he believed that humans provided the one exception to the law of zoological provinces because they ranged over the whole earth. He was not long in America before he adopted human polygenism. Two experiences during a visit to Philadelphia precipitated his conversion. One was the time he spent with Morton poring over the doctor's famous collection of skulls. The other was his first view of Africans at close range, when "men of color" served him at his hotel. "I could not take my eyes off their face in order to tell them to stay far away," he wrote with revulsion. "And when they advanced that hideous hand towards my plate in order to serve me, I wished I were able to depart in order to eat a piece of bread elsewhere, rather than dine with such service." Never again did Agassiz think that blacks and whites belonged to the same species.[21]

Agassiz saw no contradiction between his science and his Unitarian faith.

For him, the goal of science was to discover the God-ordained order in nature. Every law of nature and everything in nature, including every species of plant and animal, expressed an idea in the mind of God. Agassiz doggedly opposed all theories of evolution or development in nature because God's ideas do not change. Where Darwin detected evidence for transformation, such as in the increasing complexity of organic forms through time, Agassiz saw a sequence of catastrophes and re-creations. He believed that God successively wiped out life on Earth, only to repopulate it with more complex creatures, right up to the appearance of *Homo sapiens,* God's crowning achievement. It did not matter that Africans, for example, had not descended from Noah. All humans, no matter where they originated, were spiritually related in God's mind. Agassiz could thus simultaneously believe in racial polygenesis and the unity of humanity because he located polygenesis in nature and unity in the mind of God.

But what about the Bible? Agassiz regarded the traditional view of the origin of humans, animals, and plants as "a mere hypothesis." Adopting La Peyrère's approach, he claimed that Genesis intimated that Adam and Eve were not the first people and that Moses "never meant to say that all men originated from a single pair."[22] In his opinion, Genesis had been written from the limited and nonscientific viewpoint of the ancient Hebrews; therefore, questions such as the origin of the races should decided only by science unfettered by religious constraints. Since science so clearly revealed the evidence of design in nature, Agassiz thought that it ruled out "forever the idea of a natural development from law, and acknowledges a personal intelligent God."[23]

With Agassiz's contribution, the American school's arsenal of ideas reached completion. Morton's cranium studies gave an impressively empirical flavor to the argument that human racial differences are so physiologically and anatomically pronounced that the races must be distinct and unrelated. Gliddon's Egyptology placed racial diversity far enough back in human history to debunk the theory that racial characteristics had been produced by the effects of differing environments over time (fig. 7.5). Morton and Nott bombarded the standard view that different species cannot produce fertile offspring with a barrage of counterexamples. Agassiz's theory of zoological provinces provided the theoretical underpinning that tied the American school's racial polygenism into a broad understanding of nature.

After Morton died in 1851, Nott and Gliddon pulled all of the above work together and placed it squarely before the American public in one volume, *Types of Mankind.* Its 738 pages contained their reworking of Morton's unpublished writings, Nott's work on ethnology, Gliddon's extensive criticisms of the Bible, and an assortment of articles by friends of Morton. Most important, Agassiz contributed a piece on his zoological provinces, complete with a foldout tableau

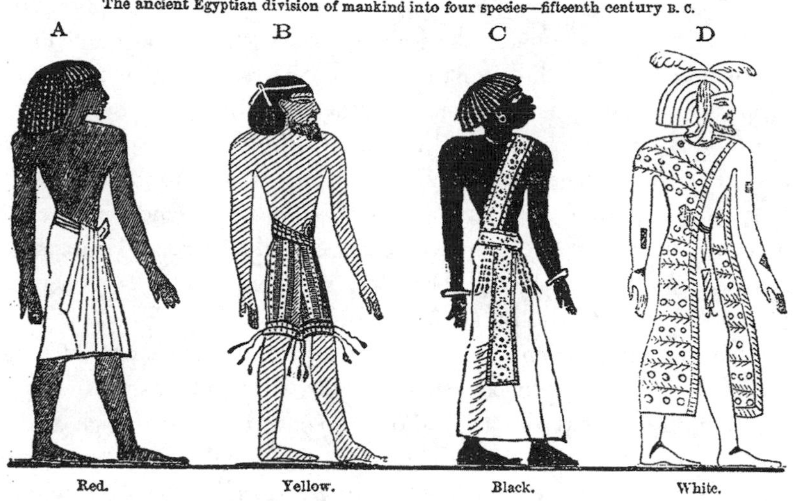

Figure 7.5. Examples of the ancient Egyptian art that Nott and Gliddon claimed proved that racial types have been fixed for all of human history. From Josiah Nott and George Gliddon, *Types of Mankind* (Philadelphia, 1854), 85.

illustrating the human and animal species that inhabited each zone and a map of the earth displaying each province in a different color. Released in 1854, *Types of Mankind* was an immediate success. The subscription list numbered almost one thousand, and the book quickly went through at least nine editions, a major achievement for an expensive book on science. Although it sold largely on Agassiz's reputation, the book fell short of the academic standards of the time. Christian leaders had been following the unity debate and had made concerned comments on it in the religious press, especially after Agassiz entered the fray in 1850. But the appearance of *Types of Mankind* pushed the defenders of orthodoxy to intensify their efforts.[24]

Christian Responses to the American School

Some Christians dismissed polygenism merely because it conflicted with "the word of God" or because it undermined the doctrines of universal sin and salvation. But most Christian leaders by the mid–nineteenth century had far too much respect for science to condemn polygenism without citing scientific reasons. Many theologians and ministers paid close attention to scientific developments, and some even practiced science in one form or another. Most Christian journals reported on key scientific literature with keen interest and appreciation. The respect for science rested on the deeply held belief that divine revelation appeared in two books, the Bible and the Book of Nature. Since nature came from the hand of God just as the Bible came from the mouth of God, both spoke truth to those who studied them properly. Antebellum Christian leaders felt close enough to science to engage it in a way that became difficult for laypersons only a few decades later. After the Civil War specialization, jargon, and the sheer complexity and volume of scientific knowledge increasingly separated scientists from the general public. The career of the American school's leading opponent, the Reverend John Bachman, illustrates this shift.

Bachman, a Lutheran minister and a professor of natural history in Charleston, South Carolina, both shepherded his congregation and pursued an extraordinary scientific career as a leading authority on quadrupeds. Although Gliddon and Nott dismissed him as a clerical ignoramus, his qualifications as a natural historian surpassed theirs. He publicly debated Morton's views on hybridity right up to the Philadelphian's death, refuting Morton's examples of fertile hybrids by the score. Bachman argued that the established anatomical criteria for identifying members of a common species—the similarity of teeth and bones—showed the human race to be a single species. Bachman clearly regarded polygenism as a heresy, but, like many Christians of his day, he believed that it should be refuted primarily on scientific rather than biblical grounds. In spite of Nott and Gliddon's jeering, he emerged as their leading scientific critic.[25]

Christian writers less scientifically qualified than Bachman also evaluated the scientific claims of the polygenists. For example, the Presbyterian theologian Charles Hodge published sophisticated critiques of the work of Morton and his disciples that foreshadowed those made in 1981 by the Harvard paleontologist Stephen Jay Gould. Both attacked Morton's findings as based on too small a group of skulls to be statistically significant, rejected the assumption that brain size correlated with intelligence, and criticized Morton's failure to distinguish between male and female skulls. Hodge also cited linguistic evidence for the Asian origins of American Indians, compiled by scholars who studied ethnology by exploring the relationships between languages rather than between physical characteristics. But Hodge did not have to borrow from outside his discipline to expose the American school's shortcomings. His theological method involved careful attention to the precise meanings of words and their logical interactions. On this basis he scrutinized Agassiz's logic, focusing particular attention on his sloppy use of the word "species." His conclusion: "Agassiz is a genius, but he is no logician."[26]

Christian writers routinely subjected new ideas to two tests: the quality of the science and the character of the scientist. Although the belief that true science could never contradict Christian doctrine remained unshakable, especially for Protestants, the hostility of some men of science to religion fueled a fear of science being used for infidel purposes. Startled by the outrages of the French Revolution and its aftermath, many Americans associated the revolutionaries' atrocities with the unbelief of their leaders and feared the infiltration of similar infidelity into American culture, particularly its science. Article after article appeared in the Christian press lamenting the "science falsely so-called" of infidels, who studied God's creation without proper respect for the Creator.

The semipopular character of *Types of Mankind* made it an easy target, since it lacked both the rigor and the restraint that critics expected of scientific writing. Even worse in the eyes of Christian readers was Gliddon and Nott's irreverent rhetoric. One critic scornfully suggested that the American's school's flagship publication had been misnamed: "It should not have been 'Types of Mankind,' but 'Types of Infidelity.'"[27] Gliddon, especially, was identified as a man too closely associated with dangerous and infidel ideas to be trusted on any topic pertaining to religion. His contribution to *Types of Mankind,* which he aimed specifically at the religious opposition, drew the most fire. In fact, several long and detailed reviews in Christian quarterlies ignored the bulk of the book and focused exclusively on Gliddon's contribution, depicting the author as a dilettantish intellectual lightweight with an inflated ego who confused an acquaintance with great scholars with scholarship itself. Even the review of *Types of Mankind* in the *Christian Examiner,* which had published Agassiz's earlier article on polygenism and had defended his right to pursue his science free of religious constraints, attacked Gliddon's work with venom.[28] In view of the many attacks on Gliddon, it is understandable why Nott believed that *Types of Mankind* would have had a much better reception if Gliddon had not been involved.[29] But the fault was not all Gliddon's; Nott himself shared the blame for the book's irreligious excesses.

Morton and Agassiz fared much better. Although Christian writers acknowledged that these two distinguished men of science had provided the scientific grounding for the offending theories, their scientific reputations largely spared them from ridicule. One critic even tried to disassociate Morton's name from *Types of Mankind,* speculating that, had he lived, "who can say that he would not indignantly disown all that Messrs. Nott and Gliddon have put forth under color of his authority?"[30] Morton, who believed that polygenism could be reconciled with Christianity, refrained from baiting Christians in the way Nott and Gliddon did. Because of Agassiz's repudiation of Noah's Flood as a geologically significant event, he may have been more theologically suspect than Morton among orthodox Christians, but his view of science as a religious pursuit—the study of God through the study of God's handiwork—endowed his work with an air of piety. In the 1860s Agassiz emerged as the foremost scientific critic of Darwinism in America, and, despite his close association with polygenism, his reputation in the Christian community soared.[31]

Although African-American Christians had a special stake in the unity debate, relatively few of them possessed the opportunity and education to participate. In 1861, however, a black Methodist newspaper, *The Christian Recorder,* presented an eighteen-part review of ethnological theory, devoting more space to the issue in a four-month period than most other newspapers did in a decade. The anonymous writer took the conventional Christian view that although the Bible does not teach science, it has enough information "to establish the doctrine of the unity of the human races." (Citing some of Gliddon's data in support, he also argued that the ancient Egyptians had been black!) In 1863 the paper reviewed Darwin's work positively, suggesting that his religious critics had misunderstood him and crediting him with undermining the case for polygenism. "One question of much dispute seems to be settled by Mr. Darwin," it observed; "thus the Caucasian, the Malay, and the Negro, according to his facts, are varieties of species, and may all have descended from a single pair, as set forth in the scriptures." Long after the publication of the *Origin of Species* (1859) many African Americans continued to view polygenism as a greater threat than Darwinism.[32]

Religious critics often deployed their scientific and theological arguments effectively, but they remained handicapped by the lack of a suitable alternative. Some Christian commentators explained the origin of the human races miraculously, associating it with the curse on Noah's son Ham or with the experience at the Tower of Babel, but many appealed to the environmental theory of Samuel Smith, which they had learned in their school days. In doing so, they were, ironically, defending the faith with a theory more naturalistic than that of their "infidel" opponents. Nature, rather than God, had produced racial differences. Whichever tack they took, the great majority of antebellum American Christians who left their views on the topic solidly opposed polygenistic race theory, as well as any notion that challenged the traditional understanding of Adam and Eve.

Christian Polygenism and Negro Suffrage

By 1850 Bachman was the only American naturalist openly opposing polygenism; yet the theory never won widespread acceptance in the American scientific community. Many of the country's most influential men of science rejected polygenism but remained silent on the issue because they had no good scientific alternative to offer and few convincing answers to the American school's arguments.[33] By the outbreak of the Civil War scientific interest in polygenism had waned considerably. Morton and Gliddon had already passed away, Agassiz had turned to fighting evolution, and Nott had become increasingly distracted by the threat of war. As the voice of the American school grew fainter and fainter, polygenism slipped from sight on the scientific stage, but for years it remained contentious in the public sphere. Leading religious journals continued to publish on the unity of man long after Appomattox, and polygenism played a role in the debate over Negro suffrage.

The Civil War ended slavery in America, but not white Americans' belief in their own racial superiority. Those who wanted to maintain some form of a racially ordered society faced a new threat in the movement in the late 1860s to gain African Americans the right to vote. Freedom from slavery was one thing, political equality quite another. In 1867 Buckner Payne, a white Nashville pamphleteer, argued that Africans were not only inferior but also subhuman. In a pamphlet titled *The Negro: What Is His Ethnological Status?,* written under the pseudonym "Ariel," Payne combined a detailed interpretation of the Hebrew text of the Bible with a smattering of scientific material in an attempt to prevent the social and political disaster that he was sure would result from any further elevation of the status of African Americans. This ignited a small pamphlet war, the so-called Ariel controversy, which continued into the twentieth century.[34]

Payne believed that a proper reading of the biblical Creation stories taught that Negroes had been created before Adam and Eve, that they belonged to the animal kingdom, and that they possessed no souls. In his opinion, they constituted an order of monkeys distinguished by the ability to talk. He reasoned that the "serpent" that tempted Eve in the Garden of Eden must have been a Negro, since no other animal could talk. Whenever true humans committed the unforgivable sin of interbreeding with Negroes, God had severely punished them. If the United States decided, contrary to nature and the command of God, to give African Americans political equality, Payne warned, then God would destroy America as surely as he had destroyed the antediluvian miscegenists at the time of Noah.[35]

Although some racist Christians found Payne's fusion of preadamism and polygenism appealing, most readers, even among those who believed in the inferiority of Africans, found his rants repulsive. One critic suspected Payne of disingenuously writing his outrageous attack on Negro suffrage in order to discredit genuine opponents of African-American rights. Agassiz, to whom Payne sent a copy of his pamphlet in hopes of receiving an endorsement, rejected the content as disgusting.[36] Payne's arguments rested primarily on an idiosyncratic reading of the Bible, but a smattering of scientific ideas also supported them. One Methodist minister readily detected the source of Payne's science: not in the work of "the great naturalists of the world" but in the writings of "Messrs. Nott and Gliddon."[37]

As marginal as they may have been in the larger context of the American debates on the relationship between religion and science, these Christian polygenists illustrate the range of opinions on the origin of races in the late nineteenth century. In a period when leading scholars were reclassifying many biblical stories as myths, Payne and his disciples, along with many other Americans, insisted on the historical and scientific accuracy of every word of biblical history. Instead of seeing themselves as scientific and theological obscurantists, they attacked their monogenist opponents as clerical despots, behaving just as the Catholic inquisitors had done in condemning Galileo in the seventeenth century.[38]

Human Antiquity and an Evolutionary Adam

The Payne party was not the only one to promote preadamism in post–Civil War America. Preadamism also benefited from two scientific developments: the discovery of the great antiquity of the human race and the growing popularity of Darwinism. Before the late 1850s little evidence suggested that humans were older than the generally accepted six thousand years. Geologists had successfully pushed back the origin of the earth by millions of years, but they had left human history relatively untouched. It was not until the careful excavation of Brixham Cave in southern England in 1858 that the evidence for human antiquity became scientifically compelling. By 1860 leading men of science were concluding that human beings had once lived among now extinct creatures. The exact age of humans remained undetermined, but it was clearly much older than the biblical six thousand years, possibly earlier than the creation of Adam and Eve.[39]

The reaction of American Christians to this development was mixed. Some, especially liberals who already believed that Adam and Eve were mythological rather than historical characters, experienced little difficulty accommodating the antiquity of humans. More conservative Christians tended to view the extension of human history as another example of science encroaching on the domain of theology and often criticized the archaeologists' methods of dating their findings. Some thoroughly orthodox Christians, however, attempted to accommodate the archaeological findings with a literal reading of Genesis, arguing that the Bible did not limit human history to a mere six thousand years. A few went even further and used preadamism to harmonize science and a literal reading of Scripture, arguing that early humans had inhabited the earth during the period between the Creation "in the beginning" and the much later creation in the Garden of Eden.[40]

One of the earliest to do so was Alexander Winchell (1824–91), a respected geologist and devout Methodist layman. Winchell published a great many articles interpreting modern science to Christians, especially his fellow Methodists (fig 7.6). He, like many pious scientists, rejected Darwin's thoroughgoing naturalism for a version of human evolution in which God acted through the laws of nature to guide the evolutionary development of organisms. Thus he adopted evolution to explain the appearance of humankind in general—while at the same

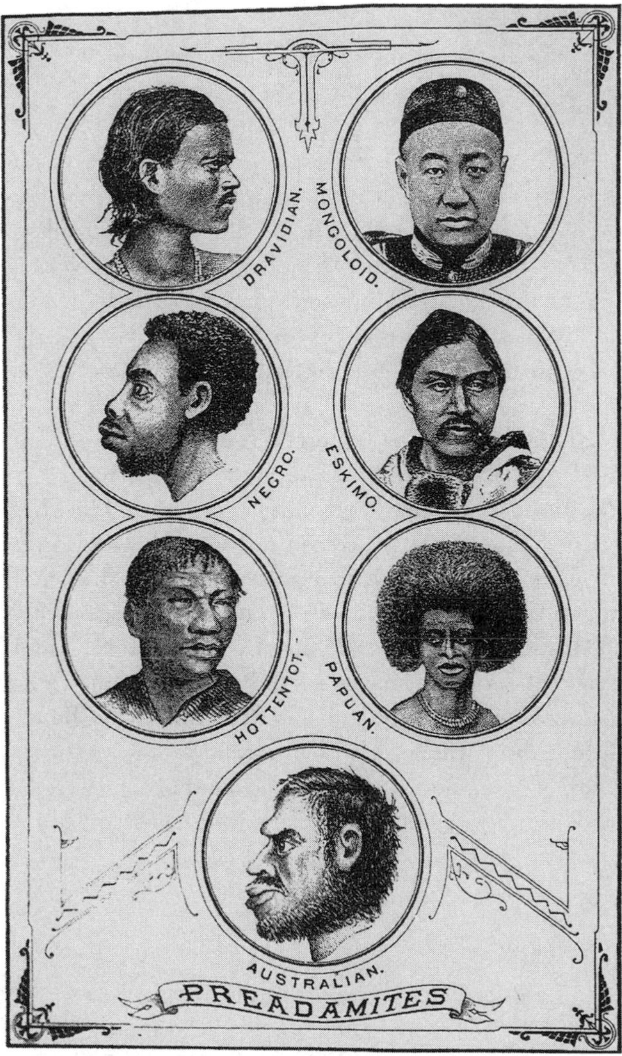

Figure 7.6. The frontispiece of Winchell's *Preadamites* depicted the races that he believed were created before Adam. Winchell, like Nott and many other ethnologists of the period, could not avoid caricature when discussing racial distinctions. From Alexander Winchell, *Preadamites; or, A Demonstration of The Existence of Men Before Adam* (Chicago, 1880).

time identifying the historical Adam as the most remote ancestor to whom ancient Jews could trace their lineage. In tracing the physical history of the human races back to an original stock, Winchell argued in *Preadamites; or, A Demonstration of The Existence of Men Before Adam* (1880) that the Negro race was preadamic in origin and—more controversial still—that the Caucasian race had developed from Negro ancestors. In his opinion, preadamism was scientifically correct, biblically warranted, and theologically sound.[41]

Unlike Payne's preadamism, Winchell's was evolutionary, monogenetic, and relatively orthodox theologically. "The biblical moral unity of mankind, whether it implies necessarily their genetic unity or not, is fully provided for by the monogenous origin of man under the present scheme of Preadamism," he insisted.[42] In spite of his protests of orthodoxy, the fledgling Vanderbilt University dismissed him from a lectureship shortly after his first preadamist publication appeared in 1878. The university's motives remain somewhat unclear, but Winchell's notion that whites descended from blacks may well have upset Vanderbilt's board of trustees more than his preadamic theory.[43] Despite losing his job, Winchell retained his position as the leading interpreter of science to American Methodists.

Biblical scholars, too, sometimes turned to preadamism as a potential tool for accommodating the evolution and antiquity of humans to a conservative reading of the Bible. For example, both Charles Woodruff Shields (1825–1904), professor of the harmony of science and revealed religion at Princeton College, and his seminary colleague Benjamin Breckinridge Warfield (1851–1921), a leading proponent of biblical inerrancy, entertained monogenetic preadamism. The fundamentalist evangelist and educator Reuben A. Torrey (1856–1928) also turned to preadamism as a possible means of harmonizing the findings of modern archaeology and anthropology with a strict reading of the Bible.[44] Such uses of preadamism in the defense of Christian orthodoxy illustrate the chameleon-like character of the idea. Born a heresy in the mind of La Peyrère, it eventually became an apologetical weapon used to defend the Christian faith.

Conclusions

Ideas about human origins changed dramatically in the course of the nineteenth century, but the shift in thinking should not be read simply as a story of escape from biblical dogma into the science of human evolution. Science did not move in a single straight line—from Adam to evolution—and Christianity proved firm yet flexible. Christians faced a complex task, dealing with the plots and subplots of a developing scientific story—an aging Earth, ancient human ancestors, and evolving hominids—while at the same time coming to grips with new biblical scholarship that emptied their holy writings of scientific content. Many found room in the text of Genesis for the untold eons of Earth history. Some kept pace with science and preserved the historical veracity of the Bible by allowing for humans before Adam; others happily accepted a mythological Adam and Eve and placed their religion safely out of the path of scientific change. Those who chose to oppose polygenism did so mainly because of its association with infidelity, not because of their antipathy toward science. When Darwinism destroyed polygenism as a scientific alternative, the earlier critics of the American school appeared to have been prescient.

The debate over human origins occurred in the larger context of America's struggle with racial diversity. At every step of the way political and social concerns interacted with scientific and religious ones. In the history of polygenism, Isaac de La Peyrère's desire to enhance the importance of Jews, Samuel Smith's use of science to undergird his political theory, Josiah Nott's defense of the Old South, and Buckner Payne's war against equal rights for blacks and whites all proved to be just as significant as more circumscribed scientific issues.

Acknowledgments

I wish to thank David N. Livingstone and James Miller for their insightful comments on a draft of this chapter and Ronald L. Numbers for his continuing guidance and encouragement in this project. This material is based on work supported by the National Science Foundation under Grant No. 9818182. Any opinions, findings, and conclusions or recommendations expressed in this material are those of the author and do not necessarily reflect the views of the National Science Foundation. The title comes from Nathan L. Frothingham, "Men before Adam," *Christian Examiner,* 50 (1851), 79–96.

8

Re-placing Darwinism and Christianity

David N. Livingstone

> He saw Darwin on his knees, and there was no difference between prayer and pulling a worm from the grass. As for Mr Covington, he prayed in the old-fashioned way.
>
> ROGER MCDONALD, *Mr Darwin's Shooter: A Novel*

The popular science writer and Oxford professor Richard Dawkins once remarked that Darwin had made it possible to be "an intellectually fulfilled atheist."[1] Indeed, the idea that evolutionary theory and Christian theology are necessarily—and thus inescapably—locked in deadly combat is pretty common. The philosopher Michael Ruse, for example, insists that Darwin made it possible, for the first time, to "confidently suspend belief in any kind of God. . . . Evolution destroyed the final foundations of traditional belief."[2] At a more popular level, the short entry on Darwin for the *Hutchinson 20th Century Encyclopedia* tells us that *The Origin of Species* "aroused bitter controversy because it did not agree with the literal sense of the Book of Genesis in the Bible" (fig. 8.1).[3]

The clarity of observations such as these, however, stands in marked contrast to the ambivalence of the historical record. To be sure, examples of conflict between Christian theology and evolutionary theory can readily be found. But to suggest that Christian believers and Darwinian evolutionists have always been at each other's throats is simply mistaken. The fact of the matter is that the links between Christianity and evolution have never been clear-cut. In what follows I hope to convey something of the complexity of that story. Indeed it has not even been possible to come to a final judgment on the religious sentiments of Charles Darwin himself. Reflecting on the theology of the Devil's Chaplain, as Darwin has been called, will

Figure 8.1. Charles Darwin (1849), at the age of forty.

therefore be time well spent. The ambiguities we find here caution against any confident assertions about exactly how the relationship between evolution and theology have been configured. Along the way we will also pause to consider one or two supposedly epic encounters that occupy a prominent place in the standard history of Darwinism and Christianity. Finally, in order to prevent the story from collapsing into chaos, I shall argue for the need to take much more seriously the local circumstances within which evolutionary theory was encountered. In other words, we will try to re-*place* debates over Darwinism and Christianity in the contexts of their own time and space. By doing so, we can begin to make some sense of the remarkable diversity of religious responses to Darwin's challenge.

Devil's Deacon or Anxious Agnostic?

Taking the measure of a life is never a straightforward matter. Biographers find themselves immersed in a world of values and commitments, and these inevitably color the narrative plots they spin. Motivations differ. Strategies vary. Interests condition. And the range of interpretative languages that are available to the biographer is immense. A psychoanalytic portrait, for example, inevitably differs from the political reading of a life. One dwells on inner fixations and neuroses; the other fastens on external social conditions. Add to this the fact that a "self" is not a simple unitary thing. All of us have the capacity to project our "selves" differently in different circumstances: as student, parent, teacher, friend, patient, tourist, or whatever. People relate differently to themselves and to others in different spaces—on the sports field, at the dinner party, on the dance floor, in the lecture hall, at home, and so on. Which, if any, is the *real* person? And which should have priority in the telling of a life?

These are critical questions when trying to understand the persona of Charles Darwin. And they are particularly vital when trying to take his spiritual temperature. Indeed we might perhaps more appropriately speak of Darwin's personae. For there were a number of different Darwins—Darwin the experimenter, Darwin the invalid, Darwin the investor, Darwin the dupe of quack medicine. And perhaps most significant of all a *private* Darwin and a *public* Darwin. In different spaces different Darwins surface. To different audiences Darwin presented himself in different religious guises. So the question immediately arises, Which—if any—should take precedence?

If it is unwise to think of Darwin's "self" as monochrome, it is also unhelpful to structure his life story in such a way that things lead remorselessly toward *The Origin of Species* (1859) or *The Descent of Man* (1871). Certainly these represent significant moments in Darwin's life. Let us recall that in these works he put forward a theory of how the multitude of living things in our world, so finely adapted to their environments, could have come into being without any recourse to a divine master plan. Given the self-evident facts of heredity and variation among organisms, and the Malthusian limits on population increase, Darwin argued that a struggle for existence must take place. It followed that those who survived were better adapted to their environments than competitors. This was essentially a theory of reproductive success in which relatively superior adaptations increase while relatively inferior ones are steadily eliminated. When applied to "man" Darwin's theory delivered a naturalistic reading of human origins. But to portray Darwin's life as a sequence of events flowing inexorably toward (or away from) these diagnostic episodes is seriously flawed. A chasm clearly opens up between what has been called "a life as it is lived" and "a life as it is told." *We* may know what happened to Darwin in, say, 1859, but *he* can't have known that when he set off on the *Beagle* voyage in 1831.

The significance of matters such as these becomes clearer when we reflect on the judgments that scientific biographers habitually make in their handling of sources. In their superlative biography *Darwin*,[4] for example, Adrian Desmond and James Moore attach much significance to a series of scribbles in Darwin's private notebooks and letters: "Oh you materialist!," "It is like confessing a murder," "What a book a Devil's Chaplain might write on the clumsy, wasteful, blundering low & horridly cruel works of nature!" These throwaway phrases are seized upon and used to give explanatory bite to their narrative strategy; they become the tool that is used to shape Darwin's project. The pertinent question is, of course, how much interpretative weight they should be made to bear. It's hard to say. But to simply privilege these *private* musings and use them to construct a unitary Darwin in which his *public* profile is cast as a series of façades is certainly not the only way to proceed. In the endeavor to take the spiritual pulse of as complex a human being as Charles Darwin, it is surely mistaken, as Frank Burch Brown writes, to underestimate "the degree to which a human being—and especially a Victorian—can hold apparently incompatible beliefs and can vacillate time and again between them."[5]

All this suggests that an unyielding chronological biography might *not* be the best way of grasping the Darwin phenomenon. It might be better to look at the different spaces of Darwin's identities—the different spheres in which he moved—and how he presented himself and his beliefs in them. Such a strategy would subvert the tendency to privilege the *private* Darwin over against his public "mask." Darwin, just like the rest of us, had the capacity to rig up diverse models of himself to suit his politics, his family, his friends, his public. Which is the *real* Darwin is clearly not the only question we can pose.

If we feel that there is any plausibility to these judgments, it will come as no surprise to learn that opinions on Darwin's spiritual state are remarkably varied. On the one hand, he has been seen as moving irresistibly from youthful orthodoxy to adult agnosticism;[6] on the other he has been described as a "muddled theist to the end."[7] Add to this the variety of experiences that have been identified as key spiritual moments. Some, for example, have fastened onto his materialistic inclinations, evident in his notebooks during the period 1837–40. In this scenario, Darwin's agnosticism, even atheism, is inferred from his quest for a naturalistic account of mind, emotion, and even the idea of God itself.[8] By the same token, it has been argued that naturalistic explanations are not of themselves incompatible with theism, and that materialism Darwin-style "represented no interest in a thoroughgoing atheistic philosophical or metaphysical materialism."[9]

By contrast to those who locate Darwin's loss of faith in the realm of intellectual struggle, James R. Moore dwells on personal and moral reasons for his spiritual decline. The death of his father and, even more importantly, of his favorite daughter—the charming ten-year old Annie—presented Darwin with insuperable moral dilemmas (fig. 8.2). Now he found that he could no longer embrace such cardinal evangelical doctrines as eternal punishment. Not that these events suddenly crushed spiritual life; they just accelerated his descent into unbelief.[10] To all this we must add that Darwin was ever sensitive to the religious

Figure 8.2. Darwin's daughter Annie Darwin, who succumbed to typhoid fever in 1851 at the age of ten. This event and the death of his father three years earlier led to his rejection of orthodox Christianity.

emotions of the women in his family circle, especially his wife, Emma, whom he did not wish to wound, and to whom he presented different versions of himself. After all, his father had warned him, prior to his engagement, to carefully conceal any religious doubts he entertained since sharing them could induce miserable suffering in a marriage partner, particularly in the face of chronic illness.

Given the irresolution that these scholarly portrayals collectively convey, I think it is a mistake to hope for closure on the final state of Darwin's soul. Clarity, precision, and certainty are inappropriate goals here. Ambiguity, anxiety, and hesitancy are more pertinent, not least because they reflect Darwin's own wavering. His changing opinions over whether or not he was wise to "truckle" to public opinion and use the word "creation" in the second edition of the *Origin of Species* is one indication of shifting sentiments. Add to this his self-confessed observations, first in a letter to Joseph Hooker in 1870 that "my theology is a simple muddle," and then in 1879 to John Fordyce that "my judgement often fluctuates."[11] These jottings surely confirm the conclusion that Darwin's religious beliefs "never entirely ceased to ebb and flow. . . . At low tide, so to speak, he was essentially an undogmatic atheist; at high tide he was a tentative theist; the rest of the time he was basically agnostic—in sympathy with theism but unable or unwilling to commit himself on such imponderable questions."[12]

Realizations such as these call attention to the profound significance of existential experience in a scientific life. As Thomas Söderqvist has recently reminded us, scientific biographers ignore senses of elation or depression at their peril. For the realities of "anguish and anxiety, despair and dread, embarrassment and fear" as well as "joy, hope and love" attest to the *lived* experience of the scientific practitioner. Scientists are people with emotions and desires, not just experimenters with thoughts and theories. In Söderqvist's telling, therefore, the daily round needs rescuing, and space created for streams of consciousness and dramatic shifts in mood.[13] A *life* is a much richer thing than a mere *career.*

Fluctuations in Darwin's spiritual moods, however, do not exhaust the scope of the impact of religion on his thinking. He had, for instance, absorbed the essentials of William Paley's *Natural Theology* during his years at Cambridge, and its flavor was to linger in his own writings in manifold ways. Paley, for example, believed that he could detect parallels between the Creator and His Creation, between divine agency and human artifice. Such correspondences provided Darwin with a captivating analogy: natural selection. Could it be, he wondered, that nature was engaged in selecting organic forms in just the way that human beings picked out the right pigeons for breeding? In the structure of Darwin's theory, then, reflected glimpses of Paley can often be caught. So potent indeed was his recasting of the Paleyan formula that some suspected he had simply transferred the characteristics of Paley's God to Nature and its laws.[14] And certainly the rhetoric in which the theory was couched displays Darwin moving casually between God and Nature:

> It may be said that natural selection is daily and hourly scrutinizing, throughout the world, every variation, even the slightest; rejecting that which is bad, preserving and adding up all that is good; silently and insensibly working, whenever and wherever opportunity offers, at the improvement of each organic being in relation to its organic and inorganic condition of life.[15]

Besides this, concepts like the adaptation of organisms to their environments and the idea of the harmony of nature were as central to Paley's cosmogony as they were to Darwin's. Yet Darwin's indebtedness to theology should not be limited to architectural echoes of Paley. Some of his profoundest convictions emerged in dialogue with—or in defiance of—conventional Christian doctrine. His critique of natural theology, it has been suggested, was tied to growing suspicions about the reliability of the human mind given its descent from animal ancestors. How, to put it bluntly, could the products of an advanced monkey mind be trusted? Again, his allergy to the miraculous was inflamed by a reading of Paley's *View of the Evidences of Christianity* while his sensitivity to suffering, as Donald Fleming speculated, may have owed its intensity to a "yearning after a better God than God."[16] And his doubts about design in the world were nurtured by an inability to reconcile the doctrine of divine providence with life's daily details.[17] All in all science and religion were thoroughly interwoven in Darwin's life and thought.

Sounds of Battle

If the image of warfare between science and religion does not adequately capture the fluctuating feelings of Charles Darwin himself, surely the sounds of battle can clearly be heard in the aftermath of the publication of the *Origin of Species.* So many commentators have assured us. And the battle cries can be heard not least in the epic struggle that took place at the June 1860 meeting of the British Association in Oxford, when Thomas Henry Huxley (fig. 8.3) squared up to Bishop Samuel Wilberforce (fig. 8.4). Reportedly Soapy Sam, as the bishop was affectionately known, had titillated his audience with the absurdities of the monkey-

Figure 8.3. T. H. Huxley (1857), Darwin's advocate, colleague, and friend. From Leonard Huxley, *Life and Letters of Thomas Henry Huxley* (New York: Appleton and Company, 1916), 160.

Figure 8.4. Samuel Wilberforce (1865), bishop in the Church of England and opponent of evolution in Victorian England.

to-man theory and concluded, in an act of unbridled folly, by asking Huxley whether he would prefer the ape on his grandfather's or grandmother's side. Huxley, so the story goes, shocked old Sir Benjamin Brodie, who was sitting nearby, by hissing, "The Lord hath delivered him into mine hands." He quietly took the podium, and in a few short minutes, smote the scientifically challenged bishop hip and thigh.

In his spicy narrative of the episode, William Irvine exulted in what he saw as Huxley's—and therefore science's—triumph over the bishop and the church. With literally deathly rhetoric, Irvine spoke of the bishop's "sudden and involuntary martyrdom" and of his "perishing in the diverted avalanches of his own blunt ridicule"; he depicted Huxley as having "committed forensic murder"; and he told of Hooker botanizing "on the grave of the Bishop's scientific reputation."[18]

Irvine, of course, is not alone in the use of such violent imagery. Darwin's son Francis, for example, spoke of "pitched battles" at Oxford and much more recently Huxley's reply has been described as "forensically fatal to Wilberforce."[19] Indeed in the memories of the participants themselves, the events of that June afternoon in the recently completed museum took on Homeric dimensions. Huxley cast himself as a slayer of the Amalekites, while Wilberforce was convinced he had emerged victorious from a battle royal. Small wonder that the BBC's television production had all the qualities of an American melodrama!

And yet . . . That some sort of rumpus took place that Saturday afternoon is hardly to be denied. But exactly what the furor was about is a different matter.[20] To be sure, it has routinely been recounted as a momentous chapter in the triumph of science over religion. But the later recollections of A. S. Farrar provide an intriguing commentary on the episode. "The speech which really left its mark *scientifically* on the meeting" he recalled, "was the short one of *Hooker*"; but Huxley "scored a victory over Wilberforce in the question of good *manners.*" Apparently the most significant thing about the debate was that the good bishop had "forgotten to behave like a gentleman" when he allegedly asked Huxley whether he "would prefer a monkey for his grandfather or his grandmother." Two months after the event Huxley recalled his response, in which he rephrased the question to a choice between an ape and the bishop himself: "If then . . . the question is put to me would I rather have a miserable ape for a grandfather or a man highly endowed by nature and possessed of great means and influence and yet who employs those faculties for the mere purpose of introducing ridicule into a grave scientific discussion—I unhesitatingly affirm my preference for the ape."[21] Vernon Jensen, a student of speech communication, uses this allusion, and others like it, to call attention to crucial *rhetorical* features of the episode. In consequence of his research, we have now come to appreciate just how unstaged and spontaneous the whole affair was. We know too that the goading of the audience had a good deal to do with the tone of both sets of remarks. And to this we must add that, however the lines of "war" were drawn, they were not simply between pro- and anti-Darwinians; age was just as important, with the younger faction tending to favor Huxley, Darwin's bulldog. Besides, it seems that in the concern to take the whole thing really seriously, the element of fun has been missed as has the simple fact that Huxley and Wilberforce each felt he had carried the day.

The rhetorical shape of the Oxford tussle invites us to consider various other circumstances that might equally throw light on the massaging, not to say manufacturing, of this most celebrated of Victorian battles. Frank Turner, for example, wants to retain the idea of conflict in the late Victorian debates over science and Christianity; but he insists that it should be transformed into a *social struggle* for cultural control between the old-fashioned parson and the new thrusting scientific professional. Victorian society, in other words, was witness to a conflict, *not* so much between science and theology, but between scientists and clergymen. Because cultural authority was progressively slipping out of the hands of one elite (the clergy), and into the hands of a newer elite (the professional scientist), scientific inquiry became one further arena in which cultural, not simply cognitive, interests were fought out.[22]

A related kind of reinterpretation is provided by Robert M. Young. He has persistently urged that it is wrong to think of the post-Darwinian debates as an encounter *between* science and religion. To think this way, he urges, is to fail to realize just how profound was the *continuity* between theological and scientific belief systems.[23] To him, the existing social order, built on hierarchy, domination, and class privilege, was justified throughout the nineteenth century, first by an appeal to religion, and then by an appeal to science. Wealth and poverty, prosperity and penury were simply the expression of either the laws of God or the laws of nature. Either way, the social order was ratified. Thus the debate about "man's place in nature" emerges as fundamentally the story of the substitution of a religious ideology by a scientific one; in both, the status quo is maintained, courtesy of natural theology or, later on, natural selection.

Something of how such forces were played out in the second half of Darwin's century can be glimpsed by taking a look at the circumstances surrounding the appearance at the time of a couple of famous accounts of the supposed "warfare" between science and religion. On 18 December 1869, the *New York Daily Tribune* carried a piece on "The Battle-Fields of Science"; it reported the details of a lecture delivered the previous day to a large audience at the Cooper Union in New York City by Andrew Dickson White, the president of Cornell University. Here White urged:

> In all modern history, interference with Science in the supposed interest of religion—no matter how conscientious such interference may have been—has resulted in the direst evils to Religion and Science, and *invariably.* And on the other hand all untrammeled scientific investigation, no matter how dangerous to religion some of its stages may have seemed, temporarily, to be, has invariably resulted in the highest good of Religion and Science.[24]

The brief diagnosis delivered here was to expand greatly over the succeeding years. In 1876 *The Warfare of Science* appeared, the British edition of which carried a preface by none other than John Tyndall, to whom we will presently return; subsequently several articles on the same subject appeared in the *Popular Science Monthly.* In turn, all of these informed White's two-volume *History of the Warfare of Science with Theology in Christendom* (1896). Throughout, White engaged in what we might call writing history backward; he kept projecting the controversies of his own day back into the past. But more; his nonsectarian Cornell University had attracted opposition from clergymen and this only strengthened his resolve to keep science at the forefront of the curriculum. At Cornell he refused to apply religious tests to staff and students alike and provocatively pronounced that he was engaged in the construction of "an asylum for *Science*—where truth shall be sought for truth's sake, not stretched or cut exactly to fit Revealed Religion."[25] Here we stand witness to a wider cultural struggle in which Darwinism played only one part.

In the meantime another "violent" history had found its way onto the bookshelves. John William Draper's *History of the Conflict between Religion and Science* (1874) is a book more often referred to than read. Certainly Draper conceived the history of science as the story of a struggle between two forces, or powers—human rationality and traditional faith. But the real object of his spite was far more specific: it was the Roman Catholic church.[26] The doctrine of papal infallibility, announced in 1870, particularly burned him, and this encouraged him to turn to science as the vehicle for attacking Catholic authority. The whole book was an exercise in overdramatization. In an altogether calculated way, Draper manipulated the recent pronouncements of Pope Pius IX to convey to his readers a profound sense of impending social crisis and cultural gloom. For to Draper, the declarations of the ecumenical council had nothing but dire implications for scientific inquiry; and these must inevitably bring the Catholic church "into collision with governments which had heretofore maintained amicable relations with it."[27] Here the intensely *political* significance that Draper discerned in the encounter between Catholic teaching and scientific inquiry is clearly exposed.

Besides this, Draper had been witness to the Wilberforce-Huxley event at Oxford back in 1860 and had addressed the selfsame audience with a paper "On the Intellectual Development of Europe, Considered with Reference to the Views of Mr Darwin and Others, that the Progression of Organisms is Determined by Law." By all accounts it was a dreary affair, and the caustic Hooker recalled it as the "paper of a Yankee donkey called Draper . . . [which] did not mend my temper, for of all the flatulent stuff and all the self-sufficient stuffers, these were the greatest: it was all a pie of Herbert Spencer and Buckle without the seasoning of either."[28] But the talk that week in Oxford would have provided Draper with further ammunition for the crusade on which he was already embarked. Moreover, his *History of the Conflict between Religion and Science* had its roots in the concerns he voiced at Oxford. Within a couple of years he would express himself more fully on the subject in *The Intellectual Development of Europe* (1863), a work cast in the mold of H. T. Buckle that read history through the twin spectacles of evolutionary theory and environmental determinism. Whatever Hooker's feelings on Draper's performance that day, his speech was one foray in a lifelong campaign to liberate scientific rationalism from the restraints of dogma.

In the meantime Draper had had his fair share of war talk. In the late 1860s he published a three-volume *History of the American Civil War.* The harsh realities he encountered here helped keep political concerns at the forefront of his imagination. With an unshakeable faith in the sovereign state as the ideal space for the preservation of scientific freedom, he announced that "intellectual enlightenment" had "imported unnumbered blessings to the human race" and "in place of the sparse dole of the monastery-gate, [had] organized charity and directed legislation to the poor."[29] The Catholic church's "Syllabus of Errors" was thus nothing less than a seditious assault on civil liberty. Small wonder that Draper's biographer, Donald Fleming, resorted to religious imagery to depict the strength of his sentiments: Draper's *History of the Conflict between Religion and Science* was nothing less that the "first draft of a New Scripture, a guide to the faithful" and "a kind of hymn of science."[30]

It is plain, then, that both White and Draper were waging war in the cause of scientific rationalism. And it thus makes sense to speak of the political origins of the conflict thesis.[31] Of crucial importance in this scenario was the role played by Thomas Henry Huxley and the X Club—a confraternity of like-minded British scientists that met for dinner during the 1860s and '70s and connived to emancipate science from established civil and theological authorities. This "sort of masonic Darwinian lodge," to borrow Desmond and Moore's words, used its own collective political clout to advance naturalistic thinking at every opportunity.[32] As for Huxley himself, his "chief aim," it has been said, was "the secularization of society through the cultural domination of science."[33] And so he was frequently to be heard uttering cries of "heresy!"; that is, telling religious believers what they could or could not believe about science. No sooner, for example, had St. George Mivart found it possible to reconcile Catholic teaching with an evolutionary stance than Huxley was rummaging through papal encyclicals to exclaim that evolution was in "complete and irreconcilable antagonism to that vigorous and consistent enemy of the highest intellectual, moral, and social life of mankind—the Catholic Church." Mivart, Huxley pronounced, simply could not be "both a true son of the Church and a loyal soldier of science."[34]

Several other factors conspire to support this cultural portrait. We might refer to the architectural styles of scientific institutions with their ecclesiastical resonances. The Victorian natural history museum, for example. was often known as "nature's cathedral."[35] Or we might recall Huxley and his comrades singing "hymns to creation," preaching "lay sermons," joining the "church scientific," and entering the "scientific priesthood." Small wonder that Huxley considered himself a "Bishop" of the new ecclesiology. Such maneuvers certainly do suggest that the image of an inescapable Victorian conflict between science and religion was a product of the cultural politics of Darwin's century.

Demilitarizing the Zone

It turns out then that taking Darwin's spiritual temperature is not an easy task and that the Huxley-Wilberforce exchange of views wasn't exactly the prizefight that many have imagined. But surely, one might think, the subsequent story of the relationship between evolution and theology is appropriately phrased in the idiom of the military dispatch. Again, however, history has been sacrificed to preconception. Now let me be clear what I am after here. Nothing I have said so far is intended to suggest that there have never been conflicts between Christians and evolutionists, or that the relationship between theological dogma and evolutionary doctrine has always been rosy. No; the point, rather, is that the model of systemic conflict does not give use a fine enough tool for slicing history. Theoretical prescription triumphs over empirical investigation.

A good sense of the shortcomings of stereotypical history can be gleaned from a remarkable review of *The Life and Letters of Darwin* that appeared in the October 1888 issue of *The Presbyterian Review.* The author was the Princeton theologian Benjamin B. Warfield, a defender of conservative Calvinism and a champion of the doctrine that the Bible is without error. Given these convictions, we might expect Warfield to judge that the theory of evolution must be rejected, if only because of Darwin's own loss of faith. Warfield's essay review is thus particularly instructive in this regard. Certainly he agreed that the word "agnosticism" best captured the spiritual state into which Darwin progressively fell. And yet he paused to commend the "unusual sweetness" of Darwin's character. "On the quiet stage of this amiable life," he mused, "there is played out before our eyes the tragedy of the death of religion out of a human soul." From Warfield's perspective all this was certainly to be lamented. But it did not imply that Darwin's theory itself was atheistic. "We raise no question as to the compatibility of the Darwinian form of the hypothesis of evolution with Christianity" he wrote. Even though Warfield was convinced that it was "his doctrine of evolution [that] directly expelled his Christian belief," he did not think that the path Darwin had traveled was inevitable. Darwin's handling of the doctrine of God's design in the world lay at the heart of the matter. But Warfield was confident that Darwin's was "a very crude notion of final cause." In any case, explanation by natural law was not incompatible with a commitment to God's supervision of nature, as such nineteenth-century scientists as W. B. Carpenter, W. H. Dallinger, and Asa Gray all revealed.

Overall then, Warfield remained deeply moved by the tragic tale of Darwin, that "great man" who drifted "from his early trust into an inextinguishable doubt. . . . No more painful spectacle can be found in all biographical literature." "We stand at the deathbed of a man whom, in common with all the world, we most deeply honor," Warfield concluded. "He has made himself a name which will live through many generations; and withal has made himself beloved by all who came into close contact with him. True, tender-hearted, and sympathetic, he has in the retirement of invalidism lived a life which has moved the world."[36]

Warfield's sensitive reading of the Darwin life, of course, cannot be taken to imply that little or no Christian opposition to evolutionary theory was registered at the time. In fact, courtesy of the superlative studies of James Moore, Ronald Numbers, and Jon Roberts we know that the landscape of the religious encounter with evolution was remarkably variegated.[37] It is impossible to survey here this vast terrain. But some sense of the range of reactions to the new biology can be gleaned from a brief perusal of two specific cases. And it will be instructive, I think, to focus on individuals routinely regarded as precursors of the American fundamentalist movement, which crystallized in the early decades of the twentieth century.

We begin with Charles Hodge, perhaps the greatest of nineteenth-century Princeton Presbyterians. The diagnosis he offered in *What Is Darwinism?* (1874) is a natural place to begin. Fundamentally, this tract for the times was an extended exercise in definition. Because he was certain that Darwin's use of the word "natural" was deliberately opposed to "supernatural," Hodge insisted that "in using the expression Natural Selection, Mr. Darwin intends to exclude design, or final causes." Here the very essence of the theory lay exposed. That "this natural selection is without design," Hodge explained, was "by far the most important and only distinctive element of his theory" and the single feature that brought "it into conflict not only with Christianity, but with the fundamental principles of natural religion." Note that Hodge's diagnosis sprang *not* from Darwin's challenge to a literal reading of Genesis, but from philosophical concerns about design. By this definitional move, Hodge could set the terms of the debate and judge who was or was not a Darwinian. It plainly meant that those like the Harvard botanist Asa Gray who considered themselves Christian Darwinians were either mistaken or just plain mixed-up; that label had no meaning. That Gray was a Christian *evolutionist* Hodge had no doubt; but that was a different matter. For to Hodge, Darwinism simply *was* "atheism."[38]

Hodge's younger contemporary, Warfield, did not see things the same way at all, even though he championed the selfsame theological system—Calvinism—to which Hodge had devoted his life. In 1916, for example, Warfield recalled that during his undergraduate days, he had been "a Darwinian of the purest water."[39] Indeed the previous year, in an article on "Calvin's Doctrine of the Creation," he urged that Calvinism had opened the door to a "naturalistic" explanation of nature, and that if Calvin had recognized that the six days of Creation "should be lengthened out into six periods . . . Calvin would have been a precursor of the modern evolutionary theorists."[40] To Warfield, it seems, Calvin was a kind of Darwin in the making.

The sweep of opinion represented here could be repeated many times over. Francis Orpen Morris, for example, an Anglican rector in Yorkshire, thought Darwinism deserved only "ineffable contempt and indignation."[41] The American Baptist theologian A. H. Strong, by contrast, insisted that "the attraction of gravitation and the principle of evolution are only other names for Christ."[42] The Scottish cleric William Miller felt that humanity's ultimate choice was—to use the title of his 1900 volume—*God, . . . or Natural Selection?* His evangelical fellow countryman Henry Drummond, on the other hand, tried vigorously in *Natural Law in the Spiritual World* (1883) to Darwinize theology and to import evolutionary laws into the supernatural realm.[43]

In the midst of such a bewildering array of responses from religious intellectuals—and we should recall that we know little of the attitudes of the vast bulk of regular churchgoers—there is a real danger that any attempt to identify broad historical patterns will fall foul of the messiness of history. Some general observations can be made, however. Roberts, for example, suggests that for the last quarter of the nineteenth century only a relatively small number of religious thinkers found it impossible to adopt evolution without abandoning what they took to be theological essentials. The majority, it seems, found ways of accommodating their theology to more or less revised versions of evolutionary theory. These very general trends are now well attested and have certainly succeeded in demilitarizing the science-religion zone in the decades around 1900. What has proven to be more difficult, if not impossible, is to identify correlations between evolutionary stance and theological system or denominational allegiance. On this issue neat boundary lines do not run between, say, Calvinists and Arminians, liberals and conservatives, Presbyterians and Baptists, Lutherans and Catholics. Because attitudes to Darwinism do not map neatly onto the landscape of conventional religious geography, it is tempting to abandon all effort at generalization and allow history to collapse into mere biography. This, however, would be too rash and I want finally to suggest one way in which we might make some sense of history's untidiness.

Placing Darwinism and Christianity: A Tale of Two Cities

In recent years philosophers, social theorists, and historians of science have begun to turn away from grand theories and general narratives toward local conditions and specific circumstances. Ideas, to put it another way, need to be located in particular contexts; they need to be literally "placed." The implications of this move for reconstructing the historical relations between Darwinism and Christianity are of considerable proportions. For example, it suggests that, rather than trading in the currency of disembodied "isms"—like Darwin*ism* and

Calvin*ism*—we would be better advised to situate *particular* encounters in *particular* places and examine the *particular* tactics that were adopted for coping with *particular* claims. In doing so we should attend not just to what scientists and theologians *said,* but also to how they were *heard,* and to how their audiences influenced what could be *both said and heard* in specific settings. What follows, then, is a brief attempt to show how being sensitive to local conditions may help us make sense of religious responses to scientific claims.

In 1874, in two cities, Presbyterians with seriously similar theologies pronounced judgment on the theory of evolution. In Edinburgh and Belfast, different circumstances prevailed and differing rhetorical stances were adopted.[44]

October 1874. Robert Rainy, the new principal of New College, Edinburgh (the theological college of the Free Church), elects "Evolution and Theology" as the subject of his inaugural address. Here he insists that Darwin's theory of evolution is simply theologically irrelevant. Indeed he is remarkably casual even about the idea of human evolution. Not that he rules out divine intervention in human origins, but by arguing that the image of God has nothing to do with the human physical form, he finds it possible to liberate Christian anthropology from detailed questions over similarities in the human and anthropoid skeletons. Thereby the door is opened to an evolutionary account of the human body. That winter in Belfast, J. L. Porter, professor of biblical criticism at the Presbyterian College, delivers his opening address. But he speaks ominously of the "evil tendencies of recent scientific theories" that threaten to "quench every virtuous thought." Accordingly he stands "prepared to show that not a single scientific fact has ever been established" from which the pernicious dogmas of Huxley and Tyndall could be "logically deduced." Darwin is not to be substituted for Paley. In the key Calvinist spaces of Belfast and Edinburgh, different attitudes to evolutionary theory are already being promoted.

Pronouncements like these, of course, are never broadcast in a social vacuum. The fact is, in these different regions different issues were facing Presbyterian communities. And these issues had a crucial role to play in shaping the rhetoric adopted by commentators on evolutionary theory. At the same time, different voices were being sounded in different ways, and their modes of expression, whether aggressive or accommodating, did much to set the *tone* of the local encounter. These conditions, I contend, had a crucial effect on the style of language that was available to theological spokesmen pronouncing on evolution. What could be said, and what could be heard, about evolution in the two localities was rather different.

In Edinburgh, Robert Rainy sought to maintain close links with the evangelical heritage of the Free Church of Scotland even while trying to keep abreast of contemporary intellectual life. He was always sensitive to his denominational constituency; thus, it is particularly significant that he felt able to endorse an evolutionary reading of human descent in his inaugural lecture. That such views could be promoted at New College, the church's seminary in Edinburgh, suggests a general lack of anxiety about evolution among Scottish Calvinists. The earlier furor surrounding Huxley's direct attack on the biblical account of Creation in a speech he gave to Edinburgh's working class at the Queen Street Hall in 1862, which had the Free Church's magazine, *The Witness,* venting its spleen, had evidently been long forgotten. Besides, already by 1866, the Free Church editor of the *Daily Review,* David Guthrie, made it clear that Darwinism was not inconsistent with faith. Why?

During the 1870s, I suggest, the Darwinian issue had paled in significance beside other intellectual currents assaulting the orthodox Scottish mind. Perhaps the most dramatic of these was the protracted heresy trial of William Robertson Smith. Matters had come to a head in 1876, when litigation began; it would end in Smith's dismissal from the Old Testament chair at the Free Church college in Aberdeen. What had sparked off the matter was Smith's entries for the ninth edition of the *Encyclopaedia Britannica,* which revealed his acceptance of biblical criticism and, in particular, the Graf-Wellhausen theory of the Pentateuchal documents. Later he would produce an immensely influential speculative account, *The Religion of the Semites* (1889). Here he urged that a primitive sense of communal unity found expression in a ceremonial meal. But it was a meal with a difference; the items on the menu were provided through ritual cannibalism. Revitalization of the tribe's sense of belonging, he believed, was secured—as George Davie pungently expresses it—through "eating the gobbets of throbbing flesh, newly-killed, of their fellow-tribesmen." In his account, Smith drew on the earlier work of his friend J. F. Maclennan who, in *Primitive Marriage* (1865), had argued for the matriarchal and polyandric origins of civilization, ultimately rooted in the unintended consequences of female infanticide. Plainly, there was more than enough sex and violence among radical Calvinists in Scotland to satisfy the likes of Freud. Indeed, they were too much even for Freud, for although he acknowledged his indebtedness to Smith, he found cannibalistic nostalgia just a bit too much.[45] In these circumstances, if the orthodox mind were to take up arms, biblical criticism, conjectural prehistory, and speculative anthropology were the arenas in which engagement was required. By contrast, natural science had long been domesticated to the needs of the Scottish theologians.

The situation was rather different across the Irish Sea. On Wednesday, 19 August 1874, the local newspaper enthusiastically announced the coming of the "Parliament of Science"—the British Association—to Belfast. Ironically, the convention was welcomed as giving temporary relief from "spinning and weaving, and Orange riots, and ecclesiastical squabbles." In fact the Belfast meeting was to be an X Club festival with Huxley, Hooker, and, crucially, Tyndall himself all speechifying. Now if an assault were to be mounted by the new scientific priesthood on the old clerical guardians of Scripture and social status, there could hardly have been a better venue for a call to arms than Presbyterian Belfast. Tyndall's performance did not fall short of expectations. All "religious theories," he thundered, "must ... submit to the control of science, and relinquish all thought of controlling it." The gauntlet had been thrown down.

Events moved quickly. The next Sunday, his address was the subject of a fierce attack by the Reverend Professor Robert Watts at Fisherwick Church in downtown Belfast. Tyndall's mention of the ancient Greek philosopher Epicurus was especially galling. The philosophy with which his name is associated taught that everything in the universe, including the soul, was material and that all human actions were determined by the original shape, speed, and direction of atoms. The human race had no creator, and no destiny, and pleasure was its highest good. To Watts, predictably, Epicurus's name had "become a synonym for sensualist," and he shuddered at the moral implications of adopting Epicurean values. That system had wrought nothing but ruin in the past "and if the people of Belfast substitute it for the holy religion of the Son of God, and practise its degrading dogmas, the moral destiny of the metropolis of Ulster may easily be forecast." In Belfast, the BA event set the agenda for the Christian response to evolution for a generation and more.

Given the different circumstances prevailing in Edinburgh and Belfast, it is now understandable why the subsequent histories of Christian responses to evolution were different in these Calvinist spaces. In Edinburgh, the names of numerous clergymen could be called in support of evolutionary theory, for example James Iverach, Robert Flint, James Orr, George Matheson, and Henry Drummond. Consider briefly now just one case—the Reverend Henry Calderwood. Supportive of the evangelistic activities of Moody and Sankey, he remained committed to the modified Calvinism embodied in the doctrinal standards of the United Presbyterian Church in Scotland. Yet in his Morse lectures delivered at Union Theological Seminary in New York in 1891 he insisted that even if "the theory of the Development of Species by Natural Selection . . . were accepted in the form in which it is at present propounded, not only would the rational basis for belief in the Divine existence and government not be affected by it, but the demand on a Sovereign Intelligence would be intensified."

In Belfast, the Tyndall event brought terror to the hearts of solid Presbyterians for years to come. During that 1874–75 winter of discontent, a set of Presbyterian intellectuals put together a series of public lectures to stem any materialist tide that Tyndall's rhetoric might trigger. Just as the villagers of medieval Europe annually beat the bounds, so the Presbyterian hierarchy needed to reestablish its theological borders. But more. Just over a year earlier the Catholic serial, *The Irish Ecclesiastical Record,* had presented an evaluation of "Darwinism" in which it castigated natural selection for its "ruthless extermination of . . . unsuccessful competitors" and for its distasteful moral implications. As for the latest clash in Belfast, the Catholic hierarchy issued a pastoral letter that November in which they repudiated the "blasphemy upon this Catholic nation" that had recently been uttered by the "professors of Materialism . . . under the name of Science." The similarity between these evaluations and those of the Presbyterian commentators we have considered is certainly marked. But the sectarian traditions in Irish religion meant that the frenzy surrounding the Tyndall event merely became yet another occasion for Ulster Protestants to disclose a siege mentality. The Watts fraternity was intent on tending to its own tradition's theological space and had no interest in cultivating ecumenical relations. Secularization and Catholicism were cast as twin allies against the inductive truths of science and the revealed truths of Scripture. It thus became possible to conflate as a single object of reproach the old enemy of popery and the new one of evolution. To Watts these were indeed the enemies of God . . . and of Ulster. In 1890, poised between Gladstone's two Home Rule Bills, which proposed that Ireland should be governed from Dublin, he wrote to B. B. Warfield. Here he denounced both the Free Church and the United Presbyterians in Scotland because their support of Gladstone's Irish policy would help "deliver their Protestant brethren into the hands of the Church of Rome, to be ruled by her through a band of unmitigated villains."

In the latter decades of the nineteenth century, the intellectual leadership of the Presbyterian citadels of Edinburgh and Belfast were involved in the reproduction of theological space. In both places, the theological boundaries that these figures had done so much to control set limits on what could be said about evolution, and on what could be heard. In Edinburgh, anxieties over biblical criticism and speculative anthropology, together with the proevolution sentiments of key leaders, made reconciliation with evolutionary theory intellectually agreeable. In Belfast, the infamous BA meeting made it exceptionally difficult to be sympathetic to those who thought that evolution was not to be equated with Tyndall's naturalistic worldview. Besides, even if this possibility had been grasped, it would have been exceedingly difficult to *express* it in the doctrinal locality they had labored so diligently to reproduce. In both cases, local circumstances are thus crucial to understanding how theologians chose to negotiate their way around the issues that Darwinism seemed to be placing on their agendas.

Conclusion

To reduce the history of the dialogue between Darwinian evolution and Christian theology to a bipartite tale of heroes and villains is to fall captive to the myths so eagerly promoted by the ideologues of both scientism and fundamentalism. Whether trying to diagnose Darwin's spiritual state or the influence of natural theology on the structure of his theory, it is clear that the language of conflict fails to capture the subtleties of the record. The idea of inevitable confrontation simply does not allow us to be sensitive enough to the rhetorical nuances and cultural tensions at work in the Wilberforce-Huxley saga and in those works of historical apologetic that stage the drama of science and religion as a theater of military operations. None of this is to claim, of course, that relations between evolution and theology have been uniformly happy. To the contrary. But we have seen that locating particular encounters in their specific circumstances allows us to ascertain not only *what was said* by interlocutors, but what *could be said and heard,* about evolution and religion in particular places. What this move makes clear is that the encounter between Darwinian science and the Christian tradition cannot be squeezed into the mold of conflict or cooperation. Accordingly we might be well advised to abandon the search for a "relationship" between such disembodied "isms" as Darwinism and Calvinism, evolutionism and evangelicalism, and their more-or-less distant cousins—materialism and theism, naturalism and deism.

9

Science, Miracles, and the Prayer-Gauge Debate

Robert Bruce Mullin

Sometimes a public debate can illuminate a deep-seated issue. Such was the case with the "prayer-gauge" controversy that swirled on both sides of the Atlantic during the early 1870s. Whether or not prayer could affect the course of natural events and aid in healing became a hotly controversial question. The debate is of both historical significance and perennial interest. On the historical level it revealed a conflict between traditional Christian ideas concerning God's activity in the world and the presuppositions of some mid-Victorian scientists. As a perennial question it is a case study in contested boundaries. When scientific understandings and religious understandings appear to come into conflict, who should set the ground rules for the debate?

Background

"In the beginning God created the heavens and the earth," begins the Book of Genesis; the picture of God reflected there, and found throughout both the Hebrew Scriptures (or the Old Testament) and the New Testament, is marked by a number of key elements. The God of the Bible is pictured as personal, active, and powerful. The scriptural God possesses those qualities of love, will, and knowledge that are usually assumed to distinguish the personal from the impersonal. This God is also described as a God who acts. "I am the Lord your God who brought you out of Egypt," begins the Ten Commandments, and the mighty acts of God fill the scriptural texts. Finally, these mighty acts are in the realm not only of spirit but also of nature. "Be exalted O Lord in thy strength," sings the Psalmist, "we will sing and praise thy power" (Psalms 21:13).

Of the many mighty acts recorded in Scripture, none have been as important in the memory of the communities of faith as the miracles. The words used in the Scriptures themselves for our word "miracle" are *semeion* (sign), *teras* (wonder), *ergon* (work), and *dunamis* (power).[1] Together, they suggest what a miracle meant in the biblical context. A miracle may be defined as an extraordinary act of God in the physical world for a religious purpose. This definition contains three important elements that occur whenever miracles as religious phenomena have been discussed. A miracle is "extraordinary"; it is not seen as part of the regular order of events. It is an action in the physical world; it is not merely a spiritual action. Finally, it is not only an extraordinary event, but one that has a religious significance. These mighty acts are associated with the histories of Israel and of the Christian church and are part of their most cherished memories. They include the splitting of the Red Sea, the bringing down by Elijah of fire from heaven to smite the priests of Baal, the many miracles associated with Jesus, and the startling accounts recorded in the Acts of the Apostles. All of these events were part of the memories of the Christian communities and informed their understandings of who their God was.

But God's activity was not limited to miracles. Both the Old and New Testaments promised that the petitions and prayers of the faithful would be heard by this personal God. Jesus instructed his followers to pray, "Give us this day our daily bread" (Matthew 6:11) and assured them of the power of such petitions by stating, "What man of you, if his son asks him for bread will give a stone? . . . How much more will your father in heaven give good things for those who ask him?" (Matthew 7:9–11). Two important types of prayer for Christians have been petitions (prayers for oneself) and intercessions (prayers for others). The biblical God was a God who watched over his people.

This confidence in both miracles and prayer undergirded Christian thought and piety for millennia. Christians viewed miracles as a continuing evidence of God's blessings. In Catholic Christianity miracles served as signs of holiness. Just as Jesus' miracles witnessed to his holiness and authority, so the continuing miracles witnessed to the holiness and authority of the church. Miracles also testified to individual sanctity, and by the year 1000 they became associated with the formal system of the canonization of saints in Western or Latin Catholicism. To this day, to be canonized in Roman Catholicism potential saints must have miracles attributed to them. Similarly, the scope of prayer covered all aspects of the way in which believers negotiated themselves through the physical world. Prayer (it was believed) could preserve one from lightning and tempest, fire and plague, as well as more ordinary sorts of suffering and sickness. The scriptural God was believed to be quite close to the world.

But here one comes to the crucial paradox in the traditional Christian view of prayer. Prayer could protect believers from lightning, fire, and tempest, but did not do so invariably. Believers, like others, suffered from these calamities. One might pray for deliverance from "plague, pestilence, and famine," but all too often such prayers were answered in the most leisurely of fashions, and the plague might not be lifted before more than a few of the petitioners were no longer alive to acknowledge the removal. In all times the belief that a personal God answered the prayers of the faithful was held in an uneasy tension with the question of how prayers were answered.

The Protestant Reformation of the sixteenth century powerfully affected these traditional understandings of miracles and prayer and contributed directly to the nineteenth-century controversy. Miracles took on a new role in the divine order for Protestant writers. Rather than reflecting either individual or group holiness, miracles (they argued) testified to the authority of divine revelation. The miracles of the biblical era manifested the divine authority of the prophets and apostles, and particularly the authority of Jesus. But when miracles had served this purpose, the argument continued, they ceased to occur, being no longer necessary. The age of miracles was restricted to periods of revelation. The reformer John Calvin easily answered Catholic charges that Protestants manifested no miracles by noting that Protestants needed no new miracles because they had no new gospel.[2] Protestants summarily dismissed all postbiblical miracles as "lying wonders"—stemming from benighted superstitions or diabolical perversions. The Protestant position centered on two distinctive claims. The first was that the truth of the Bible was objectively attested by the miracle stories found within it. Biblical testimony was not subjective, but rested on a bedrock of supernatural facts. Even when they surpassed the purview of human reason, the claims of Scripture could (and should) be accepted, since they were attested by superhuman powers. As the English philosopher John Locke explained, "to know that a revelation is from God, it is necessary to know that the messenger that delivers it is sent from God, and that cannot be known but by some sure credentials given by God himself."[3] The second was that the age of miracles was over. God no longer dramatically intervened in nature. The present world was not punctuated by acts of divine fiat in which Red Seas would mysteriously split or bread multiply; the world, on the whole, was a regular place.

If, however, God no longer intervened in the physical world through the performance of miracles, what became of prayer? Here Protestant writers made a careful distinction between miracles and acts of providence. Providential acts were actions of God *through* the natural order. Some were acts of general providence. The beauty, harmony, and fruitfulness of nature were all seen as gifts of a loving God; they were showered on all human beings and flowed immediately from Creation. But there were other blessings—known as "special providences"—that were reserved for the faithful. When storm winds scattered the forces of the Spanish Armada, frustrating their invasion of the British Isles, or when other winds directed a floating branch to a drowning man, the believer (or at least the British Protestant believer) believed that God's will was behind the winds. The difference between a "miracle" and a "special providence" was in the interconnection between God's will and the physical action. In a miracle the divine intervention was dramatic and immediate; in a special providence the divine action was remote and comparatively hidden in the regular course of nature. Storms emerge and wind moves driftwood in the regular course of nature; loaves of bread do not normally multiply to feed thousands. A miracle possessed an "objective" witness; a special providence was unrecognizable without the eyes of faith. To an outsider, the branch and the storm were random happenstance; to the believer, they came from God. Prayer belonged to the world of special providences, not that of the miraculous. Prayer could effect God's providential activity in the world, but a believer never could expect a miracle. Hence a mother might pray for her sick son and see in the lad's restoration to health the hand of God, but ought never pray for a miraculous multiplication of loaves to prepare for dinner.

Through these understandings Protestants also believed that they were forging a creative peace with a developing scientific understanding of the nature of the physical world. Law governed the natural world, and God (at present) neither broke nor altered these laws. Since these laws were discoverable by human reason, they in turn attested to the order and plan of God. But along with this there remained a place for the Bible and its miracles. God had provided two sources of knowledge. Nature, in its order and plan, revealed the glory of God in creation. Scripture, resting on miracles, offered revealed truth. The rightly ordered individual and society needed both. Thus one of the most famous Christian apologists of the eighteenth century, William Paley, explained that God's truth was seen both in the general beneficent order of nature (which he likened, in a famous analogy, to a well-running watch), and in the specific claims of Scripture, anchored by miracles.[4]

Before we leave this Protestant compromise, one more point merits emphasis, since it too will play a role in the later discussion. The ideas of miracles and of special providences had subtly different focuses. In a miracle the focus was on the divine action. Miracles were objective and evidential and pointed the individual toward the divine. A special providence focused much more on the human action or the prayer. If the divine action was more hidden, the human act of prayer or petition became more predominant and more closely connected with hope and trust.

The Question of Miracles

These long-standing issues provide the deep background to the mid–nineteenth century debate over miracles and prayer. In part the issue emerged as one aspect of the science and religion debate inaugurated with the publication of Charles Darwin's *Origin of Species* (1859), but the immediate cause was a sharp debate between two nineteenth-century scholars: the theologian James B. Mozley (1813–78) and the physicist John Tyndall (1820–93). The English-speaking religious world was abuzz over an attack on its traditional views of Scripture and the evidential value of miracles set forth in a multiauthor collection called *Essays and Reviews* (1860) which exposed various "erroneous views" in the Bible. Mozley, a former fellow of Magdalen College Oxford and future Regius Professor of Divinity at Oxford, offered a defense of the traditional view in the 1865 Bampton Lectures delivered at Oxford University, which were published under the title *Eight Lectures on Miracles Preached before the University of Oxford* (1867). He carefully defended the traditional Protestant position. Miracles anchored revelation and thus were essential to the Christian faith. Miracles differed from special providences because the latter were subjective in nature. Both were important for Christianity, however, and both stood or fell on the same issue, the plausibility of divine intervention based on the notion that God was a moral being involved in the world. "The primary difficulty of philosophy . . . [in] relating to deity is action at all," explained Mozley. Increasingly, philosophical and religious critics of miracles questioned the possibility of miracles on the basis of the observed uniformity of nature. As Baden Powell, Savilian Professor of Geometry at Oxford, argued in *Essays and Reviews,* the growing "inductive philosophy" confirmed "the grand truth of the universal order and constancy of natural causes as a primary law of belief." To defend the plausibility of God's activity in the world, Mozley challenged this assumption. One could not use an appeal to the uniformity of nature to reject miraculous claims. The uniformity of nature rested simply on the accumulation of empirical evidence; it was descriptive, not prescriptive. The principle of induction was useful in gathering information but could never exclude any alleged empirical fact. To reject a reported occurrence in the name of the uniformity of nature was to turn an intellectual principle into a "prejudice." The actions of God must be interpreted and evaluated according to their own rules.[5]

Mozley's lectures gave John Tyndall (fig. 9.1) the opportunity of drawing the

Figure 9.1. John Tyndall (1860), during his successful career at the Royal Institution of Great Britain.

issues of miracles and prayer into a broader discussion of the claims of science and religion. Mozley and Tyndall were a study in opposites. Mozley reflected the long-standing interconnections of university, church, and nation, and the heretofore careful balancing of the claims of religion and those of science. Tyndall stood outside of the religio-cultural establishment. Born in northern Ireland to a family far outside the centers of power, Tyndall epitomized both the intellectual vigor and the social longing of many of the mid-nineteenth-century scientific community.[6] From the early 1850s Tyndall had established a scientific reputation from both his experiments in the molecular components of crystals and his research into the property of diamagnetic polarity. But his social background and unorthodox religious views kept him from positions of established authority.[7] Tyndall enthusiastically supported Charles Darwin.

Tyndall attacked Mozley on a number of fronts. The first centered on Mozley's cavalier dismissal of the principle of induction. Mozley had disparaged induction to protect the possibility of miracles; Tyndall, in turn, lifted it up as the key engine for the advancement of human knowledge. The inductive principle was the backbone of the new science. It manifested that nature had no gaps and inspired the confidence that all the apparent holes in our knowledge would eventually be filled. "The inductive principle," declared Tyndall, "is founded in man's desire to know—a desire arising from his position among phenomena which are reducible to order and intellect." All scientific research rested on the orderly quality of nature and on the human mind's ability to grasp that order. "Nothing has occurred to indicate [that] the operation of th[is] has for a moment been suspended; nothing has ever intimated that nature has been crossed by spontaneous action, or that a state of things at any time existed which could not be rigorously deduced from the preceding state." Tyndall offered his readers the story of the gradual discovery of atmospheric pressure. The scientific investigators—from Galileo Galilei to Evangelista Torricelli to Blaise Pascal—each built on the others' observations and experiments as they solved the mystery of atmospheric pressure. Scientific knowledge advanced, Tyndall argued, because its theories and claims could be empirically tested. Theologians spun their wheels, he implied, because they were willing blithely to dismiss such discipline as "prejudice."[8]

Tyndall used the inductive principle to land still another blow against the miraculous. What scientific value existed in the category "miracle"? Even if an account of a "miraculous" event (such as walking on water) could be incontestably attested, how could a scientist ever know that it was caused by a supernatural being? Just because an event was temporarily inexplicable did not make it a miracle. "It appears to me that when he infers from Christ's miracles a divine and altogether superhuman energy, Mr. Mozley places himself precisely under this condemnation. For what is his logical ground for concluding that the miracles of the New Testament illustrate Divine power? May they not be the result of expanded human power?" Since human knowledge of the intricacies of nature was constantly expanding, it was inappropriate for a scientist to move from a mystery of nature to a miracle. T. H. Huxley put the issue more bluntly: "The day-fly has better grounds for calling a thunderstorm supernatural, than has a man, with his experience of an infinitesimal fraction of duration, to say that the most astonishing event that can be imagined is beyond the scope of natural causes."[9]

Tyndall originated neither of these arguments. A challenge to the objective or evidential nature of miracles had been building throughout the nineteenth century. The English philosopher John Stuart Mill had written that "if we do not already believe in supernatural agencies, no miracle can prove to us their existence. The miracle itself, considered merely as an extraordinary fact, may be satisfactorily certified by our senses or by testimony, but nothing can ever prove that it is a miracle." Religious writers had also increasingly questioned whether religious truths could ever be proved in an objective evidentialist way. But Tyndall's scientific criticism landed a crowning blow to the idea of miracles as objective evidences. Together these arguments would change the nature of the dialogue between science and Christianity concerning miracles. In the decades after 1865 many religious writers in both Britain and America began to recast their language about miracles to meet these objections. The concept of a miracle, they admitted, was ultimately a religious category. It presupposed not only a belief in God but also a distinctive understanding about the nature of God. The English writer Frederick Temple acknowledged that since science concerned itself with phenomena rather than ultimate causes (the how, not the why), "Science can never in its character of Science admit that a miracle has happened." Likewise, many English-speaking Protestants began to backpedal on the objective quality of a miraculous event and interpret miracles more like special providences. Even if one granted the general uniformity of nature, they argued, could not God operate both through nature and through his own volition?[10]

But Tyndall also questioned "special providences." And here he shifted to an increasingly cultural argument. He described a scene sure to confirm the anti-Catholic prejudices of Victorian Protestants: a Tyrolean landscape dotted with shrines to various saints, filled with gold and silver offerings given by peasants in recognition of saintly intercessions. How was this scene, Tyndall asked, different from the Protestant belief in special providences? "Each of them assumes that nature, instead of by cause and effect, is mediately ruled by the free human will." Were not both beliefs equally superstitious and reflective of an archaic worldview? Protestants had assumed that their view of divine activity in the physical world allowed a place for both science and divine activity. Tyndall rejected this assertion. Special providences reflected a superstitious worldview that was passing away: "For thousands of years witchcraft, and magic, and miracles and special providences . . . had the world to themselves," but this day was now passing. If religion was to adapt itself to this changing world, then it must eschew any attempt to justify itself by evidences of miracles or special providences.[11]

Testing the Power of Prayer

By the 1870s the debate over miracles was focusing on the cogency of petitionary and intercessory prayer. As was the case in many cultures, well into the 1860s the clergy of Britain were often called on to pray for divine intervention in the social and natural orders. When there was drought or calamity, the clergy interceded for their people. Although some liberal-leaning clergy balked at this as an anachronism, most accepted it as a duty. Such occurred in the late fall of 1871. The prince of Wales, Albert Edward, heir to the British throne, fell gravely ill from typhoid fever, a disease that had taken the life of his father, Prince Albert, ten years earlier. Many feared that the royal tragedy was about to repeat itself. The disease worsened, and the bulletins of the court physician, Sir William Jenner, grew gloomier. The Crown asked the British clergy to pray for the prince (fig. 9.2). Amazingly, the prince began to feel better—exactly on the tenth anniversary of the death of Prince Albert. An exuberant Queen Victoria called for a great service of thanksgiving to be held at the Westminster Abbey (fig. 9.3). There the queen, surrounded by leading clergy, gathered to give thanks for the wonderful divine answer to prayer. Absent from the festivity, however, were many of the leading figures of Victorian science, including John Tyndall. Tyndall responded by publishing in the July 1872 issue of the *Contemporary Review* an article entitled "The 'Prayer for the Sick': Hints towards a Serious Attempt to Estimate Its Value." The intellectual world was divided over the plausibility of acts of providence: one group (the clergy and their supporters) affirmed the "actual intrusion of supernatural power" into the course of nature, while skeptics such as Tyndall questioned any such occurrence. The only way to move forward, reasoned Tyndall, was to put such "acts of providence" to a scientific test. He then cited an anonymous letter that proposed a way of giving "quantitative pre-

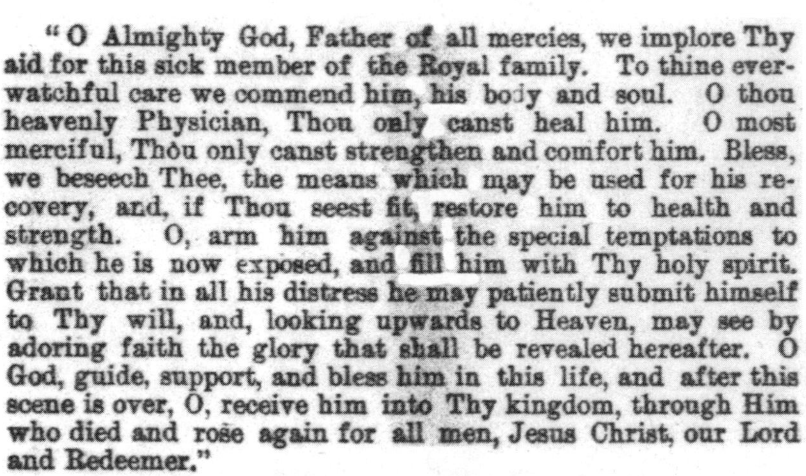

"O Almighty God, Father of all mercies, we implore Thy aid for this sick member of the Royal family. To thine ever-watchful care we commend him, his body and soul. O thou heavenly Physician, Thou only canst heal him. O most merciful, Thôu only canst strengthen and comfort him. Bless, we beseech Thee, the means which may be used for his recovery, and, if Thou seest fit, restore him to health and strength. O, arm him against the special temptations to which he is now exposed, and fill him with Thy holy spirit. Grant that in all his distress he may patiently submit himself to Thy will, and, looking upwards to Heaven, may see by adoring faith the glory that shall be revealed hereafter. O God, guide, support, and bless him in this life, and after this scene is over, O, receive him into Thy kingdom, through Him who died and rose again for all men, Jesus Christ, our Lord and Redeemer."

Figure 9.2. A prayer issued for the recovery of the prince of Wales, by the Reverend Luke Onslow, vicar of Sandringham, on Sunday, 10 December 1871. "Illness of the Prince of Wales," *The Illustrated London News Second Supplement*, 60, no. 1683, 16 December 1871, p. 597.

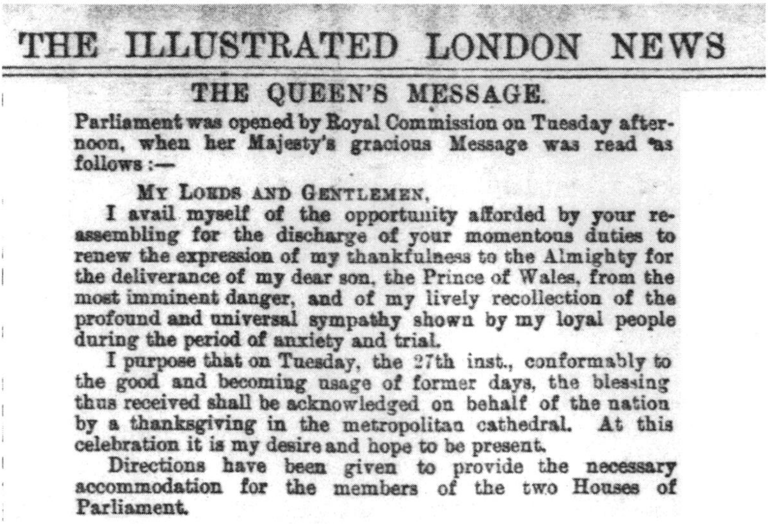

THE ILLUSTRATED LONDON NEWS

THE QUEEN'S MESSAGE.

Parliament was opened by Royal Commission on Tuesday afternoon, when her Majesty's gracious Message was read as follows :—

MY LORDS AND GENTLEMEN,

I avail myself of the opportunity afforded by your reassembling for the discharge of your momentous duties to renew the expression of my thankfulness to the Almighty for the deliverance of my dear son, the Prince of Wales, from the most imminent danger, and of my lively recollection of the profound and universal sympathy shown by my loyal people during the period of anxiety and trial.

I purpose that on Tuesday, the 27th inst., conformably to the good and becoming usage of former days, the blessing thus received shall be acknowledged on behalf of the nation by a thanksgiving in the metropolitan cathedral. At this celebration it is my desire and hope to be present.

Directions have been given to provide the necessary accommodation for the members of the two Houses of Parliament.

Figure 9.3. Queen Victoria calls for a national day of thanksgiving to give thanks to God for the prince's return to health. "The Queen's Message," *The Illustrated London News,* 60, no. 1692, 10 February 1872, p. 131.

cision" to the question. The author of the letter, it later turned out, was Sir Henry Thompson (1820–1904), a noted surgeon and professor of surgery and author of several important works in the field of medicine.[12]

Thompson stated that if prayer produced physical force, then such forces should be able to be recorded and evaluated. "I propose to examine . . . a means of demonstrating, in some tangible form, the efficacy of prayer." No form of prayer was more open to testing, he continued, than were prayers for the sick. Was there a correlation between such prayers and healing? "Such an important influence, manifestly either does, or does not exist," he wrote. "If it does a careful investigation of diseased persons by good pathologists, working with this end seriously in view, must determine the fact. The fact determined, it is simply a matter of further careful clinical observation to estimate the extent or degree in which prayer is effective." Thompson proposed that the healing rates of a prayed-for ward could be compared with those commonly obtained without prayer.

> I ask that one single ward . . . under the care of first rate physicians and surgeons, containing a certain number of patients afflicted with those diseases which have been best studied, and of which the mortality rates are best known . . . should be, during a period of not less, say, than three to five years, made the object of special prayer by the whole body of the

> faithful, and that . . . the mortality rates should be compared with the past rates, and also with that of other leading hospitals.

This procedure avoided the awkwardness of securing a ward of sick for whom no prayers could or would be offered to serve as a control group. With a touch of sarcasm, Thompson concluded: "I shrink from depriving any of . . . his natural inheritance in the prayers of Christendom." Such a test would finally answer the question of the "real power of prayer."[13]

Before examining the fracas provoked by this proposal for a "prayer-gauge," one must recognize that two issues lurked beneath the surface. The first concerned what might be called the question of scientific imagination. What would it mean as a scientific statement to claim that a spiritual or nonphysical reality could affect the course of nature? Mid-Victorian science, we have seen, presupposed a uniform nature. Throughout the nineteenth century the picture of nature had been becoming more and more regularized. The growing acceptance of the idea of the conservation of energy contributed to this picture. If energy could be neither created nor destroyed, nature was in key ways a closed system. As Tyndall opined, "No power can make its appearance in nature without an equivalent expenditure of some other power." From this perspective any intervention, not merely a miracle, was a puzzle and a threat. To believe that prayer could influence either the course of weather or human health was to substitute a universe based on will and whimsy for one based on law.[14]

The second question touched on professional expertise. Recall the language of the prayer-gauge challenge. Not only was a hospital ward to be selected as the object of prayer, but it was to be under the supervision of "first rate physicians and surgeons." One reason why people believed that the divine acted in the world was because they thought they "saw" the activity of the divine. "In the fall of the cataract the savage saw the leap of a spirit, and the echoed thunder pool was to him the hammer clang of an exasperated God." Modern clergy, in turn, saw the hand of God in the "providential" healing of a crown prince. But were clergy any better at interpreting events than had been their "savage" predecessors? By their own admission, defenders of providence viewed events of nature through the prism of faith; indeed, the very hidden nature of acts of providence required faith. Yet this mixing of proof and trust, for Tyndall, resulted in a horrible monster. "He acts the part of a Frankenstein" who does so. Only a rational, dispassionate scientist could truly understand and interpret phenomenal events. Clergy were for Tyndall dysfunctional remnants of an earlier church-directed culture. Tyndall's dismissal of clerical prerogative, as some scholars have suggested, was not merely anticlericalism, but was part of his case for the superiority of the new scientific method. Scientific knowledge advanced (as it did in the case of atmospheric pressure) because it was based on careful observation and rigorous investigation. The cultural implications were clear. If British society were to continue its upward advance, it ought to hitch its wagon to the rising star of the new science rather than to the old religion.[15]

Debating the Prayer-Gauge

Such a breaching of the traditional boundaries between a scientific worldview and a religious worldview was unacceptable for most religious commentators. The prayer-gauge touched a tender nerve. Religious writers had largely ignored an earlier questioning by Tyndall of prayers for weather. But prayers for healing had pastoral, not just theoretical, ramifications. Praying for the sick was deeply embedded in Christian piety. Hence the vigor of the defense. Some Christians merely responded with scornful disdain. "It has been proposed that we erect two hospitals, sneered one critic, "in one the patients are to be 'physicked' and the other 'prayed for.' Evidently the proposer had need of both remedies." Yet almost all of Tyndall's critics recognized the high stakes involved in the debate. As one writer admitted, when trying to defend traditional religion against its scientific detractors, the efficacy of prayer "is the key to the position; . . . if this ground is lost, all is lost." He went on to note that "of all the threatened points [it] is the one most difficult to defend."[16]

To answer Tyndall, many argued that the proposal for a prayer-gauge misunderstood the nature of God. To assume that any divine power had to be like all other regular and testable forces of nature mocked the traditional idea of God. A "visible test" would involve "the coercion of a Being when it is the first doctrine of those who believe in prayer to declare beyond the possibility of such coercion." If God were omniscient as well as all-powerful, explained the president of Princeton College, James McCosh, he would weigh all factors of a prayer request, not simply the sincerity of the petition. God answers "sometimes in one way and sometimes in another." Prayer no more bound the will of God than did a child's request bind the will of a parent. When a specific prayer was not affirmatively answered, it may be because a higher good was involved. Hence some suggested that in the celebrated case of the death of Prince Albert, his untimely demise contributed to the decision of the British government not to intervene on the side of the South during the American Civil War. Albert strongly opposed any such intervention, and his views carried greater public weight as a result of his death than they might otherwise have.[17]

Tyndall and his associates, the critics charged, also misunderstood the true nature of prayer. Prayer was most essentially the offering of the heart to God, not a series of specific petitions and intercessions. "You pray, if you pray in the spirit of Christ at all, not for a specific external end," wrote one, "but because of a deep relief to pour out your heart to God in the frankest way possible to limited human nature, and in the hope that if your *wish* is not granted, your want may be."[18] Prayer was an impulse of the religious nature, not a sophisticated divine answering machine. A prayer for healing that stemmed not from a concern for the sick individual but from a desire to examine the reality of prayer reduced prayer to merely a "barometric gauge to the providence of God."

Such arguments had a long heritage in the Christian tradition, but they carried little weight among Tyndall's defenders. Few attacked the Christian position with greater relish than Francis Galton (1822–1911), cousin of Charles Darwin and noted scientific writer, whose volume *Hereditary Genius* is still considered one of the pathbreaking studies of hereditary traits.[19] Galton shared with Victorian scientists such as Tyndall two presuppositions, which he brought to the prayer-gauge controversy. The first, reflected in his interest in inherited traits, was an overwhelming confidence in the law of physical causality. The second was a mistrust of religion in general and clergy in particular. In analyzing the attributes of all of the leading professions for *Hereditary Genius,* he found clerics both physically and emotionally weaker than average.

Like Tyndall and Thompson, Galton claimed that the scientific method was the only appropriate means for knowing the natural world: "An unscientific reasoner will be guided by a confused recollection of crude experience. A scientific reasoner will scrutinize each separate experience before he admits it as evidence, and will compare the cases he has selected on a methodical system." It was of no import that defenders of prayer saw it as a universal religious impulse; there were many such anthropological tendencies. If prayer were to be viewed any more sympathetically than were the views of "pagans" and "fetish worshippers," then it must have a "better foundation than the universal tendency of man to gross credulity."[20]

When prayer was judged by the high standard of inductive science, it failed miserably, according to Galton. No scientific evidence had ever been set forth demonstrating any relationship between prayer and healing. "The medical works of modern Europe teem with records of individual illness and of broad averages of disease; but I have been able to discern hardly any instance in which a medical man of any repute has attributed recovery to the influence of prayer," Galton declared. Employing some of his statistical findings from *Hereditary Genius,* he claimed that he could find no evidence of any correlation between prayer and health. Sovereigns were more prayed for than doctors, yet they died at an earlier age. Clergy were in all probability more regular in prayer than lawyers, but they were a more sickly lot. Foreign missionaries (another oft-prayed-for group) had a horrendous mortality rate. When one turned from health to the ability of prayer to secure temporal blessings, the same pattern emerged. "Is a bank or other commercial undertaking more secure when devout men are among its shareholders, or when the funds of pious people, or charities, or religious bodies, are deposited in its keeping, or when its proceedings are opened with prayer . . . ? It is impossible to say yes. There are too many sad experiences of the contrary." Prayer had no power to affect the physical world, he concluded. Just as lightning rods had supplanted prayer in the protection of buildings, so too should the very idea of prayer as a physical force be abandoned.[21]

Toward the end of his article, Galton offered an olive branch. A critique of prayer as a physical force did not undermine the essence of prayer, he conceded. Prayer was a spiritual reality, useful in shaping and soothing the human spirit. "There is a yearning of the heart," he acknowledged, "a craving for help, it knows not where, certainly from no source that it sees." Through such fine feelings the human heart may be comforted and the human spirit strengthened. "A confident sense of communion with God must necessarily rejoice and strengthen the heart, and divert it from petty cares." Prayer could in this way have a practical effect. Furthermore, if prayer made no claim to influence the natural order, then it could coexist with science.[22]

Galton's olive branch required religious leaders to concede the physical and natural realms to the scientist and to focus solely on the role of prayer as a spiritual or psychological power, rather than as a physical power. It called for the abandonment of the metaphor of divine intervention. Some Christians were willing to accept the challenge. For them, the prayer-gauge controversy served as a clarion call to reinterpret prayer for a scientific age. Many of these persons represented the movement of theological liberalism—often referred to as the Broad Church movement in England—which strove to reconceptualize traditional religious categories. "The doctrine of prayer as popularly held," wrote one liberal, "is one which is completely in opposition to all positive science, professing, as it does, to be based on facts, which, if true, throw doubt on all constancy in natural phenomena." Accepting the Victorian scientists' idea of a world in which "the conception of the fixity of natural law is one which the progress of science has made axiomatic," they thus agreed that the spiritual could never directly influence the physical: "A spiritual antecedent will not produce a physical consequent." Accordingly, it was archaic to ask the clergy to pray for rain, because physical forces alone caused rainfall. The role of prayer lay exclusively in the moral and spiritual arena.[23]

Prayers for health were more complicated. The liberals acknowledged that prayer could affect human existence, indeed the arena of prayer was human consciousness. God, for them, answers prayer "not by alteration of external circumstances, but by a change of the supplicant's relation to circumstance." True prayer could order the human spirit and unite it with God. But could it influence health? Here liberal writers of the 1870s tended to say no. They seemed to accept not only the axiom that the spiritual could not directly influence the physical but also Galton's assertion that no medical authorities recognized any healing value of prayer. One sees this attitude in the essay of a Scottish cleric, William Knight. After appealing to the reality of prayer as a spiritual force, he asked what practical value prayer for the sick could have? Since the spiritual could not affect the physical, he concluded that there could be no direct influence between prayer and health. But he admitted that prayer might indirectly affect the healing process by inspiring the doctor. Prayer "may bring a fresh suggestion to the mind of the physician," and in this way influence health, but it did not directly affect the patient.[24]

Who Should Judge?

The proponents of the prayer-gauge assumed that the physicist and scientist were the final judges of what was real in the phenomenal world. But was this necessarily so? Was a Victorian scientist the best judge of the fullness of reality? The British clerical writer R. F. Littledale (1833–90) did not think so, and he charged scientists such as Tyndall with having their own blind spots. "They are like Jedediah Buxton, the calculating boy of the last century, taken to see Garrick act Shakespeare, and coming away unimpressed alike by poet and actor, but being able to state with precision how many separate words Garrick uttered in the course of the drama." All of the rhetoric with which "intelligent men" have rejected the idea of special providences must be taken with care, since "'educated and sane' is used in a novel and arbitrary sense, as equivalent to holding the opinions of Professor Tyndall."[25]

The Christian system, continued Littledale, was a fact witnessed to in both history and present existence and deserved to be taken seriously. With a clear reference to a famous section of Darwin's *Origin of Species,* in which Darwin used the breeding of pigeons to establish the theory of natural selection, Littledale coyly noted that Christianity "is a fact as much as the cross-breed of pigeons." And here he turned the rhetorical tables on his adversaries by appealing to another Darwinian principle. The universal existence of a phenomenon, he suggested, probably indicated that the phenomenon fulfilled some real purpose. Useless customs and practices passed away in societal development as surely as useless traits passed away in evolution. Mixing spirits with the spiritual, he noted that just as the universal use of alcohol demonstrated that alcohol served some useful purpose, so too did the universal appeal to prayer. "The fact of anything continuing to live is a proof that it has vitality in it, and that such vitality must be as true as any other fact in the physical or moral universe, and therefore fitting matter for scientific research."[26]

The universal experience of humankind gave evidence of the reality of prayer, but why did not persons such as Tyndall recognize it? Because, answered their critics, they were so devoted to their methodological principles that they could not see any reality outside of their own categories. The very technical successes of the new scientific specialties had given to persons such as Tyndall, Thompson, and Galton a spirit of intellectual hubris. Earlier scientific figures such as Francis Bacon and Isaac Newton, although dedicated to an empirical knowledge of nature, had seen the harmony between scientific knowledge and knowledge gleaned from other aspects of human experience. Nineteenth-century scientists strove not for such integration but rather for the triumph of the so-called scientific method. "Now when a man devotes forty years of study to butterflies," Littledale argued, "he does much to enlarge the store of facts at our disposal, but he inevitably cramps his own intellect in the process, and becomes incapable of giving a valuable opinion on any subject outside his routine." No group of scientists was more liable to a reductionist methodology than were the physicists. The neatness, precision, and clarity of the world of force and matter predisposed them to reduce all reality to these categories. But other scientists recognized a more complicated picture. Biological scientists recognized that forces of life added complexity to the picture; students of humanity knew that will and mind similarly nuanced the picture. Accordingly, the notion of prayer was not an archaism, according to Littledale, but reflected the interplay between divine and human minds and wills.[27]

A similar argument was used by the small number of scientists who weighed in on the side of the defenders of prayer. George Douglas Campbell, the eighth duke of Argyll, was one of the best known amateur scientists of his time. He too argued that the relationship between material and nonmaterial reality was far more nuanced than some of the devotees of the new naturalism assumed. Their error was pride; they presumed to know what human beings do not know, "that we can define what we cannot define, and that these *words* constitute reality."[28]

If persons were truly interested in exploring the question of the reality of prayer, Littledale continued, they would have to look elsewhere than the ward proposed by Tyndall and Thompson. Prayer could get no fair trial there. Thus, like a defense attorney, he proposed a change of venue. The proper place to gather evidence of the power of prayer was not in an artificial prayer laboratory but in the places where such prayers had taken place for millennia, in religious communities. An impartial investigator could easily apply for data "to a certain number of ministers of religion, belonging to different societies, and ask them to send in details of cases." These individuals might be asked for cases that fulfilled three conditions: there was an extreme need of obtaining some benefit that could not have been gotten by ordinary means, there was devout prayer on behalf of the individual, and there was "the obtaining of the desired benefit in an unforeseen manner, subsequently to the prayer." If hundreds or thousands of such cases were reported, it would raise a strong presumption in favor of the Christian theory.[29]

Littledale's arguments were meant to turn the tables on the defenders of the prayer-gauge. It was not the defenders of prayer, but its opponents, who were afraid of a fair appeal to the evidence. The problem with the prayer-gauge was not that it was a trial of prayer, but that it was a false trial before a prejudiced judge and jury. A fair trial needed a different venue. But behind Littledale's clever argumentation lay two presuppositions, very different from those of Tyndall, Thompson, and Galton. The first concerned the value of the accumulated experience of humankind. Is it trustworthy? Littledale's argument rested on the principle that where there is smoke there is usually fire. The universal popularity of prayer suggested that some strong benefit existed in its use. The researcher sought to answer the question of how. As we have seen, Tyndall, Thompson, and Galton were far more critical of this universal human experience. The history of humanity was largely one of benighted superstition, a source of error, not wisdom. One recalls how often witchcraft and magic were invoked when these individuals spoke of the human experience. Scientific knowledge had freed the nineteenth century from the past, and the passport to freedom was something called the scientific method.

Littledale's second presupposition concerned the nature of prayer. He regarded answers to prayer as fundamentally religious phenomena, best evaluated by religious professionals. A minister could judge far better than a physicist whether prayer had been "devout." The evaluation of the efficacy of prayer was a joint responsibility of clergy and scientists, but the clergy were the obvious field agents in any evaluation. For Tyndall, if prayer were a phenomenal reality, it had to be in the form of a physical force; and if it were a physical force, it belonged in the purview of specialists who studied physical forces. Similarly, if the claim was that prayer influenced healing, any experimental investigation had to be under the control of learned physicians. There were to be no ministers or clergy in the ward reserved for prayer. The debate over the prayer-gauge thus involved a larger contest over the respective spheres of status and influence of clergy and scientists.[30]

Wider Reactions

The general public reacted to Tyndall's suggestion with shock, caused more by the impolitic nature of his suggestion than by the central thrust of his challenge. His was an offense of bad manners. Proper Victorians valued both religion and science and did not like to see them squabbling. One Methodist journal chided Tyndall for his lack of propriety:

> Tyndall belongs to that class of men who, in their pride of intellect and of science, and in the bold confidence with which they hold all scientific knowledge, condemn all other kinds of knowledge, and smile with a sort of patronizing contempt upon all matters of mere faith. . . . He has special contempt for Christian ministers whenever they venture to question any scientific positions, and on several occasions read them sharp rebukes, remanding them back "to their old theologies, traditions, and faiths." And yet this man of material things feels quite free to discuss such matters as prayer, miracles, special providences, etc.[31]

Many popular accounts offered evidences of answered prayer, but often they missed the real point of the debate, as did one journal when it announced triumphantly that the prayer of Methodist ladies had helped close a saloon. Despite the furor, the Victorian confidence in prayer seems to have been little disturbed. One might not know how prayer operated, but the practice of prayer continued unabated.

Even many of those sympathetic to Tyndall's defense of scientific naturalism distanced themselves from his views concerning prayer. In the midst of the controversy over the prayer-gauge Tyndall made a successful tour of America. Despite the protests of many clergy, Tyndall won the affections of the American scientific community (in no small part because he contributed all of the proceeds of his American lectures to a fund for advancing American science). He was feted at a farewell dinner, where leading scientific luminaries gathered to pay him honor. Yet in the speeches there was a coolness about his views on prayer, with one scientist proposing in a toast that "our learned and generous guest may rest assured that when he quits our shores many an honest heart in this land will apply the 'prayer-gauge' to him, and invite the God who creates the storm and stills the tempest, to carry him to his beautiful sea-girt home."[32]

Some religious groups took seriously the challenge of the prayer-gauge. The 1870s witnessed not only the debate over the prayer-gauge but also a renewed interest in the possibility of divine healing. One troubling factor in the traditional view of prayer was the quixotic nature of answered prayer. Sometimes God sent healing; more often he did not. By the early 1870s some Protestants had begun to argue that the New Testament offered not only spiritual blessing but physical healing as well. The Bible contained such promises about healing as that in James 5:13–14: "Is any among you suffering? Let him pray. . . . Is any among you sick? Let him call the elders of the church, and let them pray over him, anointing him with oil in the name of the Lord; and the prayer of faith will save the sick man, and the Lord will raise him up." Protestants had rarely anointed the sick since the era of the Reformation, but some began to claim that it possessed a far more powerful healing force than had ever been imagined. As one writer noted, "It seems to me that Christians are not living up to their gospel privileges when they fail to claim God's promises, not only for spiritual but for temporal blessings, and also for the healing of the body."[33]

By the end of the 1870s faith-healing movements were emerging on both sides of the Atlantic. These movements ran the gamut from Christian Science, which denied the ultimate reality of sickness, to Holiness groups, who believed that the miraculous healings of the New Testament were still available to believers.

It is easy to dismiss such movements as retrogressive, but they shared some of the cultural presuppositions of the Victorian scientists. Together they viewed the past largely as a failure; the discoveries of the nineteenth century allowed for true advance. Likewise, both the Christian healers and the scientific naturalists displayed impatience with the traditional equivocations about the efficacy of prayer. If prayer were truly efficacious, both argued, it should be regular in its operation. The Christian healers saw the fruits of divine healing as the final answer to the challenge of the prayer-gauge. As one wrote,

> twenty years ago, Professor Tyndal [*sic*] proposed . . . to make a challenge that so many sick people in a hospital should be treated with medicine, and so many in the same hospital by prayer alone—that Christian people should honestly make the test. The Christian Church was then afraid to accept the challenge, but I doubt whether he could make it to-day. I think we should have sufficient evidence to show that prayer is a power, a force, an answer—sufficient to prove that Christ is a Saviour of the body as well as the soul.[34]

Divine healing settled the question of the reality of God in the physical world.

Shifting the Debate

The prayer-gauge controversy helped to shift the debate over miracles from a Protestant-Catholic axis to a naturalist-supernaturalist axis. It also challenged the prevailing relationship between science and Christianity. Tyndall, Thompson, and Galton all but excluded the divine from the physical world and, by implication, claimed the physical realm for natural scientists. Liberal-leaning clergy supported them in their move, calling for religious individuals to rethink what they meant by prayer. Most religious respondents (and their scientific supporters) opposed the implications of the prayer-gauge and defended the reality of petitionary and intercessory prayer. They might not know how God answered prayer, but they were adamant that he did. Finally, some Christians wanted to push for a much more exalted role for divine healing. Throughout the decade of the 1870s, however, one alternative did not emerge: a serious scientific exploration of the interrelationship between religion and physical healing. The prayer-gauge seems to have been a rhetorical challenge not supposed to be taken up practically. Mid-Victorians saw healing largely as a physical or material process. Galton confidently asserted that physicians had no interest in the role of prayer in healing. But at the very time of the prayer-gauge controversy this assumption was beginning to be challenged. In 1872 Daniel Hack Tuke published *Illustrations of the Influence of the Mind on the Body in Health and Disease.* In it, he noted that although doctors had always seen evidence that the forces of mind and imagination possessed "a power which ordinary medicines have failed to exert," they had chosen "with a shrug of the shoulders to dismiss the circumstance[s] from [their] minds without further thought."[35] If the mind and imagination had such power in healing, perhaps the realms of mind and spirit and that of physical nature were not as hermetically separated as Tyndall and Galton assumed? By century's end some medical professionals were beginning to reassess the question of religion and healing.

In addition, other manifestations of interest in nonmaterial reality began to emerge. By the 1890s a scientific (or at least a quasi-scientific) interest had developed in examining psychical phenomena. The Society for Psychical Research was founded in Britain in 1892, and an American branch followed three years later. Fin-de-siècle research into psychical phenomena attracted an impressive group of intellectuals. The physicists William Barrett and Oliver Lodge, the philosophers Henry Sidgewick, Josiah Royce, and William James, and the codiscoverer of the theory of natural selection, Alfred Russel Wallace, as well as many other noteworthies, explored such psychical phenomena as the continuation of the soul after death, telepathic communication, and communications between the living and the dead. In the words of the American philosopher William James, psychical research attempted to push back the boundaries of the known and to explore that "unclassified residuum . . . usually . . . treated with contemptuous scientific regard." The identification of this mystical reality fascinated James, at least in part because most scientists remained so oblivious to it. "Facts are there only for those who have a mental affinity with them," he noted, adding that the mental predilection of many scientists was to ignore them. "If there is anything which human history demonstrates, it is the extreme slowness with which the ordinary academic and critical mind acknowledges facts which present themselves as wild facts, with no stall or pigeon-hole, or as facts which break up an accepted system."[36] Psychical research suggested a reality more complicated than that presupposed by most Victorian scientists.

An interest in nonmaterial reality also occasioned a reopening of the question of religion and healing. If mind could direct the course of healing, what of divine mind? A common psychological theory at the turn of the century posited the existence of a subliminal self in the human mind, an ultramarginal consciousness where body and spirit came together. Students of psychical research suggested that through this subliminal consciousness communication could take place between the living and the souls of those who had died. If this were true, then it could equally serve as the place and means by which the divine could influence the course of health. The conduit between God and the physical world, which had been challenged conceptually ever since the emergence of the theory of the conservation of energy, was tentatively restored. The early years of the twentieth century witnessed a new (albeit guarded) openness among both religious and scientific figures to discussion of prayer and healing. During the second decade of the century, for example, both the British Medical Association and the Church of England sponsored major studies of the role of prayer in healing. Religion and science seemed to have moved a far distance from the world of Tyndall and Galton.[37]

But had they? The new understanding of the potential powers of the human mind in healing opened a far richer view of the nature of healing, but did it allow any more place for God than had Victorian science? It remained unclear how one could move from a "mystical" or unexplained phenomenon to acceptance of the divine authorship of such events. Many of those most enthusiastic about the widening vistas of healing saw healings not as answers to prayer (at least in any traditional sense) but as the result of the mind affecting the body. Indeed, some commentators reinterpreted the great healing miracles of the Bible as illustrations of this phenomenon. Jesus' cures were not miracles but natural occurrences. Jesus' unique gift was that he was ahead of his time; he intuitively knew the interplay between mind and body and used that knowledge to heal.

Conclusion

The nineteenth-century discussion of miracles and prayer was clearly tied to the scientific and religious presuppositions of the time. The prayer-gauge controversy emerged at perhaps the high-water mark of Victorian scientific naturalism. The atomic theory and the law of the conservation of energy anchored a confidence in a mechanical and predictable universe. The rise of interest by century's end in the powers of mind and spirit profoundly shifted the discussion of prayer away from a presumed skepticism and toward a new openness. But presuppositions never remain fixed, and twentieth-century developments have further modified the nature of the discussion. Neither the mechanistic universe of the Victorians nor the idea of a subliminal consciousness open to a nonmaterial reality has found favor among most twentieth-century scientists. The triumph of modern physics has led to a questioning of the strictly deterministic model of the universe favored by Tyndall. Likewise, the failure of psychical research to establish a place for itself within the scientific community has dampened the hopes of psychical researchers.

Yet we may be able to draw some broad generalizations. The Victorian clashes over miracles and prayer occurred because cherished scientific and Christian values seemed to be in conflict. Scientists valued a knowable universe and strove to make the universe more knowable. Christians typically cherished a God who cares for believers and who is involved in the life of the world. The prayer-gauge became controversial in its day because of the difficulty of conceptualizing how God could be active in the world without disrupting the known patterns of nature. There seemed to be a fundamental conflict, requiring one to choose between divine activity and a scientific worldview. By century's end, this conflict was abating, as it was no longer seen to be inevitable. It became possible to speak of both divine activity and an orderly natural pattern. In such an environment, for example, some researchers took up the study of the mystical, which had engendered little scholarly interest earlier—a pattern confirmed by the continuing history of the question of prayer. During the middle decades of the twentieth century, such categories as subliminal consciousness lost favor, psychical research began to lose any scientific status, and scientific interest in prayer and healing waned. The last third of the century, however, saw increased cultural interest in nonmaterial and spiritual reality as well as renewed attention to alternative methods of healing. This, in turn, unleashed a new scholarly debate over the relationship between prayer and healing.[38]

With the question of miracle, providence, and prayer, the realms of science and religion touch. They come into contact because for members of both communities the world is seen as being real and important. Furthermore, both scientists and religious believers also maintain that a knowledge of how the world works informs the very essence of their worldviews. These views, as we have seen, do not always come into conflict, but they do constitute different ways of rendering reality.

10

Psychoanalysis and American Christianity, 1900–1945

Jon H. Roberts

In 1909 Sigmund Freud crossed the Atlantic in response to an invitation to present a series of lectures to an audience of notable American intellectuals who had gathered to help Clark University celebrate its twentieth anniversary (fig. 10.1). The five lectures that he delivered—hastily prepared each day during the course of a half-hour walk around Worcester—became part of an ongoing dialogue within American culture concerning the nature of mental processes. In demanding that mind be taken seriously as an agent in the cause and cure of psychological distress, Freud found himself allied with numerous American Christians in a larger movement embracing "psychotherapy."

This alliance may account for why historians have failed to appreciate just how competitive, even rancorous, the relationship between psychoanalysis and American Christianity really was. Many have acknowledged that Sigmund Freud, a self-professed "godless Jew," outspokenly belittled religion as a case of arrested psychological development akin to many of the neuroses that he analyzed and treated. Some historians have also pointed out that Freud and many of his followers regarded psychoanalysts as superior to the clergy as midwives of personal well-being. For the most part, however, they have ignored the response of Christian thinkers to the substance of psychoanalytic thought. One of the few historians who has addressed this question has claimed that liberal Christians in the

Figure 10.1. Some of the preeminent figures in psychology who spoke at Clark University (1909). Front row: Sigmund Freud, G. Stanley Hall, Carl A. Jung. Back row: A. A. Brill, Ernest Jones, Sandor Ferenczi.

United States made "efforts to mediate the gospel in light of Freud's psychological discoveries." In reality, however, in the period prior to 1945 the vast majority of Christians who discussed the claims of psychoanalysis—Protestants and Catholics, liberals and conservatives—regarded those claims as an assault on a proper understanding of the nature of mind. They also denounced the psychoanalytic interpretation of selfhood for being profoundly at odds with that espoused by Christian thought.[1]

The complex relationship that existed between Freudianism and Christianity within American culture cannot be understood without some appreciation of the cultural matrix in which psychotherapeutic perspectives emerged. During the late nineteenth century most physicians in the United States and Europe alike attributed symptoms of mental illness to pathological conditions in either the structure or the functioning of the brain and nervous system. They generally ignored questions concerning the ultimate metaphysical status of mind; in practice they treated mental processes as a function of neurological anatomy and physiology. This "somatic," or biologically oriented, view of the mind accorded with a series of advances in nineteenth-century medical research and physiological psychology. It also reflected the triumph of efforts to banish recourse to the supernatural from discussions within the medical profession. So compelling did most neurologists find the somatic model that they brought even "functional" disorders—that is, disorders lacking evidence of organic damage, such as hypochondria, neurasthenia, fears and compulsions, and the like—within its purview.[2]

The problem with the somatic model, however, is that it yielded disturbingly modest results. Neurologists enjoyed few successes in finding either physical causes or effective cures for most of the diseases with which they dealt. To make matters worse, many commentators during the late nineteenth and early twentieth centuries issued panicky warnings that the incidence of "nervousness" and other forms of mental disorder within the United States was increasing. No less important, the strains of modernity seemed to leave in their wake a growing number of instances of dis-ease: anxiety, depression, grief, and boredom. These too were conspicuously unresponsive to therapies predicated on somaticism.[3]

In the face of mounting concerns about mental health, the limitations of the somatic approach bred frustration. This frustration prompted the psychiatrist Charles G. Hill to denounce conventional therapies in his presidential address before the American Medico-Psychological Association in 1907 as "simply a pile of rubbish." Not all physicians were as dejected as Hill; the somatic approach to mental illness prevailed within medical circles long after 1907. Nevertheless, the increasing incidence of psychological distress, coupled with the seeming failure of efforts based on the somatic model to minister effectively to that distress, made Americans receptive to a host of alternative strategies: hypnosis, suggestion, and a variety of other techniques based in one way or another on persuasion and "reeducation."[4]

Underlying these approaches, characteristically grouped together under the rubric of "psychotherapy," was the conviction that mental illness was not always, or even usually, the result of physical pathology. Instead, it was rooted in psychological malfunctioning, most notably the inability of individuals to adapt effectively to their cultural and social environments. Spokespersons for this view thus affirmed that ideas and emotions and the associations that formed around them, both present and past, could determine the mental well-being and in some cases even the physical behavior of individuals.[5]

Sigmund Freud, the Birth of Psychoanalysis, and the Broader Psychotherapeutic Movement

Sigmund Freud did not begin his career as an enthusiastic champion of psychotherapy. Born in 1856 in a small Moravian town to a family headed by an impoverished Jewish merchant, Freud moved at the age of four to Vienna, the city where he spent the rest of his life. After graduating with distinction from the gymnasium (or high school) in 1873, Freud began the study of medicine at the University of Vienna. He chose this field, he later recalled, less out of desire to "help suffering humanity" than from a "need to understand something of the riddles of the world in which we live and perhaps even to contribute something to their solution."[6]

That quest for understanding led Freud into the laboratory. While pursuing his medical degree, he did neurological research for six years under the guidance of Ernst Brücke, an eminent physiologist and an ardent proponent of the idea that all life, including mind, could be reduced to physicochemical forces. So stimulating and challenging did Freud find this work that even after he took his degree in the spring of 1881, he did not leave his laboratory post until the summer of 1882, when he accepted a junior position at the General Hospital in Vienna. There, for three years while he rotated from one specialty to another, Freud undertook a variety of research projects under the direction of the neuroanatomist Theodor Meynert.

In 1885 Freud, who had opted to specialize in psychiatry, received a small grant allowing him to spend six months in Paris. There he came under the spell of the clinical neurologist Jean Martin Charcot, who introduced him to the possibilities of hypnosis and helped steer him in the direction of psychology. By that time Freud's desire to marry and set up a middle-class household had thoroughly undermined his commitment to laboratory science. On his return to Vienna, he resigned from the General Hospital and at the age of thirty began private practice as a specialist in the field of nervous disorders.

Many of the patients whom Freud encountered in his practice were afflicted with neuroses, conditions that fell within that enormously complicated region where mind and body intersected. As time went on, he experimented with hypnosis and a variety of other techniques. As late as 1895, however, he continued to subscribe at least in principle to the notion that mental processes could be described within a somatic framework. In that year he began a draft of a "Project for a Scientific Psychology." The announced goal of that project was to produce "a psychology that shall be a natural science," by which he meant a psychology that could "represent psychical processes as quantitatively determinate states of specifiable material particles." Freud abandoned this project after a little more than a month, when it became apparent to him that his goal was unattainable. Although he never repudiated in principle the idea that mental phenomena might ultimately be reduced to physiological functions, he henceforth opted to limit his own discussion of mental processes to psychological terminology. He continued to think of himself as a man of science, however, and in keeping with the goals of nineteenth-century positivism, he remained committed to providing a naturalistic account of mental processes.[7]

Toward that end, Freud developed a theory about the nature of the human mind and a set of prescriptions concerning therapy. He continued to fine-tune and in some cases even significantly to modify his system until the end of his life. This undoubtedly contributed to the almost endless diversity of psychoanalytic formulations and probably rendered public discussion of Freudianism more confusing—and confused. Still, by the time Freud presented his American lectures in 1909, he had substantially hammered out many of the major tenets of his "depth psychology."[8]

In keeping with his desire to put forth a psychological theory consistent with the canons of scientific discourse, Freud committed himself to a rigorously deterministic conception of mental processes. He insisted that the human mind was as subject to lawlike behavior as was the physical universe. In attempting to bring order out of the confusing array of symptoms characterizing mental illness and distress, Freud put forward a "dynamic" psychology that used insights he had gleaned through the use of hypnotism, free association, dream interpretation, and conversations with his patients. This psychology envisioned the mind as a turbulent assemblage of forces constantly in tension, even conflict, with one another.

Convinced that human beings had exaggerated the significance of conscious mental activity, Freud placed special emphasis on the unconscious, a region of the psyche that he thought served as the underlying source of all mental processes, the causal center of human behavior, and the repository of all memories and fantasies. Within the unconscious lay enormously powerful, amoral, irrational, instinctual forces. Those forces, or "drives"—preeminently the sexual drive, which he termed the "libido"—were products of evolution. They helped sustain the life of the human organism and thereby contributed to the survival of the species. However, in an effort to cater to their primal impulses, Freud maintained, individuals often generated ideas, desires, and wishes that were inconsistent with the "civilized" standards of behavior embraced by their conscious minds. When this occurred—as it often did during the early years of life—people resisted those unacceptable thoughts by "repressing" them within the unconscious. Unfortunately, this psychic defense mechanism was not always successful; repressed material did not always remain entirely buried. The surfacing of repressed material, Freud argued, caused hysteria, anxiety, obsessions, a bewildering variety of sexual disorders, and a number of other mental problems (fig. 10.2).

Freud believed, nevertheless, that there was hope. With the subtle guidance of their analysts, patients could attain knowledge of both the material that had been repressed and the forces of repression generated by the "civilized" impulses within their consciousness. By thus bringing to consciousness thoughts and wishes that were previously unconscious, they could free themselves of

Figure 10.2. Sigmund Freud working in his study in the later years of his life.

their neurotic conflicts, begin to choose their own goals and ambitions, and achieve "common unhappiness."[9]

Freud's work was simply one element in a considerably broader movement affirming the importance of mind as a causal agent in determining human behavior. By the first decade of the twentieth century that broader movement had become quite popular within American culture.[10]

One source of diffusion was the medical profession. During the late nineteenth and early twentieth centuries a growing number of physicians expressed doubts about the adequacy of somatic ideology. A noteworthy representative of such physicians was James Jackson Putnam, a Harvard neurologist who had begun his professional career in the 1870s as a committed proponent of somaticism. During the 1890s Putnam became fascinated with the neuroses, and in 1895 he informed his medical colleagues that he had become convinced of the "psychical origin" of many symptoms of mental illness. By 1906 Putnam had come to believe that mental processes were the "ultimate reality" in determining mental health.[11]

Between 1904 and 1909 an increasing number of physicians abandoned the somatic tradition. Several events during that period helped to swell the ranks of the defectors. Of particular importance were the visits of the noted French psychopathologist Pierre Janet to the United States and the English translation of

Paul Dubois's *Psychic Treatment of Nervous Disorders* (1905). These helped to endow the psychotherapeutic movement with the prestige of European medicine. In 1909 the United States hosted its first medical congress on psychotherapy.[12]

The Christian Conception of Mind and Mental Health

The medical profession's increased openness to psychotherapy undoubtedly helped to confer the prestige of science on that approach to psychological distress. It would be misleading, however, to ascribe the growing popularity of psychotherapeutic strategies within American culture entirely to the growing sympathy of physicians. Perhaps even more influential was the commitment of American religious thinkers to the crucial importance of psychological factors in determining personal well-being. The vast majority of Christians—Protestants and Roman Catholics alike—continued to embrace a dualistic view of mind and matter in which the mind, or spirit, constituted the essence of the self and the body was "its close ally and servant." By the late nineteenth century, use of the term "soul" had largely given way to the terms "mind," "spirit," and "personality," but this revised terminology did not mean that Christians embraced somaticism. To the contrary, they found the materialistic thrust of the somatic tradition's conception of mental processes decidedly uncongenial to Christians, for they recognized that it ignored, if it did not actually deny, precisely those elements of selfhood that defenders of Christianity found most significant and valuable.[13]

During the late nineteenth and early twentieth centuries, the mind occupied a central place within the framework of Christian theology. It constituted, for example, the source of humanity's superiority over other creatures. However much variety might have existed among Christians in their attitudes toward the evolutionary hypothesis, they continued to agree that the human species was unique and that the source of that uniqueness lay in the realm of humanity's endowments of mind. The attributes of personality, Christians asserted, constituted the source of humanity's special kinship to God. From this perspective, the stakes at issue in discussions of the status of mind included both the Christian view of humanity and the Christian view of God. William Rupp, a clergyman who wrote extensively on theology in the late nineteenth century, provided a succinct summary of this position when he wrote in 1888 that

> the fundamental idea in the Biblical conception of God is the idea of personality. God is a person, a self, a being that thinks, and wills, and loves, and is good, and holy and just. And these qualities, in a finite form,

> man has in common with God. He possesses the faculty of reason, and will, and sensibility; he has the power of self-determination, which makes him a free moral agent, responsible for his actions and capable of sin as well as righteousness.[14]

Christians also privileged the human mind because they regarded it as the primary site of humanity's interaction with the divine. During the late nineteenth century the testimony of "conscious experience, sanctified imagination, and the best Christian sentiment" became increasingly central to the defense of Christianity. Whether discussing the marvelous design of a universe sufficiently orderly in its structure and operation to be intelligible to human beings or the value of the "religious consciousness" in attesting to the reality of a divine Absolute, Christians insisted that in the divine-human encounter, minds were meeting Mind. Convinced that "spiritual things are spiritually discerned," they valued the mind as the appropriate instrument for recognizing the revelation of God within nature, the Bible, and individual human souls. From this perspective, some Christian thinkers understandably concluded that "the main religious question of our times concerns . . . the reality of our spiritual perceptions."[15]

In spite of the centrality of the mind in the theological discussions of mainline Protestants and Roman Catholics alike, the religious groups that led the way in demanding that greater attention be given to its role in the realm of health during the late nineteenth century were Christian Science and a disparate variety of New Thought sects. These groups, which combined a lively sense of the role of the spiritual with an allegiance to philosophical idealism, regarded matter as essentially illusory and reality as the emanation of a Cosmic Mind. From this perspective, they asserted that the mind was the appropriate agency to be invoked in combating psychological distress and even physical disease. "Mind cure" became central to their vision of religious experience.[16]

During the late nineteenth and early twentieth centuries the claim that the mind played a decisive role in causing and treating functional nervous diseases and other, less severe, forms of mental illness and discomfort resonated well beyond the membership of Christian Science and New Thought churches. In 1902 the eminent psychologist and philosopher William James wrote that "mind-cure principles are beginning so to pervade the air that one catches their spirit at second-hand." Even allowing for the possibility that James's own scorn for "medical materialism" led him to exaggerate a bit, it seems clear that by the first decade of the twentieth century recognition of the religious value of the principles underlying psychotherapy had begun to permeate American culture.[17]

Some American religious thinkers regarded the growing appreciation for the psychological basis of many disorders, coming as it did during a time of mounting concern about mental health, as a golden opportunity to till the vineyards of Christianity for practical fruits. The early years of the twentieth century witnessed a variety of efforts to reshape lives by combining the insights of Christian theology with the tools of modern psychotherapy. At the institutional level, the most important of these efforts was the Emmanuel movement. This movement began in 1906 at the behest of Elwood Worcester, a theological liberal serving as the rector of Boston's Emmanuel (Episcopal) Church, his associate Samuel McComb, and the future Freudian psychiatrist Isador Coriat of Worcester State Hospital. The Emmanuel movement, driven by concern that organized religion had not responded adequately to the challenges of modernity, was informed by two major theological precepts: first, that healing was central to Jesus' ministry and that anyone wishing to adhere to the "religion of Jesus" (a standard liberal catch phrase) would share his commitment to healing; second, that through the new revelations of psychotherapy, God had made opportunities for ministering to mental anguish (and other challenges to psychic well-being) more available than ever before. Armed with these principles, as well as with the conviction that the Christian religion constituted "the greatest of all therapeutic agents," the ministers at Emmanuel Church developed a number of psychotherapeutically oriented programs that during the next five years gained national attention.[18]

Many physicians criticized the Emmanuel movement for usurping the role of the medical profession, while some clergymen and theologians assailed the movement for devoting too much attention to matters of health. Nevertheless, a wide variety of Christian opinion leaders embraced the view that religion could foster mental health. Many clergy and theologians, concerned about the status of Christianity in modern culture, were convinced that one of the ways to counter secularism was to show the necessity of accepting the Gospel in order to enjoy a truly fulfilling life. Many also believed that in showing people how to cope effectively with their emotional problems, "Christian psychotherapy" was simply expressing the concern of Jesus himself with enabling individuals to live more abundant lives.[19]

Christianity and Psychoanalysis: Early Encounters

While many American Christians became quite receptive to psychotherapy, the same cannot be said of the psychoanalytic branch of the movement. On the contrary, as a result of their commitment to the idea that the resources of their faith were essential to the development of integrated personalities and fruitful lives, most Christian thinkers who addressed the issue found it impossible to accept many of the tenets of psychoanalysis.

Freud's now-famous visit to the United States in 1909 serves as a marker for the introduction of his work to the larger American intellectual community. Few religious thinkers, however, participated in discussions of psychoanalysis prior to 1920. During the preceding years, the commentary of journalists, psychologists, and medical specialists had convinced them that while Freud had made a number of claims about the human mind that were either true or innovative, the claims that he had made that were true were not particularly innovative, and the claims that he had made that were innovative were too outlandish to be true. As a result, while Christians in the United States extolled the value of psychotherapy, they tended to ignore Freud's contributions to that endeavor until the 1920s.[20]

By contrast, a number of the early proponents of Freudian views expressed little hesitation in attacking religion. In 1912, for example, the psychiatrist Trigant Burrow declared that "psychoanalysis is subversive of those forms of religious belief upon which society is founded." Anticipating Freud himself, Burrow wrote that from the perspective of psychoanalysis, "the Heavenly Father becomes an unconscious 'projection' mechanism whereby the childhood of the race seeks to perpetuate its human progenitor." Perhaps the most blatant use of Freud's name (though not always of his actual views) to assail religious belief was made by Theodore Schroeder, a lawyer and outspoken advocate of what he called an "erotogenetic" interpretation of religion. Although Schroeder began developing this interpretation before he encountered Freud's work, he found ideas gained in the course of a seven-month analysis with the eminent Freudian psychiatrist William Alanson White (superintendent of the Saint Elizabeth Hospital in Washington, D.C.) useful in defending the proposition that "religion is but misinterpreted or perverted sexuality."[21]

By 1920 the assaults on religion initiated by professed Freudians, coupled with the attention that Freud's views had received from avant-garde intellectuals and physicians, had convinced many Christian thinkers in the United States that psychoanalysis had become a force to be reckoned with. Accordingly, although most American theologians and members of the clergy continued to insist that Freud's work was at best "half noonday sense and half nocturnal nonsense," they embarked on a sustained effort to evaluate the merits of psychoanalysis. Their critique centered on four areas: Freud's view of the unconscious and its role in mental life; his diagnosis and treatment of personality problems; the goals he envisioned for psychotherapy; and his interpretation of religion.[22]

Christianity, Psychoanalysis, and the Unconscious

Although Christian thinkers recognized that Freud had not actually discovered the unconscious, they acknowledged that he did deserve credit for confirming its existence and for heightening awareness of its significance for human psychology. Most of them concluded, however, that the grim Freudian vision of the unconscious as the abode of primitive, even savage, instinctual drives was "extremely far fetched and improbable"—the virtually inevitable result of Freud's tendency to "fit all our psychic life into abnormal moulds."[23]

In order to understand why Christians responded so harshly to Freud's view of the unconscious, it is necessary to remember that they believed that the "fringe" region lying beyond consciousness contained insights concerning "the spiritual reason and conscience and the deeper faculties of the soul." Although few were inclined to deny that the unconscious served as the source of sexual and other "primitive" instinctual drives, they were convinced that it was much more than this. Some Christians regarded the unconscious as an arena in which the Holy Spirit "moves in His mysterious way" in effecting regeneration. Even more popular was the idea that the unconscious constituted a repository of ideas and values—a "wonderful subterranean vault"—that were constantly being produced by the conscious mind. This idea helps account for why Christians held pure thoughts and spiritually enriching experiences in great esteem; such thoughts and experiences were more likely to provide people with appropriate objects of cosmic meaning and devotion around which to integrate or reintegrate their personalities. Hence, Cecil V. Crabb, a theologically conservative Mississippi Presbyterian clergyman who authored two books on the relationship between Christianity and psychology, asserted that human beings must deposit within the "great subconscious chamber" of the mind "only good, true, pure, and noble thoughts." Crabb declared that

> if we plant there seed thoughts of anger, malice, lust, failure, fear and selfishness, then . . . we should expect a harvest of failure, licentiousness, sickness, and death in our bodily and moral beings. On the other hand, if we have stored our subconscious chamber with thoughts of love, unselfishness, and Godliness, then we can readily expect peace and happiness in our whole personality.[24]

Given this conception of the unconscious, widely embraced by American Christians, claims that Freud's depiction of the unconscious constituted "a menace to the innocent and credulous" were rooted in important theological considerations; they did not simply represent instances of mindless Victorian prudery. Far from serving as a temple of the Holy Spirit, the unconscious envisioned by Freud seemed to base human personality on ignoble impulses that dragged human nature itself into the "moral mire." William Hallock Johnson, a professor of Greek and New Testament literature at Lincoln University (Pennsylvania), declared that if Freud's interpretation of the human mind were correct, the proper inference would be that "every man is possessed by 'the spirit of an unclean devil.'" As a conservative Christian who believed in the doctrine of original sin, Johnson allowed that "the demon in us is bad enough"; it was surely not, however, "as black, in the direction of gross obscenity, as it is painted in the Freud-Jung psycho-analysis and dream interpretation." The liberal Protestant Charles T. Holman of the University of Chicago was somewhat more charitable. He acknowledged that by showing that the unconscious contained impulses and motives that were undesirable, even demonic, Freud had succeeded in confirming the scriptural view that "the heart is deceitful above all things." Holman also insisted, however, that it was unnecessary to "go all the way with Freud" in describing the unconscious. Human beings also possessed the ability to appreciate "higher satisfactions" and, with God's help, could attain mastery over their "lower" instinctual drives.[25]

Freud's emphasis on the irrational, self-interested nature of the unconscious reinforced Christians' hostility toward his view of the dynamics of the human mind. Protestants and Catholics alike had long rejected views of psychology that denied free will as an assault on the very foundations of moral responsibility. Thus they quickly challenged what they took to be Freud's assumption that conscious thought was little more than a tissue of rationalization for unconscious impulses that were largely irrational and animalistic. This view, they charged, was a blatant effort to discount the significance of human reason and will.[26]

Although Christians frequently emphasized the theological significance of the unconscious and acknowledged its "influence," most continued to regard consciousness as the site within the self where human beings would "be most in touch with God" and, consequently, as the most valuable component of human personality. They emphasized that the conscious mind served as the "source of the content of our subconscious" and the location of the will, which was charged with the task of "harness[ing] the unconscious to the service of the good." From this perspective, many Christians concluded that Freud had unduly minimized the significance of consciousness while exaggerating the power of the unconscious.[27]

Christianity, Psychoanalysis, and Mental Illness

Proponents of the Christian worldview also devoted a good deal of attention to Freud's approach to the diagnosis and treatment of personality disorders. Some, intent on legitimizing the use of psychoanalysis as a psychotherapeutic strategy, attempted to make Freudianism more theologically acceptable by emphasizing that it was intended not as a set of philosophical doctrines but simply as "a system of mental hygiene, ridding the mind of encrustations out of the past and preparing it to face the problems of the future." Those who reasoned in this manner concluded that clergy could avail themselves of Freudian therapeutic insights in their pastoral roles or—even better—collaborate with psychoanalysts as they dealt with parishioners' problems.[28]

Many Christians observed that both Christianity and psychoanalysis dealt with conflicts arising out of the human condition that prevented people from living emotionally healthy lives. This prompted some commentators to call attention to the similarity between Freud's discussion of conflicts between the libido and the superego and the apostle Paul's view of conflicts between one's good intentions and one's sinful impulses. Christians expressed no more enthusiasm than did Freudians for repression as a means of dealing with unacceptable impulses. Some, in fact, regarded the idea that repression was "a dangerous and generally ineffectual solution" to emotional problems as the real "kernel of truth" in psychoanalysis.[29]

In the end, however, Christian thinkers could not bring themselves to believe that the afflictions besetting human personality were always—or even mainly—the result of repression. The problems were rather more generally the result of sin and unbelief: wrong decisions, sordid motives, and unworthy objects of devotion that people had at one time or another consciously chosen. Christians believed that these difficulties should be dealt with in a spirit of love rather than harsh condemnation, but they insisted that it was shameful to abdicate judgment altogether.[30]

American Protestant clergymen and theologians characteristically evaluated the Freudian diagnosis of psychological problems from this perspective. To be sure, a few of those thinkers agreed with John Moore, a clergyman from New Haven, Connecticut, when he suggested that "psycho-analysis will help the modern minister to understand sin more fully, it will give him a vivid sense of its intense reality, and finally it will enable him to lay his gospel deeper into its root and resource." More typically, however, Christian commentators criticized Freudians for ascribing sinfulness to "some form of psychic abnormality." Those commentators warned that to hear partisans of psychoanalysis tell it, "our churches should be transformed into hospitals." Christians dissented from that view. The tendency of Freudians and others to use the metaphor of illness when describing people's capitulation to baser desires and instincts, Christian critics warned, attested all too eloquently to the fact that "the conviction of sin rests very lightly on this generation."[31]

Christians frequently charged that psychoanalytic therapy reinforced the prevailing tendency to belittle sin. Indeed, most of them were convinced that the very essence of such therapy lay in destroying personality disorders by eliminating the inhibitions, or restraints, that were causing them. Although this was not an entirely accurate interpretation of Freud's own not-always-consistent position, one can see how Christians came to view Freud as an exponent of the unbridled pursuit of pleasure. Popularizations of Freud's work in American journals of public opinion frequently espoused this interpretation. From the vantage point of Christians of widely different theological perspectives, Freudianism seemed to embrace "the morals of the barnyard." Freud's views were utterly irreconcilable, they argued, with the emphasis on self-control and the concentration on "higher" impulses commended in Christian ethics. The ability—and the responsibility—to bring sinful tendencies under control, the Lutheran theologian Sverre Norborg maintained, distinguished human beings from other animals. Self-control was an inherent element in the existential burden of being human. Christians were not to absolve individuals of that responsibility but to preach the need to adhere to it and, just as important, to show individuals how to obtain forgiveness when they failed to do so.[32]

Closely connected to the issue of forgiveness was the subject of confession. During the period between 1920 and 1940 a number of Christians called attention to the existence of certain similarities between the classic Christian doctrine of confession and psychoanalysis. One lesson that could be gleaned from psychoanalysis, they believed, was that open discussion of sin could reduce the incidence of repressed conflicts.

In the end, however, most Christians who closely analyzed the relationship between confession and psychoanalysis concluded that the differences were more profound than the similarities. Not only did the confession of sins involve knowledge present to the consciousness rather than the release of repressed material from the unconscious, but the very goals of confession—forgiveness and absolution—radically distinguished the process from psychoanalytic therapy, which sought to eliminate neuroses. However useful psychoanalysis might be in treating some personalities under the yoke of mental illness, Christians reasoned, it could not serve as a substitute for repentance. Just as important, they insisted that confession, repentance, and forgiveness involved a divine element absent from psychoanalysis. Indeed, for many Christians the relative significance of the human and the divine in changing lives was a paramount distinction between the Freudian and the Christian worldviews: whereas proponents of psychoanalysis held that human beings, in consultation with an analyst, were ultimately able to heal themselves through the attainment of "insight," Christianity taught that release from sin—spiritual healing—required the infusion of divine grace. As one theologian put it,

> Christians willingly admit that they have been driven to God because they cannot master themselves. . . . They had to face the bitter defeat of their

"ideals," "super-ego," "psychic power," or whatever grandiose name one likes to give to the ego. . . . [S]elf and sex, again and again, showed their demonic power over their lives.[33]

Christianity, Psychoanalysis, and the Achievement of Abundant Life

Underlying much of the Christian assault on Freudian therapy was concern that some people viewed psychoanalysis as an alternative to the Christian faith. Seeking to counter this view, defenders of Christianity emphasized that no system of psychotherapy that ignored the need for an encounter with God could help people achieve "a healthy, full human life." In attempting to justify this claim, they noted that only religion was capable of bringing about "the integration and unification of personality" that served as psychotherapy's goal. Psychoanalysis, after all, promised only a "release from inner discord." The removal of "unhealthy complexes," however, was insufficient; without the higher purposes and the appropriate objects of devotion and commitment that only the Christian worldview could provide, human beings would remain unable to lead triumphant lives. This conviction prompted Harry Emerson Fosdick, the best-known liberal Protestant clergyman of his generation, to complain that Freudians were guilty of "taking people apart, and finding it utterly impossible to put them back together again." By contrast, Christianity specialized in refurbishing peoples' lives. Convinced that "sooner or later the object of our faith affects vitally the emotional state that is aroused in our hearts," Cecil V. Crabb insisted that "the only faith that will give real psychic peace, power, and plenty is a rational faith in a historic Jesus who died for our sins, and who is abundantly able to deliver us not only from the guilt, but also from the present power of sin as it works in the entire range of our personality." By putting individuals in touch with "the ultimate spiritual forces of the universe," Charles T. Holman wrote, Christianity brought people into contact with an "organizing center for their higher loyalties." Lives thus grounded experienced freedom from much of the fear and anxiety associated with the human condition and possessed an "inner power" enabling them to live morally, even victoriously, "in the heat of every day." Hence, systems of psychotherapy that included a strong "religious element" had been most successful. As a psychotherapist, Christians concluded, Jesus was preferable to Sigmund Freud.[34]

Given their belief in the therapeutic value of their faith, it is hardly surprising that some Christians emphasized that the suppression of religious impulses was a major source of unhappiness in the modern world. Others noted that human sinfulness served as the cause of many of the "maladies of the soul." And because sin was a spiritual rather than a psychological affliction, psychology was powerless to deal with it. Only "a spiritual ministry" that went well beyond the practice of the psychiatrist could effectively address the issue.[35]

Christians also expressed their conviction that in their analysis of the relationship between "feelings" and therapy, psychoanalysis failed to comprehend the human condition. A number of defenders of the Christian worldview conceded that conflicts and their attendant guilt, shame, and feelings of inferiority produced unhappiness, but they emphasized that those unpleasant feelings often impelled people to make the leap of faith necessary to reorganize their lives on a higher spiritual plane. Convinced that faith in God and commitment to a gospel of love and forgiveness provided human beings with the only bases for a truly abundant life, Christians decried the efforts of psychoanalysis to seek peace at any price.[36]

Implicit in such criticism was the major indictment that Christian apologists advanced against the sufficiency of the Freudian gospel: that mere "health" was a woefully modest object of human aspiration. Granville Mercer Williams, the pastor of Saint Paul's Episcopal Church in Brooklyn, reminded his readers that Christians "are called not only to make men happy, which is the goal of psychology; but to go further, and make men saints, which is the goal of religion." This conviction prompted a number of religious thinkers to point out that the integration of personality was not an end in itself; after all, personality could be integrated in accordance with a host of "unworthy purposes." The appropriate goal was the achievement of integration on a moral level representative of a person's "highest self." In the judgment of exponents of Christian psychotherapy, "new birth" was the phrase that best described the kind of "fundamental reorganization and redirection of the total forces of personality" that they had in mind. From this perspective, it seemed obvious that Christians were not to force their faith into molds created by secular psychology but rather to make their "psychology and . . . psychiatry more religious and employ them effectively in the service of God and of the human soul."[37]

Christianity, Psychoanalysis, and the Nature of Religion

In contrast to Christian commentators on psychoanalysis, the vast majority of committed Freudians during the early twentieth century assumed that religion was not the solution to problems associated with being human but was itself one of the problems. In 1927 Freud added to an already formidable body of psychoanalytic literature assailing religious belief with *The Future of an Illusion,* a work that presented a series of speculations concerning the origin and nature of religious beliefs that he had been thinking about for some time (fig. 10.3). In effect, Freud

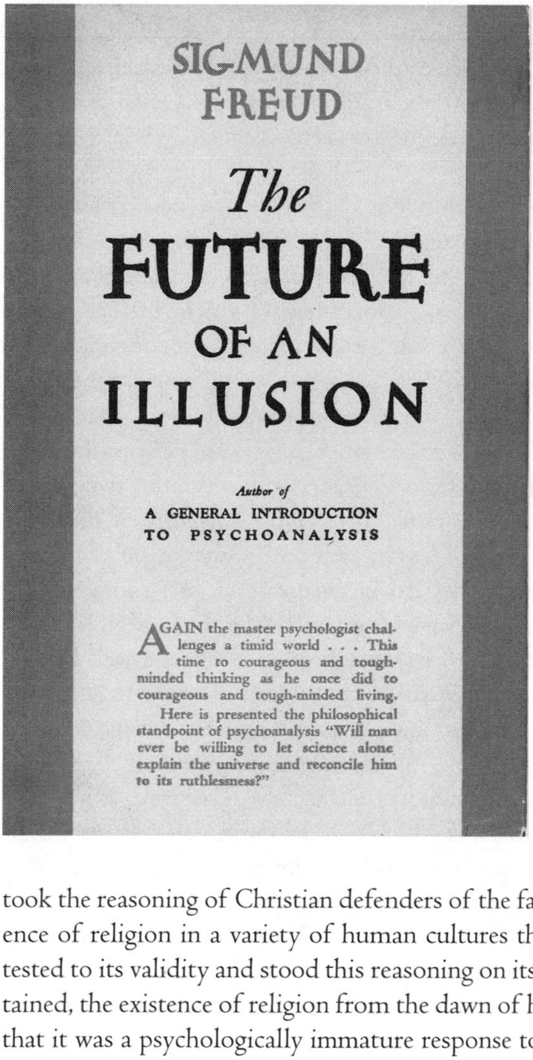

Figure 10.3. Front cover of the fourth printing (1949) of Freud's *Future of an Illusion* (New York: Liverwright Publishing, 1949).

took the reasoning of Christian defenders of the faith who argued that the presence of religion in a variety of human cultures throughout human history attested to its validity and stood this reasoning on its head. In reality, Freud maintained, the existence of religion from the dawn of human history suggested only that it was a psychologically immature response to life that the species had developed during the childhood stage of its development. In Freud's hands, gods—including the Christian God—became projections of human fathers, developed in the face of an awareness of superior powers within nature and the larger culture and the reality of impending death. Although Freud acknowledged that the origin of religious beliefs did not logically speak to the truth or falsity of those beliefs, he refused to concede the reality of the objects of religious devotion. Convinced that religion was an essentially unhealthful response to human experience, he both hoped and expected that human beings would eventually abandon the "burden of religious doctrines" and replace them with the more mature, more "objective" instruments of scientific knowledge. Freud exempted his views concerning religion from the list of tenets that he regarded as essential to psychoanalysis. Nevertheless, those views, coupled with similarly hostile discussions of religion in the work of many of his supporters, further weakened prospects for the establishment of a friendly dialogue between psychoanalysis and Christian theology.[38]

As the historian Peter Gay has rightly noted, the persuasiveness of *The Future of an Illusion* largely depended on the convictions that readers brought to the text. It is therefore hardly surprising that many Christians who evaluated that work dismissed it as "puerile." One theologian (possibly with tongue in cheek) submitted that in discussing religion Freud "suffers from a subconscious self-deception of a non-scientific origin." That theologian's verdict on the merits of Freudianism as "a philosophy of life," however, was emphatically not facetious; the psychoanalytic view of the human condition, he said, was "perhaps the most dangerous enemy of modern youth." Some Christian commentators denounced Freud's views of religion as simply another form of the age-old view that "the cosmic process is . . . essentially a soul-less mechanism." Others assailed Freud's effort "to reduce all religious thinking to rationalization" as an implicit endorsement of the psychogenetic fallacy—that is, that the truth of an idea depended on the nature of its origin. Edgar Sheffield Brightman, a well-known philosopher at Boston University, accused psychoanalysts of trying to poach on the preserve of philosophers of religion, who were, he insisted, far more qualified than analysts to judge the validity of arguments for and against religion. Brightman found Freud's reductionist approach to religion absurd. It was one thing, he asserted, to suggest that some kind of relationship existed within the subconscious between one's love of father and one's love of God; it was quite another to reduce the belief in God to a "father-complex" and declare the Deity nonexistent. Still other Christian thinkers took the more pragmatic position that the validity of religion could be evaluated by the fruits it produced. In reality, they maintained, religion was, "and always has been, a source of motivation greater in scope and more enduring in value than any other type of ideational and emotional experience."[39]

Defenders of Christianity acknowledged that some forms of religion were escapist and immature. They denied, however, that this description applied to Christianity itself. To the contrary, they maintained, the Christian faith, rightly interpreted, provided a more tough-minded, unflinching confrontation with reality than did schemes of psychotherapy committed to eliminating unhappiness and conflicts at all costs. Professor Samuel Nowell Stevens of Northwestern University thus maintained that

> the peculiarly significant contribution of the Christian religion has been that it has placed a premium on that attitude of mind which enabled the

individual to find the supreme value in facing life as it is rather than denying it or in fleeing away from it. It has placed a tremendous imperative in the lives of thousands of people to live richly and abundantly.[40]

Conclusion

For all of the reservations Christians expressed about psychoanalysis, Freudianism, almost in spite of itself, played a role in fostering changes in Christian theology. In particular, during the 1930s Freud's work helped to reinforce, if it did not actually provoke, a reaction against the tendency to interpret Christian theology in psychological terms that had been taking place since the late nineteenth century. Freud's work, by showing that the existence of religious beliefs over space and through time could be used to challenge as well as defend the legitimacy of religion, helped disclose the essential weakness of arguments predicated on the mere prevalence of belief. In this way, Freudianism contributed to the effort to generate the development of a "realistic" Christian theology that centered its attention more on the object of religious devotion—God—than on human beings, the subjects of such devotion.[41]

Christian realism, however, remained a minority movement. It would therefore be wrongheaded to use that movement as a springboard for claiming that psychoanalysis played a large role in shaping the history of Christianity in the United States. In fact, the most notable aspect of the interaction of American Christians with psychoanalysis was not how much but how little they drew on Freudian thought. As the foregoing discussion has shown, much of their discussion was critical rather than accommodating.

The history of the relationship between psychoanalysis and American Christianity should make students of American culture mindful of the need to make more careful distinctions between the position that Christians took toward the general enterprise of psychotherapy and the views they held of Freudianism and other particular expressions of that enterprise. Failure to make such distinctions has led to distorted images of both the views embraced by Christians and the nature of their role in creating and sustaining a "therapeutic culture."

Perhaps even more important, the response of American Christians to Freudian psychoanalysis should serve as a cautionary tale that will aid us in thinking about the relationship between science and religion in the twentieth century. One of the reigning views of modern scholarship is that it is inappropriate to employ martial metaphors such as "warfare" in discussing the historical relationship between science and religion. To be sure, scholars have convincingly demonstrated that such metaphors do not always accurately describe the complex relationship between these two approaches to human experience. Nevertheless, as the encounter of American Christians with Freudianism suggests, even liberal religious thinkers have not always approached views put forward in the name of science in the spirit of appeasement. Rather, in the ongoing dialogue between people speaking in the name of science and proponents of the Christian worldview, the human mind remains contested turf even today.

Acknowledgments

During the course of writing this paper, I have incurred a number of intellectual debts, and I am happy to have the opportunity to acknowledge them. Deborah Coon, Nathan G. Hale Jr., Robert Bruce Mullin, and James Turner read this manuscript in various stages and offered invaluable criticisms and suggestions. I also benefited enormously from the comments that an earlier version of the manuscript generated at a conference dedicated to this book project at Berkeley, California. David C. Lindberg's superb editorial eye has sharpened my thinking and tightened my prose. Ronald L. Numbers has read several versions of this piece and has provided me with his usual superb advice. Finally, I am delighted to express my appreciation to my wife, Sharon (ILYS), and my son, Jeffrey, for their love and support.

11

The Scopes Trial in History and Legend

Edward J. Larson

The Scopes trial may be the most famous encounter between science and religion to have occurred on American soil. It took place in 1925, soon after Tennessee enacted a statute forbidding public school teachers to instruct students about the theory of human evolution. The new law attracted national attention, with some conservative religious leaders praising it but many mainstream cultural leaders scorning it. Interest soon focused on the small Tennessee town of Dayton, where a local science teacher named John T. Scopes accepted the invitation of the American Civil Liberties Union to challenge the new law in court. As this young teacher (backed by the nation's scientific, educational, and cultural establishment) stood against the forces of fundamentalist religious lawmaking, the news media promptly proclaimed the encounter "the trial of the century."

The trial gained attention in part because it captured so many of the contradictions that characterize America. In the eighteenth century, the United States had become the first modern nation officially to separate church and state—and yet this trial involved a state statute that linked church and state by seeking to conform public education to religious belief. During the nineteenth century, American cultural leaders had generally embraced a civil religion that endorsed Protestant Christian norms and values for society and public education—and yet in this case the cultural elite turned against Tennessee for placing the biblical story of Creation above the established scientific theory of evolution. By the twentieth century, the United States had risen to the forefront in scientific research, especially in the biological sciences—and yet the enactment of the Tennessee statute suggested that significant political support existed for accepting religion over science on a matter central to modern biology. For many Americans at the time and ever after, the Scopes trial represented a landmark in the seemingly inevitable conflict between modern scientific thought and traditional religious belief.

In time, the Scopes trial became part of American folklore. As with any legend, the story evolved with each retelling. In its most enduring version, which appeared in the 1955 Broadway play and 1960 Hollywood movie *Inherit the Wind,* a mob of townspeople stirred by the preaching of Reverend Jeremiah Brown—whose name evokes images of the fanaticism of the biblical Jeremiah and the abolitionist John Brown—drags a young science teacher from his public school classroom and throws him in jail for telling his students about the Darwinian theory of human evolution. Such teaching supposedly undermines their faith in the biblical account of human creation. "Do we curse the man who denies the Word?" Brown rhetorically asked the assembled townspeople at one point. "Yes," they reply in unison. "Do we cast out this sinner in our midst?" he adds, prompting a mightier affirmation from the crowd. "Do we call down hellfire on the man who has sinned against the word?" Brown shouts. The mob roars its assent.[1]

Although real names and dates are not used, viewers of this scene presumably know that the teacher was John Scopes and that his trial had occurred during the summer of 1925 in Dayton, Tennessee. In both the play and movie, the main characters (except Brown and Scopes's fiancée, who had no parallels in Dayton) are given soundalike pseudonyms; the time is set as "summer, not too long ago." The stage directions begin, "It is important to the concept of the play that the town is always visible, looming there, as much on trial as the individual defendant." The movie version opens with the mob hauling Scopes out of his classroom for teaching evolution. Limited to a few sets, the play begins with a jailed Scopes explaining, "You know why I did it. I had the book in my hand, Hunter's *Civic Biology.* I opened it up, and read to my sophomore science class Chapter 17, Darwin's *Origin of Species.*" For innocently doing his job, Scopes "is threatened with fine and imprisonment," according to the script.[2] It makes for powerful drama—a scene seared into the national consciousness—but it never happened that way. Indeed, the movie prompted the great American journalist Joseph Wood Krutch, who had covered the Scopes trial for *The Nation,* to rebut, "The little town of Dayton behaved on the whole quite well. The atmosphere was so far from being sinister that it suggested a circus day."[3]

Dayton in 1925

Krutch remembered the episode well. It was no witch hunt led by a fundamentalist firebrand, but a bizarre publicity stunt concocted by secular civic leaders. Earlier in 1925, the Tennessee state legislature had passed a statute making it a misdemeanor, punishable by a maximum fine of $500, for a public school teacher "to teach any theory that denies the story of the Divine Creation of man as taught in the Bible, and to teach instead that man had descended from a lower order of animal."[4] Upon learning of the restriction, the fledgling American Civil Liberties Union (ACLU) in New York City issued a press release offering to assist any teacher willing to challenge the constitutionality of the new law in court. The ACLU statement appeared in its entirety on 4 May in the *Chattanooga Times*, which had opposed enactment of the antievolution statute. "Our lawyers think a friendly test case can be arranged without costing a teacher his or her job," the ACLU release stated. "Distinguished counsel have volunteered their services. All we need now is a willing client."[5] John Scopes became that willing client at the invitation of Dayton school officials. Although technically under arrest, Scopes was neither jailed nor threatened with imprisonment; he spent much of the time until his trial traveling and talking with reporters.

"Why Dayton, of all places?" a *Saint Louis Post Dispatch* editorial asked shortly after Scopes's arrest. Civic boosters adopted this question as the title for a promotional booklet sold during the trial. "Of all places, why not Dayton?" the booklet asked back.[6] Midway between Knoxville and Chattanooga, in the valley carved by the Tennessee River in the rising foothills of east Tennessee, Dayton lacked both a sense of tradition and confidence in the future. Only a few farmhouses existed in the area at the time of the Civil War, which, in 1925, remained a vivid memory for many Tennesseans. The town sprang up in the late nineteenth century with the coming of the railroads and became the commercial and governmental center for Rhea County. It was part of the so-called New South. Northern money financed laying the rail lines, digging nearby coal and iron mines, and building a blast furnace that attracted hundreds of Scottish immigrants and underemployed Southerners to the new town. Optimistic county officials erected a handsome, three-story courthouse on a spacious downtown square. By linking Dayton to northern markets, the rail lines facilitated the development of commercial farming in the surrounding valley, with Rhea County becoming a major center for strawberry production by the 1920s. But though the berry crop flourished and mining continued, the blast furnace went cold. A hosiery mill opened early in the new century, but it could not offset the loss of jobs at the furnace. New commercial construction slowed, leaving the downtown with three blocks of one- and two-story storefronts and two sides of the courthouse square undeveloped. Concerned civic leaders actively courted new industry as they watched their town's population cut almost in half, from a peak of about 3,000 around the turn of the century to fewer than 1,800 by the time of the Scopes trial.[7]

A native New Yorker with some training in civil engineering at Ohio Northern College, George W. Rappleyea managed the largely dormant local coal and iron facilities for their northern owners in 1925. Only thirty-one years old, Rappleyea had drifted from religion while in college, accepted the theory of human evolution, and written letters to the *Chattanooga Times* opposing enactment of Tennessee's antievolution law. Upon reading in that newspaper about the ACLU's offer to help any Tennessee schoolteacher challenge the new law in court, Rappleyea saw a chance to strike the statute, and he set about drawing other townspeople into his scheme.[8]

Rappleyea hurried down to Fred E. Robinson's drugstore with newspaper in hand, or at least that is how the most credible version of this legend goes. Robinson chaired the Rhea County school board, and the soda fountain at his downtown drugstore served as the watering hole for the town's business and professional elite during those days of national prohibition. "Mr. Robinson, you and [local attorney] John Godsey are always looking for something that will get Dayton a little publicity. I wonder if you have seen the morning paper?" Robinson later recalled Rappleyea asking.[9] Robinson had seen the morning paper but had not noted the ACLU's offer. Rappleyea then related his scheme of staging a test case in Dayton and boasted of having connections to the ACLU in New York. Robinson slowly warmed to the idea, as did school superintendent Walter White, a former Republican state senator who liked the antievolution law but loved publicity for his town even more. Godsey agreed to assist the defense. Before long Rappleyea was confident enough of local support to place his initial call to New York, asking whether the ACLU would make good on its offer if Dayton indicted one of its own schoolteachers. Other key participants signed on by the next day, when the ACLU accepted the arrangement.

Then the drugstore conspirators summoned the high school's twenty-four-year-old general science instructor and part-time football coach, John T. Scopes. "Robinson offered me a chair and the boy who worked as a soda jerk brought me a fountain drink," Scopes later wrote. "'John, we've been arguing,' said Rappleyea, 'and I said that nobody could teach biology without teaching evolution.' 'That's right,' I said, not sure what he was leading up to." Scopes then pulled down a copy of George W. Hunter's *Civic Biology* from a sales shelf (the enterprising Robinson also sold public school textbooks) and opened it to the section on human evolution. This was the state-approved text, prescribed for use in all Tennessee high schools. "'You have been teaching 'em this book?' Rappleyea said. 'Yes,' I said. I explained that I had got the book out of storage and had used it for review purposes while filling in for the principal during his illness. He [the principal] was the regular biology teacher," Scopes recalled. "'Then you've been violating the law,' Robinson said." The school board official then told Scopes about the ACLU offer. Scopes remembered the fateful question: "'John, would you be willing to stand for a test case?' Robinson said. 'Would you be willing to let your name be used?' I realized that the best time to scotch the snake is when it starts to wiggle. The snake already had been wiggling a good long time."[10]

Scopes presented an ideal defendant for the test case. Single, easygoing, and without any fixed intention of staying in Dayton, he had little to lose from a summertime caper—unlike the regular biology teacher, who had a family and administrative responsibilities. Scopes also looked the part of an earnest young teacher, complete with horn-rimmed glasses and a boyish face that made him appear academic but not threatening. Cooperative, well liked, and naturally shy, he would not alienate parents or taxpayers with soapbox speeches on evolution or give the appearance of a radical or ungrateful public employee. Yet his friends knew that Scopes disapproved of the new law and accepted an evolutionary view of human origins. He knew little about the issue—he coached football and taught physics and math, not biology—but he had been a student at the University of Kentucky when that institution's president led the fight against antievolution legislation in the Bluegrass State, and he admired the president's courage. Further, Scopes's father, an immigrant railroad mechanic and labor organizer, was an avowed socialist and agnostic who, as the *Chattanooga Times* reported, "could talk long and loud against the political and religious system of America."[11] John Scopes inclined toward his father's views about government and religion, but in an easygoing way. Indeed, he liked to talk about sports more than politics and occasionally attended Dayton's northern Methodist church as a way of meeting people. The defendant was an establishment's rebel who would test the law without causing trouble. "Had we sought to find a defendant to present the issue," ACLU counsel Arthur Garfield Hays later confided, "we could not have improved on the individual."[12]

Despite Scopes's ideal casting for his role, he played a bit part in *Inherit the Wind* and at the actual trial. In both, the prosecutors and defense attorneys assumed the leading roles. William Jennings Bryan (fig. 11.1) and Clarence Darrow (fig. 11.2) stole the show.

William Jennings Bryan

In *Inherit the Wind,* Bryan appears almost out of thin air. The audience learns simply that he ran for president three times, remained an influential political orator ("the biggest man in the country—next to the President, maybe," Scopes's fiancée observes),[13] and retained a fundamentalist faith in the Bible. Only at the end of the play does the famed journalist H. L. Mencken explain, "Something happens to an Also-Ran. . . . He becomes a national unloved child . . . [and] unloved children, of all ages, insinuate themselves into spotlights."[14] In this instance, however, the real-life Bryan did much more than insinuate himself into the Scopes trial; he all but ignited that spotlight by focusing public attention on the social implications of Darwinism.

Ever since Charles Darwin published his theory of evolution in 1859, some conservative Christians objected to what they regarded as the atheistic implications of its naturalistic explanation for the origin of biological species, particularly of humans. Early in the twentieth century, these objections intensified with the spread of fundamentalism as a reaction by some traditional American Protestants to increased religious liberalism ("modernism," they called it) within the mainline denominations. Fighting for the fundamentals of biblical orthodoxy, many fundamentalist leaders denounced evolutionary thinking as the heart of the modernist heresy.

Figure 11.1. William Jennings Bryan. Newspapers throughout the United States published this photograph of Bryan scowling during the Scopes trial. Bryan suffered greatly from the oppressive heat in the courtroom during the trial, and had removed his collar on the day this picture was taken.

Bryan, a political liberal with decidedly conservative religious beliefs, added his voice to this chorus following the First World War, as he came to see Darwinian survival-of-the-fittest thinking (known as social Darwinism when applied to human society) behind excessive militarism, imperialism, and laissez-faire capitalism, the three greatest sins in Bryan's political theology. With his Progressive political instinct of seeking legislative solutions to social problems, Bryan called for the enactment of state restrictions against teaching the Darwinian theory of human evolution in public schools, campaigning for such laws across America during the early 1920s. Following a near miss in Kentucky during 1922, antievolution bills inspired by Bryan appeared in state legislatures throughout the country, culminating in passage of the Tennessee law in 1925.

Bryan's antievolutionism was compatible with his progressive politics because both involved reform, appealed to majority rule, and sprang from Christian convictions. Bryan alluded to these issues in his first public address dealing with Darwinism, which he composed at the height of his political career, in 1904. From this earliest point, he described Darwinism as "dangerous" for both

Figure 11.2. Clarence Darrow. A newspaper photographer took this picture during a break in the action of the Scopes trial. With piercing eyes looking out from a rumpled visage, this photograph captures the public image of the famed litigator.

religious and social reasons. "I object to the Darwinian theory," Bryan said with respect to the religious implications of a purely naturalistic explanation for human development, "because I fear we shall lose the consciousness of God's presence in our daily life, if we must accept the theory that through all the ages no spiritual force has touched the life of man and shaped the destiny of nations." Turning to the social consequences of the theory, Bryan added, "But there is another objection. The Darwinian theory represents man as reaching his present perfection by the operation of the law of hate—the merciless law by which the strong crowd out and kill off the weak."[15] These abiding convictions drew Bryan to the Scopes trial to defend Tennessee's antievolution statute, a law that he had inspired and endorsed.

In *Inherit the Wind,* Darrow accounts for the late-in-life antievolutionism of his once-liberal adversary with the words, "A giant once lived in that body. But [Bryan] got lost. Because he was looking for God too high up and too far away."[16] Bryan's best biographer, Lawrence Levine, rejected this view. "In William Jennings Bryan, reform and reaction lived happily, if somewhat incongruously, side by side," Levine concluded. "The Bryan of the 1920s was essentially the Bryan of the 1890s: older in years but no less vigorous, no less optimistic, no less certain."[17]

Clarence Darrow

Inherit the Wind provides a fuller rationale for Darrow's participation in the small-town misdemeanor trial than it does for Bryan's appearance in it, but even here the drama diverges sharply from the historical record. In the actual historical episode, Darrow volunteered his services directly to Scopes. In the play, however, a jailed Scopes appeals to Mencken's newspaper for help in getting a lawyer. Mencken then announces that his paper was sending "the most agile legal mind of the Twentieth Century," Clarence Darrow. "The agnostic?" the Reverend Brown gasps. "A vicious, godless man!"[18] Both descriptions capture historically valid perceptions of Darrow.

By the twenties, Darrow unquestionably stood out as the most famous—some would say infamous—trial lawyer in America. Born into an educated working-class family in rural Ohio, Darrow first gained public notice in the 1890s as a Chicago city attorney and popular speaker for liberal causes. He secured the Democratic nomination to Congress in 1896, but spent most of his time campaigning for the party ticket, headed by presidential nominee William Jennings Bryan, and lost by about one hundred votes. About this time Darrow took up the cause of labor, beginning with the defense of the famed socialist labor leader Eugene V. Debs against criminal charges growing out of the 1894 Pullman strike. "For the next fifteen years Clarence Darrow was the country's outstanding defender of labor, at a time when labor was more militant and idealistic and employers more hardened and desperate than ever before or since," an article in *The Nation* observed during the Scopes litigation. "The cases he was called upon to defend were almost invariably criminal prosecutions in bitterly hostile communities."[19] The final such case, a dramatic 1911 murder trial involving two union leaders accused of blowing up the *Los Angeles Times* building, tarnished Darrow's reputation with labor when the lionized defendants confessed their guilt to avoid the death penalty.

Thereafter, Darrow gradually shifted his practice to criminal law, defending an odd mix of political radicals and wealthy murderers. These activities kept Darrow's name in newspaper headlines, such as during the 1924 Loeb–Leopold case, one of the most sensational trials in American history. In it, Darrow used arguments of psychological determinism (informed by notions of evolutionary naturalism) to save two wealthy and intelligent Chicago teenagers from execution for their cold-blooded murder of an unpopular schoolmate, a crime that the defendants apparently committed for no other reason than to see if they could get away with it. Although Darrow's defense outraged many Americans who believed in individual responsibility, it reflected his longstanding and oft-proclaimed belief in materialistic determinism.[20]

Not content with simply questioning popular notions of criminal responsibility, Darrow delighted in challenging traditional concepts of morality and religion. One historian has described Darrow as "the last of the 'village atheists' on a national scale," and in this role he performed for America the same part that his father once played in his hometown.[21] "He rebelled, just as his father had rebelled, against the narrow preachments of 'do gooders,'" Darrow biographer Kevin Tierney concluded. "He regarded Christianity as a 'slave religion,' encouraging acquiescence in injustice, a willingness to make do with the mediocre, and complacency in the face of the intolerable."[22] In the courtroom, on the lecture circuit, in public debates, and through dozens of popular books and articles, Darrow spent a lifetime ridiculing traditional Christian beliefs. He called himself an agnostic, but he sounded like an atheist. In this, he imitated his intellectual mentor, nineteenth-century American social critic Robert G. Ingersoll. According to Ingersoll, "The Agnostic does not simply say, 'I do not know [if God exists].' He goes another step, and he says, with great emphasis, that you do not know. . . . He is not satisfied with saying that you do not know—he demonstrates that you do not know, and he drives you from the field of fact."[23]

Darrow welcomed the hullabaloo surrounding the antievolution crusade. It rekindled interest in his legalistic attacks on the Bible, which had begun to appear hopelessly out-of-date in light of modern developments in mainline Christianity. In response to comments about evolution made by William Jennings Bryan in 1923, for example, Darrow again made front-page headlines in the *Chicago Daily Tribune* by simply asking Bryan such questions as: "Did Noah build the ark?" and, if so, "how did Noah gather [animals] from all the continents?"[24] Leading Chicago ministers complained that both Bryan's comments and Darrow's questions missed the point, but the public loved it.[25] And so did Darrow. When the Scopes trial began two years later, Darrow volunteered his service for the defense, the only time he ever offered free legal aid. He saw a chance to grab the limelight and debunk Christianity. "My object," Darrow later wrote, "was to focus the attention of the country on the programme of Mr. Bryan and the other fundamentalists in America."[26] Thus, contrary to *Inherit the Wind,* Darrow did not need to be—nor was he—sent to Dayton by Mencken's newspaper. (Coincidentally, Mencken and Darrow had been together when word of the pending trial first broke, and they had discussed whether Darrow should defend Scopes. The sixty-eight-year-old attorney had just announced his retirement, however, and let the matter pass.) The ACLU would not want his help anyway, Darrow surmised, because his zealous agnosticism might transform the trial from a narrow appeal for free speech to a broad assault on religion. As the ACLU later assured its many liberal religious supporters, it did not want Darrow anywhere near Dayton.

Darrow's initial surmise about the ACLU, and his later machinations to insinuate himself into the trial by appealing directly to Scopes and reporters (who welcomed the colorful litigator) rather than going through the ACLU, highlight a telling omission from the version of events in *Inherit the Wind.* Scopes need not have (and did not) ask Mencken's newspaper to obtain defense counsel for him because he agreed to test the law only after the ACLU offered to defend him. It was the ACLU's case from the start, and Darrow would not have joined it if Bryan had stayed out. Once he did, however, Darrow demanded in.[27] Despite assuming a leading role at trial, Darrow served merely as one member of a star-studded defense team assembled by the ACLU that included Arthur Garfield Hays, a wealthy Park Avenue attorney who regularly defended free-speech cases for the ACLU, and Dudley Field Malone, an even wealthier New York divorce lawyer, drawn to the spotlight of the Scopes litigation.

These other attorneys, who actually argued more of the case than Darrow, and the ACLU itself are never mentioned in *Inherit the Wind.* The omission simplifies the script, but it also serves another purpose. As they later explained, Jerome Lawrence and Robert E. Lee wrote the play during the mid-1950s as a means of awakening public concern about the innocent victims of the mass hysteria then feeding McCarthy-era assaults on alleged communists and leftists.[28]

Because the ACLU remained a much maligned target of those assaults, identifying its role in the Scopes trial would have made it harder to portray Scopes as an innocent victim. Thus the ACLU disappeared from the legend of its own most famous trial.

National Interest

The prospect of the two renowned orators Bryan and Darrow actually litigating the profound issues of science versus religion and academic freedom versus popular control over public education turned the trial into a media sensation then and the stuff of legend thereafter. News of the trial dominated the headlines during the weeks leading up to it and pushed nearly everything else off American front pages throughout the eight-day event. Two hundred reporters covered the story in Dayton, including some of the country's best correspondents, representing many of the major newspapers and magazines. Thousands of miles of telegraph wires were hung to transmit every word spoken in court, and pioneering live radio broadcasts carried the oratory to the listening public. Newsreel cameras recorded the encounter, with the film flown directly to major northern cities for projection in movie houses. The media billed it as "the trial of the century" before it even began, and it lived up to its billing. By its end, Dayton civic leaders could only marvel at the success of their publicity stunt and dream of ways to capitalize on their town's new notoriety.

Because the judge permitted radio microphones, newsreel cameras, and telegraph tickers in the courtroom, the encounter looked more like a popular debate before a national audience than a criminal prosecution tried before local jurors. "I have set the date when all universities and schools will be through their terms of school in order that scientists, theologians and other school men will be able to act as expert witnesses," he commented before trial. "My suggestion is that a roof be built over a large vacant lot . . . and seats be built in tiers. At the very least, the place should seat twenty thousand people."[29] At the time, Dayton's population stood at less than one tenth of that figure. Ultimately, most of the trial occurred in the town's oversized courtroom, with inside seats reserved for reporters and loudspeakers carrying the proceedings outside and to the high school auditorium (fig. 11.3).

The courtroom arguments and speeches by both sides addressed the nation rather than the jurors (who missed most of the oratory anyway, because it had so little to do with the facts of the case that it was delivered with the jury excused). The defense divided its arguments among its three principal attorneys. Hays presented the standard ACLU argument that Tennessee's antievolution statute

Figure 11.3. Clarence Darrow defends his client before the courtroom and jurors, who are seated in the front of the crowd with ties and black coats.

violated the individual rights of teachers. Malone, a liberal Catholic, mostly argued that the scientific theory of evolution did not necessarily conflict with an open-minded reading of Genesis. Darrow, for his part, concentrated on debunking fundamentalist reliance on revealed Scripture as a source of knowledge about nature suitable for setting education standards. Their common goal, as Hays stated at the time, was to make it "possible that laws of this kind will hereafter meet the opposition of an aroused public opinion."[30] Elements from all three lines of argument appear in the words attributed to Darrow in *Inherit the Wind,* which makes him sound unduly tolerant of Christian beliefs when he gives voice to arguments originally articulated by Hays and Malone.

The prosecution countered with a half dozen local attorneys, led by the state's able prosecutor and future U.S. senator, Tom Stewart, plus Bryan and his son, William Jennings Jr., a California lawyer. In court, they focused on proving that Scopes had broken the law and objected to any attempt to litigate the merits of that statute, mainly because they could not find expert witnesses opposed to the theory of evolution capable of matching those assembled by the defense in support of that theory. "Mr. Scopes might have taken his stand on the street corners and expounded until he became hoarse," Stewart maintained in a typical statement, "but he cannot go into the public schools . . . and teach his theory." Legislators, "who are responsible to their constituents, to the citizens of Tennessee," should control public education, he asserted.[31] *Inherit the Wind* gives due billing to Stewart, but the other local prosecutors and the younger Bryan disappear in that account. The elder Bryan, who had not practiced law for three decades, stayed uncharacteristically quiet in court, saving his oratory for lecturing the assembled press and public outside the courtroom about the vices of teaching evolution and the virtues of majority rule. He also prepared a thunderous three-hour-long address on these points that he planned to deliver as the prosecution's closing argument. As the actual trial played itself out, however, Darrow managed to frustrate Bryan's plan by waiving his own close because, under Tennessee practice, the defense determined whether there were to be closing statements. In this and other major scenes from the trial, events unfolded roughly in the order presented in *Inherit the Wind.*

The Trial Unfolds

First came jury selection. Darrow typically stressed this part of a trial as critical for the defense and often spent weeks going through hundreds of persons summoned for jury duty (called veniremen) before settling on twelve jurors who just might be open to his arguments and acquit his typically notorious defendant. Darrow had a different objective at the Scopes trial, however. He wanted to convict the statute rather than acquit the defendant, and only judges could do this. Jurors simply applied the law to the facts of the case. Darrow could have won an acquittal by arguing that Scopes (who, after all, was not even a biology teacher) never violated the statute, but that would have left the law intact. Instead, the defense sought either to have the trial judge strike the statute, which was all but beyond his role, or to have Scopes convicted. If the latter happened, the defense could appeal to a higher court, which could review the statute.

Rather than spend weeks seeking sophisticated jurors open to acquitting Scopes despite the law, Darrow quickly accepted the simplest of veniremen. Their presence on the jury, the defense reasoned, would dramatize to the watching nation the spectacle of nonexperts sitting in judgment on a scientific theory, which constituted a key objection to the antievolution law. This strategy paid off when one of the veniremen turned out to be illiterate. His exchange with Darrow became the subject of countless news stories and needed little exaggeration in later accounts. "Ever read anything in a book about evolution?" Darrow asks in *Inherit the Wind.* "Nope," answers the man, giving a typical reply at the trial. Darrow follows up, "I'll bet you read your Bible." Again: "Nope." When a puzzled

Darrow asks why, the venireman simply replies, "Can't read." To ensure that viewers got the point, the playwrights called for "a few titters through the courtroom," before Darrow declares "He'll do."[32] After relating several such exchanges stressing the jury's ignorance about the theory of evolution, Mencken presented the situation to newspaper readers across America just as the defense wanted: "Such a jury, in a legal sense, may be fair, but it would certainly be spitting in the eye of reason to call it impartial."[33]

No sooner was the jury selected than it was excused from the courtroom—for days—as the parties wrangled over defense motions to strike the statute as unconstitutional. Although these arguments occasionally soared into dramatic pleas for either individual freedom or majority rule, their technical nature makes them difficult to follow by nonlawyers. The trial judge denied the motions.

Inherit the Wind skips ahead to the prosecution's case. The scene begins with a schoolboy named Howard (Howard Morgan in the actual trial) testifying that Scopes taught his class about evolution. On cross-examination, Darrow asks Howard the famous question, "Did it do you any harm?" and coaches out a denial.[34] Morgan remains an all-American boy. Dropped from the legend is Bryan's telling retort: "Mr. Darrow asked Howard Morgan, 'Did it hurt you?' Why did he not ask the boy's mother?"[35] Instead, the play inserts testimony from Scopes's fictional fiancée, whom Bryan mercilessly grills about Scopes's agnosticism. In 1960, when paid to tout the movie for its makers, Scopes dismissed this female part: "They had to invent romance for the balcony set."[36] More likely, the writers created the scene to discredit McCarthyite inquisitions of the families and friends of alleged communists. It gave a dark taint to the trial wholly missing from the actual episode, in which Scopes had to instruct his students to testify that he had violated the law.

Following the prosecution's brief presentation, the defense offered the testimony of fifteen experts in science and religion, all prepared to testify against the statute. The prosecution immediately objected. The admissibility of such testimony was the key issue at trial. It passes quickly in *Inherit the Wind* but consumed days at trial. The script captures the heart of the matter, however. "Their testimony is basic to the defense of my case," Darrow pleads. "For it is my intent to show this court that what [John Scopes] spoke quietly one spring afternoon in the [Dayton] High School is no crime! It is incontrovertible as geometry in every enlightened community of minds!" Bryan formally objects to the testimony as irrelevant to the question of whether Scopes violated the law. He then adds his real complaint: "And I refuse to allow these agnostic scientists to employ this courtroom as a sounding board, as a platform from which they can shout their heresies into the headlines!" Not surprisingly given legal limits on expert testimony, the judge sides with the prosecution. Stage directions for *Inherit the Wind*, however, instruct Darrow to appear "flabbergasted" and to look around "helplessly" until "there's a glint of an idea in his eye."[37] This sets up the most dramatic moment of *Inherit the Wind,* when the defense calls Bryan to the stand as a expert on religion. It was equally dramatic at the actual trial, although Darrow had quietly planned it for days. It has become the most famous scene in the folklore of American legal history. And it really happened, though not quite as portrayed in *Inherit the Wind.*

Bryan's Testimony

It was Hays, not Darrow, who called Bryan as the defense's expert on the Bible. The volunteer prosecutor proved cooperative. Up to this point, Stewart had masterfully controlled the proceedings and confounded his wily opponents. Indeed, Tennessee's governor had just wired the young prosecutor, "You are handling the case like a veteran and I am proud of you."[38] But Stewart could not control his impetuous cocounsel, especially because the judge seemed eager to hear Bryan defend the faith. "All the lawyers leaped to their feet at once," Scopes recalled.[39] At least one of the local prosecutors formally objected. Stewart seethed with anger. But Bryan vainly welcomed the opportunity to face his adversaries. "They did not come here to try this case," Bryan explained early in his testimony. "They came here to try revealed religion. I am here to defend it, and they can ask me any questions they please."[40] Darrow did just that.

Thinking the trial all but over, except for the much awaited closing oratory, and hearing that cracks had appeared in the ceiling below the overcrowded second-floor courtroom, the judge moved the afternoon session outside, onto the courthouse lawn. The crowd swelled as word of the encounter spread. From the five hundred persons that evacuated the courtroom, the number rose to an estimated three thousand people spread over the lawn—nearly twice the town's normal population. The participants appeared on a crude wooden platform erected for the proceedings, looking much like Punch and Judy puppets performing at an outdoor festival. Enterprising youngsters passed through the crowd hawking refreshments. "Then began an examination which has few, if any, parallels in court history," the *Nashville Banner* reported. "In reality, it was a debate between Darrow and Bryan on Biblical history, on agnosticism and belief in revealed religion."[41] Darrow posed the well-worn questions of the village skeptic, much like his father would have asked fifty years before: Did Jonah live inside a whale for three days? How could Joshua lengthen the day by making the Sun (rather than the earth) stand still? Where did Cain get his wife? In a narrow sense, as Stewart persistently complained, Darrow's questions had nothing to do with the case because they never inquired about human evolution. In a broad sense, as Hays repeatedly countered, they had everything to do with the case because they challenged biblical literalism. Best of all for Darrow, no good answers existed. The queries compelled Bryan "to choose between his crude beliefs and the common intelligence of modern times" or to admit ignorance, as Darrow later observed.[42] Bryan tried all three approaches at different times during the afternoon, without appreciable success.

Darrow questioned Bryan as a hostile witness, peppering him with queries and giving him little chance for explanation. At times it seemed like a firing line:

Darrow: "You claim that everything in the Bible should be literally interpreted?"

Bryan: "I believe everything in the Bible should be accepted as it is given there; some of the Bible is given illustratively. . . ."

Darrow: "But when you read that . . . the whale swallowed Jonah . . . how do you literally interpret that?"

Bryan: ". . . I believe in a God who can make a whale and can make a man and make both of them do what he pleases. . . ."

Darrow: "But do you believe he made them—that he made such a fish and it was big enough to swallow Jonah?"

Bryan: "Yes sir. Let me add: One miracle is just as easy to believe as another."

Darrow: "It is for me . . . just as hard."

Bryan: "It is hard to believe for you, but easy for me. . . . When you get beyond what man can do, you get within the realm of miracles; and it is just as easy to believe the miracle of Jonah as any other miracle in the Bible."[43]

Such affirmations undercut the appeal of fundamentalism. On the stump, Bryan had effectively championed the cause of biblical faith by addressing the great questions of life; the special creation of humans in God's image gave purpose to every person, and the bodily resurrection of Christ gave hope to believers for eternal life. But Darrow did not inquire about these grand miracles. For many Americans, laudable simple faith became laughable crude belief when applied to Jonah's whale, Noah's Flood, and Adam's rib. Yet Bryan acknowledged accepting each of these biblical miracles on faith and professed that all miracles were equally easy to believe.

Bryan fared little better when he tried to explain two of the biblical passages raised by Darrow. In an apparent concession to modern astronomy, Bryan suggested that God had extended the day for Joshua by stopping the earth rather than the Sun. Similarly, in line with nineteenth-century evangelical scholarship, Bryan affirmed his understanding that the Genesis days of Creation represented periods of time. This led to the following exchange:

Darrow: "Have you any idea of the length of these periods?"
Bryan: "No; I don't."
Darrow: "Do you think the sun was made on the fourth day?"
Bryan: "Yes."
Darrow: "And they had evening and morning without the sun?"
Bryan: "I am simply saying it is a period."
Darrow: "They had evening and morning for four periods without the sun, do you think?"
Bryan: "I believe in creation as there told, and if I am not able to explain it I will accept it."[44]

Even in saying that Creation might have taken as long as six hundred million years, Bryan did not venture beyond the accepted bounds of biblical inerrancy. Nevertheless, the defense made the most of it, perhaps unaware that fundamentalists routinely interpreted Scripture but more likely simply seeing a good argument in the battle for general public opinion. "Bryan had conceded that he interpreted the Bible," Hays gloated. "He must have agreed that others have the same right."[45] Scopes later recalled, with questionable accuracy, that "the Bible literalists who came to cheer Bryan were surprised, ill content, and disappointed that Bryan gave ground."[46]

As Darrow pushed his various lines of questioning, Bryan increasingly admitted that he simply did not know the answers. He had no idea what would have happened to the earth if it had stopped for Joshua, or about the antiquity of human civilization or the age of the earth. "Did you ever discover where Cain got his wife?" Darrow asked. "No sir; I leave the agnostics to hunt for her," came the bittersweet reply.[47] "Mr. Bryan's complete lack of interest in many of the things closely connected with such religious questions as he had been supporting for many years was strikingly shown again and again by Mr. Darrow," the *New York Times* reported.[48] Stewart tried to end the two-hour-long interrogation at least a dozen times, but Bryan refused to stop. "I am simply trying to protect the word of God against the greatest atheist or agnostic in the United States," he shouted, pounding his fist in rage. "I want the papers to know I am not afraid to get on the stand in front of him and let him do his worst."[49] The crowd cheered this outburst and every counterthrust he attempted. Darrow received little applause but inflicted the most jabs. "The only purpose Mr. Darrow has is to slur the Bible, but I will answer his questions," Bryan exclaimed. "I object to your statement," Darrow shouted back, both men now standing and shaking their fists at each other. "I am examining your fool ideas that no intelligent Christian on earth believes."[50] The judge, having finally heard enough, abruptly adjourned court for the day. He never let the interrogation resume.

Inherit the Wind grossly oversimplifies this encounter and in doing so transforms Bryan into a mindless, reactionary creature of the mob. In the play and movie he assails evolution solely on narrow biblical grounds (never hinting at the broad social concerns that largely motivated him) and denounces all science as "Godless," rather than only the so-called false science of evolution.[51] Rather than accepting the popular "day-age" interpretation of the Genesis account of Creation week, which he did in real life, the Bryan of *Inherit the Wind* follows Bishop Ussher's traditional computation that God created the universe in six twenty-four-hour days beginning "on the 23rd of October in the Year 4,004 B.C. at—uh, at 9 A.M.!" The crowd gradually slips away from him as he babbles on, reciting the names of the books of the Old Testament. "Mother. They're laughing at me, Mother!" Bryan cries to his wife at the close of his testimony in the play. "I can't stand it when they laugh at me!" Even though Bryan opposed including a penalty provision in antievolution laws, *Inherit the Wind* ends with his character ranting against the small size of the fine imposed by the judge before fatally collapsing in the courtroom while the now-hostile crowd ignores his closing speech. "The mighty Evolution Law explodes with a pale puff of a wet firecracker," the stage directions explain, just as McCarthyism died from ridicule and disgust.[52]

As Lawrence and Lee debunk Bryan in the eyes of the audience, they uplift Darrow. In *Inherit the Wind,* Darrow makes his entrance in a "long, ominous shadow," the stage directions instruct, "hunched over, head jutting forward." A young girl screams, "It's the Devil!" But he softens as the play proceeds. "All I want is to prevent the clock-stoppers from dumping a load of medieval nonsense in the United States Constitution," he explains at one point, "You've got to stop 'em somewhere."[53] Thus Darrow remains a self-proclaimed agnostic but loses his crusading materialism. At the end, it is Mencken who delivers Darrow's famous slur of Bryan's "fool religion," and it is Darrow who reacts with anger: "You smart-aleck! You have no more right to spit on his religion than you have a right to spit on my religion! Or lack of it!" The writers have Darrow issue the liberal's McCarthy-era plea for tolerance, saying that everyone has the "right to be wrong"! The cynical Mencken then calls the defense lawyer "more religious" than Bryan, and storms from the scene. Left alone in the courtroom, Darrow picks up the defendant's copy of the *Origin of Species* and the judge's Bible. After "balancing them thoughtfully, as if his hands were scales," the staging directions state, the attorney "jams them in his briefcase, side by side," and slowly walks off the now-empty stage.[54] "A bit of religious disinfectant is added to the agnostic legend for audiences whose evolutionary stage is not yet very high," the *Village Voice* sneered in its review of the movie.[55]

Lawrence and Lee thus reduced Bryan and Darrow to one-dimensional caricatures of themselves and simplified the trial into a three-act morality play in which tolerance happily triumphs over bigotry. To ensure that the audience appreciate the point, the writers have Scopes ask Darrow after the jury convicts him, "Did I win or did I lose?" Darrow answers, "You won. . . . Millions of people will say you won. They'll read in their papers tonight that you smashed a bad law. You made it a joke!"[56] It is fine theater but hobbles efforts to understand the trial's complex historical legacy.

Legacy

Although partisans on both sides claimed victory, at the time most neutral observers viewed the Scopes trial as a draw, and none saw it as decisive. America's adversarial legal system tends to drive parties apart rather than to reconcile them. That was certainly the result in this case. Despite Bryan's stumbling on the witness stand (which his supporters attributed to his notorious interrogator's wiles), each side effectively communicated its message from Dayton—maybe not well enough to win converts but at least sufficiently to energize those already predisposed toward its viewpoint. If, as *Inherit the Wind* declares, millions of people thereafter dismissed antievolutionism as a joke, millions more saw it as a cause. When Bryan died a week later in Dayton, the fundamentalist movement acquired a martyr. Huge crowds turned out to watch as a special train carried Bryan's body to Washington for a state funeral in Arlington National Cemetery. Three months after the trial, Mencken could only sneer about the man he so despised: "His place in the Tennessee hagiocracy is secure. If the village barber saved any of his hair, then it is curing gallstones down there today."[57] Darrow's followers lionized him too, and granted him an elevated place in agnostic hagiography because of his deeds at Dayton.

The pace of antievolution activism actually quickened following the trial, but it encountered heightened popular resistance. Two states, Mississippi and Arkansas, soon enacted antievolution statutes modeled on the Tennessee law. An anticipated victory in the Minnesota legislature turned into a demoralizing defeat, however. When one Rhode Island legislator introduced such a proposal in 1927, his bemused colleagues referred it to the Committee on Fish and Game, where it died without a hearing or a vote.

With time and countless retellings, the Scopes trial became part of the fabric of American culture. It grew to symbolize not simply antievolutionism, but religiously motivated intrusions into public policy generally. *Inherit the Wind*'s dramatic plea for tolerance may have targeted McCarthyites, using fundamentalists only as straw men. But the straw men proved to be more durable than the real targets, and the threat to individual liberty attributed to them became increasingly ominous for some Americans as the power of government grew over the ensuing years. Indeed, the issues raised by the Scopes trial and legend endure precisely because they embody the characteristically American struggle between individual liberty and majority rule and cast it in the timeless debate over science and religion. For many twentieth-century Americans, the Scopes trial became both the yardstick by which the former battle was measured and the glass through which the latter debate was seen. In its review of *Inherit the Wind*'s 1996 Broadway revival, the *New York Times* described the original courtroom confrontation as "one of the most colorful and briefly riveting of the *trials of the century* that seemed to be especially abundant in the sensation-loving 1920's."[58] Dozens of prosecutions have received that designation over the years, but only the Scopes trial fully lives up to its billing by continuing to echo throughout the century.

12

Science without God: Natural Laws and Christian Beliefs

Ronald L. Numbers

Nothing has come to characterize modern science more than its rejection of appeals to God in explaining the workings of nature. Numerous scientists, philosophers of science, and science educators have made this claim. In 1982 a United States federal judge, eager to distinguish science from other forms of knowledge, especially religion, spelled out "the essential characteristics of science." At the top of his list appeared the notion that science must be "guided by natural law." No statement, declared the judge, could count as science if it depended on "a supernatural intervention." Five years later the U.S. Supreme Court affirmed the judge's reasoning.[1]

Students of nature have not always shunned the supernatural. It took centuries, indeed millennia, for naturalism to dominate the study of nature, and even at the beginning of the twenty-first century, as we shall see, a tiny but vocal group of "theistic scientists" is challenging what they regard as the arbitrary exclusion of the supernatural from science. In exploring how naturalism came to control the practice of science, I hope to answer some basic questions about the identity and motives of those who advocated it. In particular I want to illuminate the reasons why naturalism, described by some scholars as the great engine driving the secularization of Western society, attracted so much support from devout Christians, who often eagerly embraced it as the method of choice for understanding nature. Naturalization, as we shall see, did not lead inevitably to secularization.

First, however, we need some clarification about terms. Historians have employed the word "naturalism" to designate a broad range of views: from a purely methodological commitment to explaining the workings of nature without recourse to the supernatural, largely devoid of metaphysical implications about God, to a philosophical embracement of materialism, tantamount to atheism. When Thomas H. Huxley (1825–95) coined the term "scientific naturalism" in 1892, he used it to describe a philosophical outlook that shunned the supernatural and adopted empirical science as the only reliable basis of knowledge about the physical, social, and moral worlds. Although such metaphysical naturalism, rooted in the findings of science, has played an important role in the history of philosophy and religion, its significance in the history of scientific practice has remained small compared to what has recently come to be called methodological naturalism, the focus of this chapter.[2]

Naturalism and Natural Philosophy

Recorded efforts to explain naturally what had previously been attributed to the whimsy of gods date back to the Milesian philosophers of the ancient Greek world, who, six centuries before the birth of Christianity, declared such phenomena as earthquakes, lightning, and thunder to be the result of natural causes. A little later Hippocratic physicians expanded the realm of the natural to include most diseases, including epilepsy, "the sacred disease." As one Hippocratic writer insisted, "Each disease has a natural cause and nothing happens without a natural cause." The first-century Roman philosopher Lucius Annaeus Seneca, ever suspicious of supernatural causation, calmed the fears of fellow citizens by assuring them that "angry deities" had nothing to do with most meteorological or astronomical events: "Those phenomena have causes of their own."[3]

As these scattered examples show, belief in natural causes and the regularity of nature antedated the appearance of Christianity, with its Judaic notion of God as creator and sustainer of the universe. Although inspired by a man regarded as divine and developed in a milieu of miracles, Christianity could, and sometimes did, encourage the quest for natural explanations. Long before the birth of modern science and the appearance of "scientists" in the nineteenth century, the study of nature in the West was carried out primarily by Christian scholars known as natural philosophers, who typically expressed a preference for natural explanations over divine mysteries. During the philosophical awakening of the twelfth century, for instance, Adelard of Bath (ca. 1080–ca. 1150), a much-traveled Englishman familiar with the views of Seneca, instructed his nephew on the virtues of natural explanations:

> I will take nothing away from God: for whatever exists is from Him and because of Him. But the natural order does not exist confusedly and without rational arrangement, and human reason should be listened to concerning those things it treats of. But when it completely fails, then the matter should be referred to God.

A number of other medieval churchmen expressed similar views, on occasion extending the search for natural explanations to such biblical events as Noah's Flood, previously regarded as a miracle.[4]

By the late Middle Ages the search for natural causes had come to typify the work of Christian natural philosophers. Although characteristically leaving the door open for the possibility of direct divine intervention, they frequently expressed contempt for soft-minded contemporaries who invoked miracles rather than searching for natural explanations. The University of Paris cleric Jean Buridan (ca. 1295–ca. 1358), described as "perhaps the most brilliant arts master of the Middle Ages," contrasted the philosopher's search for "appropriate natural causes" with the common folk's erroneous habit of attributing unusual astronomical phenomena to the supernatural. In the fourteenth century the natural philosopher Nicole Oresme (ca. 1320–82), who went on to become a Roman Catholic bishop, admonished that, in discussing various marvels of nature, "there is no reason to take recourse to the heavens, the last refuge of the weak, or demons, or to our glorious God as if He would produce these effects directly, more so than those effects whose causes we believe are well known to us."[5]

Enthusiasm for the naturalistic study of nature picked up in the sixteenth and seventeenth centuries as more and more Christians turned their attention to discovering the so-called secondary causes that God employed in operating the world. The Italian Catholic Galileo Galilei (1564–1642), one of the foremost promoters of the new philosophy, insisted that nature "never violates the terms of the laws imposed upon her." In a widely circulated letter to the Grand Duchess Christina, written in 1615, Galileo, as a good Christian, acknowledged the divine inspiration of both Holy Scripture and the Book of Nature—but insisted that interpreters of the former should have no say in determining the meaning of the latter. Declaring the independence of natural philosophy from theology, he asserted "that in disputes about natural phenomena one must begin not with the authority of scriptural passages but with sensory experience and necessary demonstrations."[6]

Far to the west, in England, the Anglican philosopher and statesman Francis Bacon (1561–1626) was preaching a similar message of independence, warning of "the extreme prejudice which both religion and philosophy hath received and may receive by being commixed together; as that which undoubtedly will make an heretical religion, and an imaginary and fabulous philosophy." Christians, he advised, should welcome rather than fear the truth that God operates the world largely, though not exclusively, through natural laws discoverable by human effort. Although conceding that too great an emphasis on natural law might undermine belief in God, he remained confident that further reflection would "bring the mind back again to religion."[7]

The danger Bacon perceived did not take long to materialize. As natural philosophers came to view nature as "a law-bound system of matter in motion," a vast machine running with little or no divine intervention, they increasingly focused on the regularities of nature and the laws of motion rather than on God's intrusions. When the French Catholic natural philosopher René Descartes (1596–1650) boldly constructed a universe of whirling ethereal fluids and speculated how the solar system could have been formed by the action of these vortices operating according to the God-ordained laws of nature, he acquired considerable notoriety for nearly pushing God out of the cosmos altogether. His pious fellow countryman Blaise Pascal (1623–62) accused Descartes, somewhat unfairly, of trying to dispense with God altogether, according Him only "a flip of the finger in order to set the world in motion." Fearing clerical retribution in the years after Galileo's trial, Descartes disingenuously declared his own cosmogony to be "absolutely false."[8]

The English chemist Robert Boyle (1627–91)—as ardent an advocate of the mechanical philosophy as Descartes yet as pious as Pascal—viewed the discovery of the divinely established laws of nature as a religious act. A devout Protestant with great reverence for the Bible, Boyle regarded revelation as "a foreign principle" to the study of the "laws or rules" of nature. He sought to explain natural phenomena in terms of matter in motion as a means of combating pagan notions that granted nature quasi-divine powers, not as a way to eliminate divine purpose from the world. According to the historians Edward B. Davis and Michael Hunter, viewing the cosmos as a "compounded machine" run according to natural laws struck Boyle as being "more consistent with biblical statements of divine sovereignty than older, non-mechanistic views" of an intelligent nature. "By denying 'Nature' any wisdom of its own, the mechanical conception of nature located purpose where Boyle believed it belonged: over and behind nature, in the mind of a personal God, rather than in an impersonal semi-deity immanent within the world." God's customary reliance on natural laws (or secondary causes) did not, in Boyle's opinion, rule out the possibility of occasional supernatural interventions, when God (the primary cause) might act "in special ways to achieve particular ends." This view became common among Christian men of science, as well as among clerics.[9]

No one contributed more to the popular image of the solar system as a giant mechanical device than the University of Cambridge professor of mathematics Isaac Newton (1642–1727), a man of deep, if unorthodox, religious conviction, who unblushingly attributed the perfections of the solar system to "the counsel and dominion of an intelligent and powerful Being." Widely recognized as the greatest natural philosopher of all time for his discovery of the role of gravity in the operation of the universe, he insisted that natural knowledge should be based on observations and experiments, not hypotheses. Although he chided Descartes for his attempt to explain the solar system by "mere Laws of Nature," he himself believed that God typically used them "as instruments in his works." In private correspondence he even speculated in Cartesian fashion about how God might have used natural laws to construct the solar system from a "common Chaos."[10]

Endorsed by such publicly religious natural philosophers as Boyle and Newton, the search for natural laws and mechanical explanations became a veritable Christian vocation, especially in Protestant countries, where miraculous signs and wonders were often associated with Catholic superstition. As one Anglican divine complained in 1635, some Protestants were even beginning to question of the efficacy of prayer, believing that God, working through second causes, "hath set a constant course in nature."[11]

For ordinary folk the most compelling instances of supernaturalism giving way to naturalism occurred not in physics or chemistry but in such areas as meteorology and medicine, in explanations of epidemics, eclipses, and earthquakes. Already by the sixteenth century, supernatural explanations of disease had largely disappeared from medical literature except in discussions of epidemics and insanity, which remained etiological mysteries, and venereal diseases, the wages of sin. In writing about the common afflictions of humanity—fractures, tumors, endemic diseases, and such—physicians seldom mentioned God or the devil. Even when discussing the plague, the most dreaded disease of all, they tended merely to acknowledge its supernatural origin before passing quickly to its more mundane aspects. The great French surgeon Ambroise Paré (1510–90), for example, explicitly confined himself to "the natural causes of the plague," saying that he would let divines deal with its ultimate causes. Priests and theologians may have placed greater emphasis on supernatural causes and cures, but in general they too easily accommodated new medical knowledge by maintaining that God usually effected His will through natural agencies rather than by direct intervention. Theological interests thus seldom precluded searching for natural causes or using natural therapies.[12]

The most dramatic, and in some ways revealing, episode in the naturalization of disease occurred in the British colonies of North America in the early 1720s. Christians had long regarded smallpox, a frighteningly deadly and disfiguring disease, as God's ultimate scourge to punish sinners and bring them to their knees in contrition. Thus when an epidemic threatened to strike New England in 1721, the governor of Massachusetts called for a day of fasting and repenting of the sins that had "stirred up the Anger of Heaven against us." However, the Puritan Cotton Mather (1663–1728), one of the town of Boston's leading ministerial lights, offered an alternative to repentance—inoculation with an attenuated but live form of smallpox—in hopes of preventing the disease by natural means. Having heard rumors of successful inoculations against smallpox in Africa and the Middle East, Mather, a fellow of the Royal Society of London and a natural philosopher in his own right, proposed that the untested, potentially lethal, procedure be tried in Boston. The best trained physician in town, William Douglass (1691–1752), fearing that inoculation would spread rather than prevent the disease and resenting the meddling of ministers in medical matters, urged Mather to rely instead on "the *all-wise Providence* of God Almighty" and quit trying to thwart God's will. Mather and five other clerics countered that such reasoning would rule out all medical intervention. Cannot pious persons, they asked,

> give into the method or practice without having their devotion and subjection to the All-wise Providence of God Almighty call'd in question? . . . Do we not in the use of all means depend on GOD's blessing? . . . For, what hand or art of Man is there in this Operation more than in bleeding, blistering and a Score more things in Medical Use? which are all consistent with a humble trust in our Great preserver, and a due Subjection to His All-wise Providence.

Besides, added Mather, Dr. Douglass risked violating the biblical commandment against killing by refusing to use inoculation to save lives. After postepidemic calculations demonstrated the efficacy of inoculation, smallpox, previously a divine judgment, became a preventable disease. Few Christians lamented the metamorphosis. And in generations to come their descendents would give thanks to God as medical science brought cholera, diphtheria, yellow fever, and even venereal diseases under natural control.[13]

The same process occurred in meteorology. Benjamin Franklin (1706–90), who as a teenager in Boston had backed Douglass in his quarrel with Mather over smallpox inoculation, found himself on the opposite side of a similar debate a few decades later, after announcing the invention of a device to prevent another of God's judgments on erring humanity: lightning. When a French cleric denounced lightning rods as an inappropriate means of thwarting God's will, the American printer turned scientific celebrity scornfully replied: "he speaks as if he thought it presumption in man to propose guarding himself against the *Thun-*

ders of Heaven! Surely the Thunder of Heaven is no more supernatural than the Rain, hail or Sunshine of heaven, against the Inconvenience of which we guard by Roofs & Shades without Scruple." Reflective Christians quickly accepted Franklin's logic, and before long lightning rods were adorning the steeples of churches throughout Europe and North America, protecting them not from God's wrath but from a dangerous and capricious natural occurrence.[14]

Reactions to the great earthquakes of 1727 and 1755 further illustrate the inroads of scientific naturalism on popular culture. On the night of 29 October 1727 a violent earthquake shook the northern colonies of America, producing widespread damage to property. Terrified residents, humbled by this apparent display of divine anger, set aside fast days and begged God to forgive them for such sins as Sabbath breaking, pride, and drunkenness. To promote repentance among his parishioners, Thomas Prince (1687–1758), the Puritan pastor of Boston's Old South Church, preached a sermon entitled *Earthquakes the Works of God and Tokens of His Just Displeasure*. In it he conceded that the ignition of gases in the earth's interior might have touched off the tremors—but then argued that such secondary explanations only demonstrated "how the mighty GOD works invisibly by sensible Causes, and even by those that are extremely little and weak, produces the greatest and most terrible Effects in the World." Cotton Mather similarly insisted on God's active role in producing such catastrophes. "Let the *Natural Causes of Earthquakes* be what the *Wise Men of Enquiry* please," he wrote. "*They* and their *Causes* are still under the government of HIM that is the *GOD of Nature*." Twenty-eight years later, on 18 November 1755, a second earthquake jolted New England. This time nearly two months passed before community leaders called for a public fast. When the aging Reverend Prince reissued his earlier sermon on earthquakes as tokens of God's displeasure, Professor John Winthrop IV (1714–79) of Harvard College calmed the timorous with the assurance that, although God bore ultimate responsibility for the shaking, natural causes had produced the tremors. "I think Mr. Winthrop has laid Mr. Prince flat on [his] back, and seems to take some pleasure in his mortification," gloated one of the professor's admirers.[15]

Years ago the historian Keith Thomas claimed that as the mechanical philosophy pushed God further and further into the distance, it "killed the concept of miracles, weakened the belief in the physical efficacy of prayer, and diminished faith in the possibility of direct divine inspiration." Undoubtedly, some people experienced this effect. The revelations of natural philosophy helped to convince the liberal Boston minister Charles Chauncy (1705–87), for example, that

> God does not communicate either being or happiness to his creatures, at least on this earth, by an immediate act of power, but by concurring with

> an established course of nature. What I mean is, he brings creatures into existence, and makes them happy, by the intervention of second causes, operating under his direction and influence, in a stated, regular uniform manner.

But for every liberal such as Chauncy there were scores of Christians who continued to believe in miracles, prayer, and divine inspiration—while at the same time welcoming the evidence that epidemics, earthquakes, and lightening bolts derived from natural causes. And for every natural philosopher who lost his faith to science, many more found their beliefs untouched by the search for natural causes of physical events. For them, the search for natural laws led to a fuller understanding of God, not disbelief.[16]

The Decline of Natural Philosophy and the Beginnings of Modern Science

No single event marks the transition from godly natural philosophy to naturalistic modern science, but sometime between roughly the mid–eighteenth and mid–nineteenth centuries students of nature in one discipline after another reached the conclusion that, regardless of one's personal beliefs, supernatural explanations had no place in the practice of science. As we have seen, natural philosophers had often expressed a preference for natural causes, but few, if any, had ruled out appeals to God. In contrast, virtually all scientists (a term coined in the 1830s but not widely used until the late nineteenth century), whether Christians or non-Christians, came by the latter nineteenth century to agree that God talk lay beyond the boundaries of science.[17]

The roots of secular science can be traced most clearly to Enlightenment France, where the spirit of Descartes lingered. Although not a materialist—he believed in God and the existence of immaterial souls—Descartes had pushed naturalism to the point of regarding animals as mere machines. This extreme form of naturalism scarcely influenced the course of scientific investigation, especially outside of France, but it did spur some French Cartesians to go even further than Descartes. The French physician Julien Offray de La Mettrie (1709–51), for example, suggested that humans are nothing but "perpendicularly crawling machines," a claim that even the French found sensational. While acknowledging God to be the author of the Book of Nature, La Mettrie insisted that "experience and observation," not revelation, should be "our only guides." Such methods might not lead to absolute truth about human nature, but they provided "the greatest degree of probability possible on this subject." Like many men of science to follow, he was willing to trade theological certainty for such scientific probability.[18]

Much more influential on scientific practice was La Mettrie's countryman Georges-Louis Leclerc de Buffon (1707–88), an ardent admirer of Newton and one of the most prominent natural historians in the eighteenth century. Buffon called for an emphasis on the regularities of nature and a renunciation of all appeals to the supernatural. Those studying physical subjects, he argued, "ought as much as possible, to avoid having recourse to supernatural causes." Philosophers "ought not to be affected by causes which seldom act, and whose action is always sudden and violent. These have no place in the ordinary course of nature. But operations uniformly repeated, motions which succeed one another without interruption, are the causes which alone ought to be the foundation of our reasoning." Buffon professed not to care whether such explanations were true—so long as they appeared probable. A theist, though not a practicing Christian, Buffon acknowledged that the Creator had originally set the planets in motion, but considered the fact of no value to the natural philosopher. Buffon's methodological convictions inspired him to propose a natural history of the solar system, based on the notion that a passing comet, "falling obliquely on the sun," had detached the matter that later formed the planets and their satellites (fig. 12.1). Although his speculations never caught on, some Christian critics justifiably criticized him for trying "to exclude the agency of a divine Architect, and to represent a world begun and perfected merely by the operation of natural, undesigning causes."[19]

A far more successful account of the origin of the solar system came from the Frenchman Pierre-Simon de Laplace (1749–1827), who had abandoned training as a cleric for a career in mathematics and astronomy and eventually became one of the leading men of science in Europe (fig. 12.2). Finding Buffon's methodology attractive but his theory physically implausible, Laplace in 1796 proposed that the planets had been formed from the revolving atmosphere of the primitive Sun, which, as it cooled and contracted, had abandoned a succession of Saturn-like rings, which had coalesced to form the planets. On the occasion of a visit in 1802 to the country estate of Napoleon Bonaparte, Laplace entertained his host with an account of his so-called nebular hypothesis. When the French leader asked why he had heard no mention of God, Laplace supposedly uttered the much-quoted words "Sire, I have no need of that hypothesis." The only firsthand account of the exchange simply reports Napoleon's disappointment with Laplace's explanation that "a chain of natural causes would account for the construction and preservation of the wonderful system." Either way, there was no mistaking the impious message.[20]

Laplace's thoroughly naturalistic hypothesis, authored by a notorious un-

Figure 12.1. The primeval collision of a comet with the Sun, according to Buffon. From George-Louis Leclerc de Buffon, *Histoire naturelle, générale et particuliere* (Deux-Ponts: Sanson, 1795–91), 1:153. Courtesy of William Ashworth and the Linda Hall Library of Science, Engineering and Technology.

believer, represented the secularization of natural philosophy at its baldest. Not surprisingly, some Christians denounced Laplace for his transparently atheistic science. This was especially true in the English-speaking world, where the tradition of natural theology remained strong and where French science was widely viewed as tending toward godless materialism. It seemed clear to the Scottish divine Thomas Chalmers (1780–1847), for example, that if "all the beauties and benefits of the astronomical system [could] be referred to the single law of gravitation, it would greatly reduce the strength of the argument of a designing cause." One of the classic arguments for the existence of God rested on the observation that an object appearing to have been made for a particular purpose had been produced by an intelligent and purposeful designer. If the solar system looked as if it were not the result of necessity or of accident, if it appeared to have been made with a special end in mind, then it must have had a designer,

EXPOSITION

DU

SYSTÈME DU MONDE;

PAR M. LE COMTE LAPLACE,

Chancelier du Sénat-Conservateur, Grand-Officier de la Légion d'Honneur; Grand'Croix de l'ordre de la Réunion; Membre de l'Institut impérial et du Bureau des Longitudes de France; des Sociétés royales de Londres et de Gottingue; des Académies des Sciences de Russie, de Danemarck, de Suède, de Prusse, d'Italie, etc.

QUATRIÈME ÉDITION,

revue et augmentée par l'Auteur.

PARIS,

MME V^E COURCIER, Imprimeur-Libraire pour les Mathématiques, quai des Augustins, n° 57.

1813.

Figure 12.2. Left: Pierre-Simon de Laplace. Right: Title page of Laplace's *Exposition du Système du monde,* in which he develops his nebular hypothesis. From Pierre-Simon de Laplace, *Exposition du Système du monde,* 4th ed. (Paris: Mme. Ve. Courcier, 1813).

namely God. But what happened to the argument if the arrangement had resulted simply from the laws of nature operating on inert matter?[21]

John Pringle Nichol (1804–59), a minister-turned-astronomer at the University of Glasgow and an avid popularizer of the nebular hypothesis, offered one plausible answer. He dismissed Chalmers's argument that "we can demonstrate the existence of a Deity more emphatically from that portion of creation of which we remain ignorant, than from any portion whose processes we have explored," as downright dangerous to Christianity. Such fears, Nichol believed, stemmed from a misunderstanding of the term "law." To him, laws simply designated divine order: "LAW of itself is no substantive or independent power; no causal influence sprung of blind necessity, which carries on events of its own will and energies without command."[22]

As more and more of the artifacts of nature, such as the solar system, came to be seen as products of natural law rather than divine miracle, defenders of design increasingly shifted their attention to the origin of the laws that had proved capable of such wondrous things. Many Christians concluded that these laws had been instituted by God and were evidence of His existence and wisdom. In this way, as John Le Conte (1818–91) of the University of California pointed out in the early 1870s, the cosmogony of Laplace helped to bring about a transformation in the application of the principle of design "from the region of facts to that of laws." The nebular hypothesis thus strengthened, rather than weakened, the argument from design, opening "before the mind a stupendous and glorious field for meditation upon the works and character of the Great Architect of the Universe."[23]

Christian apologists proved equally adept at modifying the doctrine of divine providence to accommodate the nebular hypothesis. Instead of pointing to the miraculous creation of the world by divine fiat, a "special" providential act, they emphasized God's "general" providence in creating the world by means of natural laws and secondary causes. While the Creator's role in the formation of the solar system thus changed, it neither declined nor disappeared, as some timid believers had feared. Daniel Kirkwood (1814–95), a Presbyterian astronomer who contributed more to the acceptance of the nebular hypothesis in America than anyone else, argued that if God's power is demonstrated in sustaining and governing the world through the agency of secondary causes, then it should not "be regarded as derogating from his perfections, to suppose the same power to have been exerted in a similar way in the process of its formation."[24]

God's reliance on secondary causes in the daily operation of the world made it seem only reasonable to suppose that He had at least sometimes used the same means in creating it. "God generally effects his purposes . . . by intermediate agencies, and this is especially the case in dead, unorganized matter," wrote one author:

> If, then, the rains of heaven, and the gentle dew, and the floods, and storms, and volcanoes, and earthquakes, are all products of material forces, exhibiting no evidence of miraculous intervention, there is nothing profane or impious in supposing that the planets and satellites of our system are not the immediate workmanship of the great First Cause. . . . God is still present; but it is in the operation of unchangeable laws; in the sustaining of efficient energies which he has imposed on the material world that he has created; in the preservation of powers, properties, and affinities, which he has transferred out of himself and given to the matter he has made.[25]

To at least one observer, Laplace's cosmogony offered an even more convincing demonstration of divine providence than did the traditional view:

> How much more sublime and exalted a view does it give us of the work of creation and of the Great Architect, to contemplate him evolving a system of worlds from a diffused mass of matter, by the establishment of certain

> laws and properties, than to consider him as taking a portion of that matter in his hand and moulding it as it were into a sphere, and then imparting to it an impulse of motion.[26]

So eager were many Christians to baptize the nebular hypothesis that they even read it back into the first chapter of Genesis. The Swiss-American Arnold Guyot (1807–84), a highly respected evangelical geographer at Princeton College, took the lead in educating fellow Christians on the close correspondence between the nebular hypothesis and the Mosaic narrative. Assuming the "days" of Genesis 1 to represent great epochs of creative activity, he argued that if the formless "waters" created by God at the beginning actually symbolized gaseous matter, then the light of the first day undoubtedly had been produced by the chemical action resulting from the concentration of this matter into nebulae. The dividing of the waters on the second day symbolized the breaking up of the nebulae into various planetary systems, of which ours was one. During the third epoch the earth had condensed to form a solid globe, and during the fourth, the nebulous vapors surrounding our planet had dispersed to allow the light of the Sun to shine on the earth. During the fifth and sixth epochs God had populated the earth with living creatures. "Such is the grand cosmogonic week described by Moses," declared the Christian professor. "To a sincere and unprejudiced mind it must be evident that these great outlines are the same as those which modern science enables us to trace, however imperfect and unsettled the data afforded by scientific researchers may appear on many points."[27]

During the second half of the nineteenth century Guyot's harmonization of the nebular hypothesis and the Bible became a favorite of Christian apologists, especially in America, even winning the endorsement of such staunchly orthodox theologians as Charles Hodge (1797–1878) of Princeton Theological Seminary. "The best views we have met with on the harmony between science and the Bible," Hodge wrote in his immensely influential *Systematic Theology*, "are those of Professor Arnold Guyot, a philosopher of enlarged comprehension of nature and a truly Christian spirit."[28] In Christian classrooms across the United States and Canada, at least, students learned that Laplace's nebular hypothesis, despite its author's own intention, testified to the wisdom and power of God and spoke to the truth of Scripture.

A Preliminary Discourse on the Study of Natural Philosophy (1830) by English astronomer John Herschel (1792–1871) was described by one scholar as "the first attempt by an eminent man of science to make the methods of science explicit." Frequently extrapolating from astronomy, the paradigmatic science of the time, Herschel asserted that sound scientific knowledge derived exclusively from *experience*—"the great, and indeed only ultimate source of our knowledge of nature and its laws"—which was gained by *observation* and *experiment*, "the fountains of all natural science." Natural philosophy and science (he used the terms interchangeably) recognized only those causes "having a real existence in nature, and not being mere hypotheses or figments of the mind." Although this stricture ruled out supernatural causes, Herschel adamantly denied that the pursuit of science fostered unbelief. To the contrary, he insisted that science "places the existence and principal attributes of a Deity on such grounds as to render doubt absurd and atheism ridiculous." Natural laws testified to God's existence; they did not make him superfluous.[29]

Efforts to naturalize the history of the earth followed closely on the naturalization of the skies—and produced similar results. When students of Earth history, many of them Protestant ministers, created the new discipline of geology in the early nineteenth century, they consciously sought to reconstruct Earth history using natural means alone. By the 1820s virtually all geologists, even those who invoked catastrophic events, were eschewing appeals to the supernatural. When the British geologist Charles Lyell (1797–1875) set about in the early 1830s to "free the science from Moses," the emancipation had already largely occurred. Nevertheless, his landmark *Principles of Geology* (1830–33) conveniently summed up the accepted methods of doing geology, with the subtitle, *Being an Attempt to Explain the Former Changes of the Earth's Surface, by Reference to Causes Now in Operation,* conveying the main point. As Lyell described his project to a friend, it would in good Hershelian fashion "endeavour to establish the principles of reasoning in the science," most notably the idea "that *no causes whatever* have from the earliest time to which we can look back to the present ever acted but those *now* acting & that they never acted with different degrees of energy from that which they now exert."[30]

Lyell applauded his geological colleagues for following the lead of astronomers in substituting "fixed and invariable laws" for "immaterial and supernatural agents":

> Many appearances, which for a long time were regarded as indicating mysterious and extraordinary agency, are finally recognized as the necessary result of the laws now governing the material world; and the discovery of this unlooked for conformity has induced some geologists to infer that there has never been any interruption to the same uniform order of physical events. The same assemblage of general causes, they conceive, may have been sufficient to produce, by their various combinations, the endless diversity of effects, of which the shell of the earth has preserved the memorials, and, consistently with these principles, the recurrence of analogous changes is expected by them in time to come.

The community of geologists, comprising mostly Christian men of science, thus embraced "the undeviating uniformity of secondary causes"—with one troubling exception.[31]

Like so many of his contemporaries, Lyell, a communicant of the Church of England, for years stopped short of extending the domain of natural law to the origin of species, especially of humans. At times he leaned toward attributing new species to "the intervention of intermediate causes"; on other occasions he appealed to "the direct intervention of the First Cause," thus transferring the issue from the jurisdiction of science to that of religion. He used his *Principles of Geology* as a platform to oppose organic evolution, particularly the theories of the late French zoologist Jean-Baptiste Lamarck (1744–1829), and he professed not to be "acquainted with any physical evidence by which a geologist could shake the opinion generally entertained of the creation of man within the period generally assigned."[32]

The person most responsible for naturalizing the origin of species—and thereby making the problem a scientific matter—was Lyell's younger friend Charles Darwin (1809–82). As early at 1838 Darwin had concluded that attributing the structure of animals to "the *will* of the Deity" was "no explanation—it *has not the character of a physical* law & is therefore utterly useless." Within a couple of decades many other students of natural history (or naturalists, as they were commonly called) had reached the same conclusion. The British zoologist Thomas H. Huxley, one of the most outspoken critics of the supernatural origin of species, came to see references to special creation as representing little more than a "specious mask for our ignorance." If the advocates of special creation hoped to win a hearing for their views as science, he argued, then they had an obligation to provide "some particle of evidence that the existing species of animals and plants did originate in that way." Of course, they could not. "We have not the slightest scientific evidence of such unconditional creative acts; nor, indeed, could we have such evidence," Huxley noted in 1856 lecture; "for, if a species were to originate under one's very eyes, I know of no amount of evidence which would justify one in admitting it to be a special creative act independent of the whole vast chain of causes and events in the universe." To highlight the scientific vacuity of special creation, Darwin, Huxley, and other naturalists took to asking provocatively whether "elemental atoms flash[ed] into living tissues? Was there vacant space one moment and an elephant apparent the next? Or did a laborious God mould out of gathered earth a body to then endue with life?" Creationists did their best to ignore such taunts.[33]

In his revolutionary *Origin of Species* (1859) Darwin aimed primarily "to overthrow the dogma of separate creations" and extend the domain of natural law throughout the organic world. He succeeded spectacularly—not because of his clever theory of natural selection (which few biologists thought sufficient to account for evolution) nor because of the voluminous evidence of organic development that he presented, but because, as one Christian reader bluntly put it, there was "literally nothing deserving the name of Science to put in its place." The American geologist William North Rice (1845–1928), an active Methodist, made much the same point. "The great strength of the Darwinian theory," he wrote in 1867, "lies in its coincidence with the general spirit and tendency of science. It is the aim of science to narrow the domain of the supernatural, by bringing all phenomena within the scope of natural laws and secondary causes."[34]

In reviewing the *Origin of Species* for the *Atlantic Monthly,* the Harvard botanist Asa Gray (1810–88) forthrightly addressed the question of how he and his colleagues had come to feel so uncomfortable with a "supernatural" account of speciation. "Sufficient answer," he explained, "may be found in the activity of the human intellect, 'the delirious yet divine desire to know,' stimulated as it has been by its own success in unveiling the laws and processes of inorganic Nature." Minds that had witnessed the dramatic progress of the physical sciences in recent years simply could not "be expected to let the old belief about species pass unquestioned." Besides, he later explained, "the business of science is with the course of Nature, not with interruptions of it, which must rest on their own special evidence." Organic evolution, echoed his friend George Frederick Wright (1838–1921), a geologist and ordained Congregational minister, accorded with the fundamental principle of science, which states that

> we are to press known secondary causes as far as they will go in explanation of facts. We are not to resort to an unknown (i.e., supernatural) cause for explanation of phenomena till the power of known causes has been exhausted. If we cease to observe this rule there is an end to all science and all sound sense.[35]

All of the above statements welcoming Darwinism as a legitimate extension of natural law into the biological world came from Christian scientists of impeccable religious standing: Rice, a Methodist; Gray, a Presbyterian; Wright, a Congregationalist. Naturalism appealed to them, and to a host of other Christians, in part because it served as a reliable means of discovering God's laws. As the duke of Argyll, George Douglas Campbell (1823–1910), so passionately argued in his widely read book *The Reign of Law* (1867), the natural laws of science represented nothing less than manifestations of God's will. Christians could thus celebrate the rule of natural law as "the delight, the reward, the goal of Science." Even the evangelical theologian Benjamin B. Warfield (1851–1921), a leading defender of biblical inerrancy in turn-of-the-century America, argued that teleology (that is, belief in a divinely designed world) "is in no way inconsistent with . . . a complete system of natural causation."[36]

The adoption of naturalistic methods did not drive most nineteenth-century scientists into the arms of agnosticism or atheism. Long after God talk had disappeared from the heartland of science, the vast majority of scientists, at least in the United States, remained Christians or theists. Their acceptance of naturalistic science sometimes prompted them, as Jon H. Roberts has pointed out, "to reassess the relationship between nature and the supernatural." For example, when the American naturalist Joseph Le Conte (1821–1901), John Le Conte's brother, moved from seeing species as being "introduced by the miraculous interference of a personal intelligence" to viewing them as the products of divinely ordained natural laws, he rejected all "anthropomorphic notions of Deity" for a God "ever-present, all-pervading, ever-acting." Nevertheless, Le Conte remained an active churchgoing Christian.[37]

The relatively smooth passage of naturalism turned nasty during the last third of the nineteenth century, when a noisy group of British scientists and philosophers, led by Huxley and the Irish physicist John Tyndall (1820–93), began insisting that empirical, naturalistic science provided the *only* reliable knowledge of nature, humans, and society. Their anticlerical project, aimed at undermining the authority of the established Anglican church and dubbed "scientific naturalism" by Huxley, had little to do with naturalizing the practice of science but a lot to do with creating positions and influence for men such as themselves. They sought, as the historian Frank M. Turner has phrased it, "to expand the influence of scientific ideas for the purpose of secularizing society rather than for the goal of advancing science internally. Secularization was their goal; science, their weapon."[38]

For centuries men of science had typically gone out of their way to assure the religious of their peaceful intentions. In 1874, however, during his presidential address to the British Association for the Advancement of Science, Tyndall declared war on theology in the name of science. Men of science, he threatened, would

> wrest from theology, the entire domain of cosmological theory. All schemes and systems which thus infringe upon the domain of science must, in so far as they do this submit to its control, and relinquish all thought of controlling it. Acting otherwise proved always disastrous in the past, and it is simply fatuous today.

In contrast to most earlier naturalists, who had aspired simply to eliminate the supernatural from science while leaving religion alone, Tyndall and his crowd sought to root out supernaturalism from all phases of life and to replace traditional religion with a rational "religion of science." As described by one devotee, this secular substitute rested on "an implicit faith that by the methods of physical science, and by these methods alone, could be solved all the problems arising out of the relation of man to man and of man towards the universe." Despite the protests of Christians that the scientific naturalists were illegitimately trying to "associate naturalism and science in a kind of joint supremacy over the thoughts and consciences of mankind," the linkage of science and secularization colored the popular image of science for decades to come.[39]

The rise of the social sciences in the late nineteenth century in many ways reflected these imperialistic aims of the scientific naturalists. As moral philosophy fragmented into such new disciplines as psychology and sociology, many social scientists, insecure about their scientific standing, loudly pledged their allegiance not only to the naturalistic methods of science but to the philosophy of scientific naturalism as well. Most damaging of all, they turned religion itself into object of scientific scrutiny. Having "conquered one field after another," noted an American psychologist at the time, science now entered "the most complex, the most inaccessible, and, of all, the most sacred domain—that of religion." Under the naturalistic gaze of social scientists the soul dissolved into nothingness, God faded into an illusion, and spirituality became, in the words of a British psychologist, "'epiphenomenal,' a merely incidental phosphorescence, so to say, that regularly accompanies physical processes of a certain type and complexity." Here, at last, Christians felt compelled to draw the line.[40]

Reclaiming Science in the Name of God

By the closing years of the twentieth century naturalistic methods reigned supreme within the scientific community, and even devout Christian scientists scarcely dreamed of appealing to the supernatural when actually doing science. "Naturalism rules the secular academic world absolutely, which is bad enough," lamented one concerned layman. "What is far worse is that it rules much of the Christian world as well." Even the founders of scientific creationism, who brazenly rejected so much of the content of modern science, commonly acknowledged naturalism as the legitimate method of science. Because they narrowed the scope of science to exclude questions of origins, they typically limited it to the study of "present and reproducible phenomena" and left God and miracles to religion. Given the consensus on naturalism, it came as something of a surprise in the late 1980s and 1990s when a small group of so-called theistic scientists and camp followers unveiled plans "to reclaim science in the name of God." They launched their offensive by attacking methodological naturalism as atheistic—

or, as one partisan put it, "absolute rubbish"—and by asserting the presence of "intelligent design" (ID) in the universe.[41]

The roots of the intelligent design argument run deep in the soil of natural theology, but its recent flowering dates from the mid-1980s. The guru of ID, Berkeley law professor Phillip E. Johnson (b. 1940), initially sought to discredit evolution by demonstrating that it rested on the unwarranted assumption that naturalism was the only legitimate way of doing science. This bias, argued the Presbyterian layman, unfairly limited the range of possible explanations and ruled out any consideration of theistic factors. Johnson's writings inspired a Catholic biochemist at Lehigh University, Michael J. Behe (b. 1952), to speak out on the inadequacy of naturalistic evolution for explaining molecular life. In *Darwin's Black Box* (1996), Behe maintained that biochemistry had "pushed Darwin's theory to the limit . . . by opening the ultimate black box, the cell, thereby making possible our understanding of how life works." The "astonishing complexity of subcellular organic structure" led him to conclude—on the basis of scientific data, he asserted, "not from sacred books or sectarian beliefs"—that intelligent design had been at work. "The result is so unambiguous and so significant that it must be ranked as one of the greatest achievements in the history of science," he gushed. "The discovery [of intelligent design] rivals those of Newton and Einstein, Lavoisier and Schroedinger, Pasteur and Darwin"—and by implication elevated *its* discoverer to the pantheon of modern science (fig. 12.3).[42]

The partisans of ID hoped to spark "an intellectual revolution" that would rewrite the ground rules of science to allow the inclusion of supernatural explanations of phenomena. If Carl Sagan (1934–96) and other reputable researchers could undertake a Search for Extra-Terrestrial Intelligence (SETI) in the name of science, they reasoned, why should they be dismissed as unscientific for searching for evidence of intelligence in the biomolecular world? Should logical analogies fail to impress, ID advocates hoped that concerns for cultural diversity might win them a hearing. "In so pluralistic a society as ours," pleaded one spokesman, "why don't alternative views about life's origin and development have a legitimate place in academic discourse?"[43]

This quixotic attempt to foment a methodological revolution in science created little stir within the mainstream scientific community. Most scientists either ignored it or dismissed it as "the same old creationist bullshit dressed up in new clothes." The British evolutionary biologist Richard Dawkins (b. 1941) wrote it off as "a pathetic cop-out of [one's] responsibilities as a scientist." Significantly, the most spirited debate over intelligent design and scientific naturalism took place among conservative Christian scholars. Having long since come to terms with doing science naturalistically, reported the editor of the evangelical journal

Figure 12.3. Personalities in the ongoing debates over evolution, creation, and intelligent design, as depicted in a painting by Jody Nilsen that first appeared in the 1997 book issue of *Christianity Today* (28 April 1997). Seated: Michael Behe and Charles Darwin. Standing: evolutionist Stephen Jay Gould, Richard Dawkins, the artist's father, the artist, and Phillip Johnson. Courtesy of Jody Nilsen.

Perspectives on Science and Christian Faith, "most evangelical observers—especially working scientists—[remained] deeply skeptical." Though supportive of a theistic worldview, they balked at being "asked to add 'divine agency' to their list of scientific working tools."[44]

As the editor's response illustrates, scientific naturalism of the methodological kind could—and did—coexist with orthodox Christianity. Despite the occasional efforts of unbelievers to use scientific naturalism to construct a world without God, it has retained strong Christian support down to the present. And well it might, for, as we have seen, scientific naturalism was largely made in Christendom by pious Christians. Although it possessed the potential to corrode religious beliefs—and sometimes did so—it flourished among Christian scientists who believed that God customarily achieved his ends through natural means.[45]

I began this chapter by asserting that nothing characterizes modern science more than its rejection of God in explaining the workings of nature. That statement is, I believe, true. It would be wrong, however, to conclude that the naturalization of science has secularized society generally. As late as the 1990s nearly

40 percent of American scientists continued to believe in a personal God, and, despite the immense cultural authority of naturalistic science, the overwhelming majority of American maintained an active belief in the supernatural. During the waning years of the twentieth century, 47 percent of Americans affirmed the statement that "God created man pretty much in his present form at one time within the last 10,000 years," and an additional 35 percent thought that God had guided the process of evolution. Only 11 percent subscribed to purely naturalistic evolution. A whopping 82 percent of Americans trusted "in the healing power of personal prayer," with 77 percent asserting that "God sometimes intervenes to cure people who have a serious illness." Science may have become godless, but the masses—and many scientists—privately clung tenaciously to the supernatural.[46]

Acknowledgments

I am indebted to Louise Robbins and Richard Davidson for their research assistance in the preparation of this chapter and to Edward B. Davis, Bernard Lightman, David C. Lindberg, David N. Livingstone, Robert Bruce Mullin, Margaret J. Osler, Jon H. Roberts, Michael H. Shank, and John Stenhouse for their criticisms and suggestions.

NOTES

INTRODUCTION

1. See the excellent essay "Draper, White, and the Military Metaphor" in James R. Moore, *The Post-Darwinian Controversies: A Study of the Protestant Struggle to Come to Terms with Darwin in Great Britain and America, 1870–1900* (Cambridge: Cambridge University Press, 1979), 19–49. See also David C. Lindberg and Ronald L. Numbers, "Beyond War and Peace: A Reappraisal of the Encounter between Christianity and Science," *Church History,* 55 (1986), 338–54; and Ronald L. Numbers, "Science and Religion," *Osiris,* 2d ser., vol. 1 (1985), 59–80.

2. Edward J. Larson and Larry Witham, "Scientists Are Still Keeping the Faith," *Nature,* 386 (1997), 435–36.

3. See, for example, Stanley L. Jaki, *The Road of Science and the Ways to God* (Chicago: University of Chicago Press, 1978).

4. John L. Heilbron, *The Sun in the Church: Cathedrals as Solar Observatories* (Cambridge: Harvard University Press, 1999), 3.

5. Introduction to David C. Lindberg and Ronald L. Numbers, eds., *God and Nature: Historical Essays on the Encounter between Christianity and Science* (Berkeley and Los Angeles: University of California Press, 1986), 10.

6. For example, David A. Hollinger identifies us with a "neo-harmonist" historiography in "Justification by Verification: The Scientific Challenge to the Moral Authority of Christianity in Modern America," in Michael J. Lacey, ed., *Religion and Twentieth-Century American Intellectual Life* (Cambridge: Cambridge University Press, 1989), 116.

7. John Hedley Brooke, *Science and Religion: Some Historical Perspectives* (Cambridge: Cambridge University Press, 1991), 33.

8. Moore, *Post-Darwinian Controversies;* David N. Livingstone, "Science, Region, and Religion: The Reception of Darwinism in Princeton, Belfast, and Edinburgh," in Ronald L. Numbers and John Stenhouse, eds., *Disseminating Darwinism: The Role of Place, Race, Religion, and Gender* (Cambridge: Cambridge University Press, 1999), 7–38, quotation on 13–14.

9. See James R. Moore, "Religion and Science," in Peter J. Bowler and John V. Pickstone, eds., *The Modern Biological and Earth Sciences* (Cambridge: Cambridge University Press, forthcoming).

10. For a rare look at the history of Pentecostalism and science, see Ronald L. Numbers, *Darwinism Comes to America* (Cambridge: Harvard University Press, 1998), 111–35.

CHAPTER 1

1. Scholars will recognize this as the famous "warfare thesis" of Andrew Dickson White (*A History of the Warfare of Science with Theology in Christendom,* 2 vols. [New York: D. Appleton, 1896]) and others. Though now generally discredited by scholars specializing in the historical study of science and religion, those of us who lecture on science and religion know firsthand how vigorously it thrives among the educated public.

2. For the broader picture, see David C. Lindberg, "Science and the Medieval Church," in Lindberg and Michael H. Shank, eds., *The Cambridge History of Science,* vol. 2 (Cambridge: Cambridge University Press, forthcoming); Lindberg, "Early Christian Attitudes toward Nature," in Gary B. Ferngren et al., eds., *The History of Science and Religion in the Western Tradition: An Encyclopedia* (New York: Garland, 2000), 243–47; and Lindberg, "Medieval Science and Religion," in ibid., 259–67.

3. For most modern scientists, of course, religion has no legitimate role at all; but exceptions are ample. And for the educated public, the consumers of modern science, religion has quite a bit to say about the content of scientific beliefs in areas such as biological evolution and cosmology.

4. On the classical tradition, see John Edwin Sandys, *A History of Classical Scholarship,* 2 vols. (Cambridge: Cambridge University Press, 1903–8); R. R. Bolgar, *The Classical Heritage and Its Beneficiaries from the Carolingian Age to the End of the Renaissance* (Cambridge: Cambridge University Press, 1954).

5. For a short account of these events, see David C. Lindberg, *The Beginnings of Western Science: The European Scientific Tradition in Philosophical, Religious, and Institutional Context, 600 B.C. to A.D. 1450* (Chicago: University of Chicago Press, 1992), 133–51. For a longer account (marred by its harsh, judgmental tone), see William H. Stahl, *Roman Science: Origins, Development, and Influence to the Later Middle Ages* (Madison: University of Wisconsin Press, 1962).

6. The Christian Europe dealt with in this essay is the Western portion, which employed Latin as its language of learning and liturgy. Eastern Europe (eventually the Byzantine Empire, with its capital in Constantinople) retained Greek as its scholarly and liturgical language. Western and Eastern Christendom continued to interact throughout the period covered by this essay, but our subject is Western Christendom alone.

7. See Lindberg, *Beginnings of Western Science,* 163–82, 203–6.

8. On science and the early church, see David C. Lindberg, "Science and the Early Church," in Lindberg and Ronald L. Numbers, eds., *God and Nature: Historical Essays on the Encounter between Christianity and Science* (Berkeley and Los Angeles: University of California Press, 1986), 19–48; Lindberg, "Early Christian Attitudes toward Nature," in Ferngren, *The History of Science and Religion in the Western Tradition,* 243–47.

9. Latin text in Tertullian, *De praescriptione haereticorum,* VII, *Tertulliani Opera,* ed. Nic. Rigaltius (Paris, 1664), 204–5.

10. On Tertullian's attitude toward the learning of the classical tradition, see especially Eric Osborn, *Tertullian, First Theologian of the West* (Cambridge: Cambridge University Press, 1997), chaps. 2–3; and Timothy David Barnes, *Tertullian: A Historical and Literary Study* (Oxford: Clarendon Press, 1971).

11. Athenagoras, *A Plea for the Christians,* VI, and Clement, *Stromata,* I.17, both trans. B. P.

Pratten, in Alexander Roberts and James Donaldson, eds., *The Ante-Nicene Fathers* (Grand Rapids: Eerdmans, 1986), 2: 131–32, 320.

12. The best biography of Augustine is Peter Brown, *Augustine of Hippo* (Berkeley and Los Angeles: University of California Press, 1967). For a list of Augustine's writings and commentary on their contents, see Eugène Portalié, *A Guide to the Thought of St. Augustine* (Chicago: Regnery, 1960). On Augustine's attitude toward the natural sciences, see Kenneth J. Howell, "Augustine of Hippo," in Ferngren, *The History of Science and Religion in the Western Tradition*, 134–36, and Peter Harrison, *The Bible, Protestantism, and the Rise of Natural Science* (Cambridge: Cambridge University Press, 1998), 25–33. Harrison's interpretation, though in many ways complementary to mine, portrays Augustine as less welcoming to the sciences.

13. *Confessions*, trans. J. G. Pilkington, in *The Basic Writings of Saint Augustine*, ed. Whitney J. Oates (New York: Random House, 1948), X.35, vol. 1, p. 174 (with one change of wording), and IV.16, p. 56.

14. *Epistolae*, 120, ed. A. Goldbacher (Corpus Scriptorum Ecclesiasticorum Latinorum, 34) (Vienna: Tempsky, 1895), p. 708. On the life of faith and the rational understanding to which Augustine thinks it ought to lead, see Norman Kretzmann, "Faith Seeks, Understanding Finds: Augustine's Charter for Christian Philosophy," in Thomas P. Flint, ed., *Christian Philosophy*, University of Notre Dame Studies in the Philosophy of Religion, 6 (Notre Dame: University of Notre Dame Press, 1990), 1–36.

15. *The Literal Meaning of Genesis*, trans. John Hammond Taylor (Ancient Christian Writers, 41–42) (New York: Newman Press, 1982), II.9, p. 59 [with one change of wording]. All page citations to *Literal Commentary* refer to volume 1 of Taylor's translation.

16. *Enchiridion*, X, trans. J. F. Shaw, in *Basic Writings*, 661–62.

17. *On Christian Doctrine*, trans. D. W. Robertson Jr. (Indianapolis: Bobbs-Merrill, 1958), II.39, p. 74 with minor changes). All page citations to *On Christian Doctrine* refer to Robertson's translation.

18. *Literal Commentary*, IV.23, p. 131; I.19, pp. 42–43; see also II.1, pp. 47–48.

19. *On Christian Doctrine*, II.40, p. 75; II.18, p. 54.

20. Ibid., I.4, p. 10.

21. *Literal Commentary*, II.9, vol. 1, pp. 58–60. A parallel passage in Augustine's *City of God*, XVI.9 (cited in this chapter of the *Literal Commentary on Genesis*) refers unambiguously, and with apparent approval, to the theory of a spherical Earth.

22. *Literal Commentary*, II.2, pp. 48–49.

23. Ibid., II.2, pp. 48–49; II.4–5, pp. 50–52.

24. Ibid., II.17, p. 71. See also Augustine's *City of God*, V.1–10.

25. For more, see Lindberg, *Beginnings of Western Science*, 152–59, 183–213.

26. For five prominent examples, one each from the sixth through tenth centuries, see Lindberg, "Medieval Science and Religion," 261–62.

27. The mendicant orders (Franciscan and Dominican) were dedicated to a life of poverty but, unlike monastic orders, were active in the world rather than withdrawn from it.

28. There is no good recent biography of Bacon. The best older study is Stewart C. Easton, *Roger Bacon and His Search for a Universal Science* (Oxford: Basil Blackwell, 1952). For a more recent biographical sketch, see David C. Lindberg, *Roger Bacon's Philosophy of Nature*

(Oxford: Clarendon Press, 1983), xv–xxvi; and Jeremiah Hackett, "Roger Bacon: His Life, Career, and Works," in Hackett, ed., *Roger Bacon and the Sciences: Commemorative Essays* (Leiden: Brill, 1997), 9–23. On Bacon's scientific achievements see A. C. Crombie and J. D. North, "Bacon, Roger," in *Dictionary of Scientific Biography;* and Hackett, *Bacon and the Sciences.*

29. See David C. Lindberg, "Science as Handmaiden: Roger Bacon and the Patristic Tradition," *Isis,* 78 (1987), 518–36.

30. For additional discussion, see Lindberg, *Beginnings of Western Science,* 215–23. It is important to understand that the rationalistic tendencies in question were not created, but rather were seriously exacerbated, by acquisition of the full Aristotelian corpus and the thick version of the classical tradition. Peter Abelard is a revealing representative of twelfth-century rationalism in theology.

31. Lindberg, *Beginnings of Western Science,* 236–41; Edward Grant, "The Condemnations of 1270 and 1277, God's Absolute Power, and Physical Thought in the Late Middle Ages," *Viator,* 10 (1979), 211–44.

32. Lindberg, "Science as Handmaiden," 527–28.

33. *The Opus Majus of Roger Bacon,* ed. John Henry Bridges, 3 vols. (London: Williams & Norgate, 1900), II.1, vol. 1, p. 34. The only full English translation of Bacon's *Opus maius,* by Robert B. Burke (Philadelphia: University of Pennsylvania Press, 1928), is useful as a rough guide to Bacon's thought, but it is not reliable in matters of detail. All volume and page citations to the *Opus maius* refer to the Bridges edition.

34. Ibid., II.1, vol. 1, pp. 33, 35–36; II.3, vol. 3, p. 39.

35. Ibid., III.14, vol. 3, p. 123.

36. Ibid., III.14, vol. 3, pp. 124–25.

37. Ibid., IV [*mathematicae in divinis utilitas*] and IV.1.1, vol. 1, pp. 176, 97.

38. Ibid., IV, vol. 1, pp. 180–210.

39. This association grows out of multiple meanings of the Greek word *mathesis,* which included both astrology and mathematical science more generally. Affinities, in turn, between astrology and magic led to an association of mathematics with magic.

40. *Opus maius,* IV, vol. 1, pp. 238–48 (p. 240 for the quotation).

41. Ibid., IV, vol. 1, pp. 254–69. On astrology, see also *Opus maius,* IV, vol. 1, pp. 376–404. For more on the universal radiation of force, see David C. Lindberg, "Roger Bacon on Light, Vision, and the Universal Emanation of Force," in Hackett, *Bacon and the Sciences,* 243–75.

42. For biblical references to the Antichrist (or so judged), see especially I John 2:18–23; Daniel 7:7–24, 8:8–25; and Revelation 13:1–18. On medieval concern over the possible coming of the Antichrist, see Marjorie Reeves, *Joachim of Fiore and the Prophetic Future* (New York: Harper & Row, 1977).

43. *Opus maius,* IV, vol. 1, p. 399.

44. See David Woodward and Herbert M. Howe, "Roger Bacon on Geography and Cartography," in Hackett, *Bacon and the Sciences,* 199–222; Lindberg, "Bacon on Light, Vision, and the Universal Emanation of Force," in ibid., 243–75; and Jeremiah Hackett, "Roger Bacon on *Scientia Experimentalis,*" in ibid., 277–315.

45. Bacon, *Opus maius,* II.3 and II.8, vol. 3, p. 41 and vol. 1, p. 43. Bacon, *Perspectiva,* III.3.1, in David C. Lindberg, *Roger Bacon and the Origins of Perspectiva in the Middle Ages* (Oxford: Clarendon Press, 1986), 323. Cf. Augustine, *On Christian Doctrine,* II.16, p. 50.

46. *Opus maius,* IV, vol. 1, p. 213.

47. Ibid., II.1, vol. 3, p. 36.

48. If my purpose in this article were to tell the whole story of medieval acceptance and domestication of the classical tradition, the contributions of Albert and Aquinas would merit significant attention. Their low profile in this article is explained by my decision to tell a lean story, focused narrowly on one theologian and one natural philosopher; reinforcing this decision is my belief that the existing scholarly literature tends to blow the significance of Albert and Aquinas in the faith/reason controversy out of proportion to their actual medieval influence and that it is high time to focus attention on other actors. The literature on Albert and Aquinas is enormous; the following two sources will provide an introductory account: Ralph McInerny, *A First Glance at St. Thomas Aquinas: A Handbook for Peeping Thomists* (Notre Dame: University of Notre Dame Press, 1990); and James A. Weisheipl, ed., *Albertus Magnus and the Sciences: Commemorative Essays 1980* (Toronto: Pontifical Institute of Mediaeval Studies, 1980).

49. On these examples, see Edward Grant, "Cosmology," in David C. Lindberg, ed., *Science in the Middle Ages* (Chicago: University of Chicago Press, 1978), 275–80; John E. Murdoch and Edith D. Sylla, "The Science of Motion," in ibid., 217–18; Stefano Caroti, "Nicole Oresme's Polemic against Astrology in His 'Quodlibeta,'" in Patrick Curry, ed., *Astrology, Science, and Society* (Bury Saint Edmunds: Boydell, 1987), 75–93.

CHAPTER 2

1. Andrew Dickson White, *A History of the Warfare of Science with Theology in Christendom,* 2 vols. (New York: Appleton, 1896), 1:130–70. White's account contains more than a little fiction.

2. Predecessors in the attempt to examine the local and human aspects of the Galileo affair include Jerome J. Langford, *Galileo, Science, and the Church,* rev. ed. (Ann Arbor: University of Michigan Press, 1966); Richard S. Westfall, *Essays on the Trial of Galileo* (Vatican City: Vatican Observatory, 1989); Annibale Fantoli, *Galileo: For Copernicanism and for the Church,* trans. George V. Coyne, 2d ed. (Vatican City: Vatican Observatory, 1996); and (though I disagree with parts of his analysis) Giorgio de Santillana, *The Crime of Galileo* (Chicago: University of Chicago Press, 1955).

3. Convenient sources on Galileo's early life and career are Stillman Drake, "Galilei, Galileo," *Dictionary of Scientific Biography;* and Drake, *Galileo at Work: His Scientific Biography* (Chicago: University of Chicago Press, 1978).

4. Contrary to legend, virtually all educated Europeans since Aristotle have accepted the sphericity of the earth and, at least since the third century B.C., have had a very good idea of its actual circumference. See Jeffrey Burton Russell, *Inventing the Flat Earth: Columbus and Modern Historians* (New York: Praeger, 1991).

5. For a good introduction to the heliocentric system of Copernicus, see Thomas S. Kuhn, *The Copernican Revolution* (Cambridge: Harvard University Press, 1957). For an advanced analysis, see Noel M. Swerdlow and Otto Neugebauer, *Mathematical Astronomy in Copernicus's De Revolutionibus,* 2 parts (New York: Springer-Verlag, 1984).

6. Erasmus Reinhold's *Prutenic Tables,* published in 1551, modified Copernicus's heliocentric models in such a way as to bring about closer agreement between theory and data. But this improvement in accuracy of heliocentric predictions did not speak to the question of cosmological reality.

7. Kuhn, *Copernican Revolution,* 164–80; Robert S. Westman, "The Melanchthon Circle, Rheticus, and the Wittenberg Interpretation of the Copernican Theory," *Isis,* 66 (1975), 164–93.

8. Copernicus attempted to evade this conclusion by proposing that the radius of the earth's orbit (which he believed to be roughly five million miles) was "imperceptible" by comparison with the radius of the sphere of fixed stars. Copernicus's figure for the diameter of the earth's orbit was approximately one-twentieth of the modern figure. On estimates of cosmic dimensions, see Albert Van Helden, *Measuring the Universe: Cosmic Dimensions from Aristarchus to Halley* (Chicago: University of Chicago Press, 1985).

9. Westman, "Melanchthon Circle."

10. Albert Van Helden, *The Invention of the Telescope* (Philadelphia: American Philosophical Society, 1977).

11. For a handy source on Galileo's telescopic observations, see Stillman Drake, *Discoveries and Opinions of Galileo* (Garden City: Doubleday Anchor, 1957), which contains partial translations of Galileo's *Starry Messenger* and *Letters on Sunspots.* For a full translation of the *Starry Messenger,* see Galileo Galilei, *Sidereus Nuncius or The Sidereal Messenger,* trans. Albert Van Helden (Chicago: University of Chicago Press, 1989).

12. *Letters on Sunspots,* in Drake, *Discoveries and Opinions,* 101.

13. Galileo, *Sidereus Nuncius,* 47.

14. Kuhn, *Copernican Revolution,* 222–24.

15. Galileo was not aware of the phases of Venus and their implications for heliocentrism until after publication of the *Starry Messenger.* On the means by which the phases of Venus were brought to his attention, see the fascinating article by Richard S. Westfall, "Science and Patronage: Galileo and the Telescope," *Isis,* 76 (1985), 11–30.

16. On this model, about which Galileo chose not to inform his readers, preferring to portray the cosmological debate as a simple contest between Ptolemy and Copernicus, see Kuhn, *Copernican Revolution,* 200–206.

17. *Letters on Sunspots,* 94.

18. Jean Dietz Moss, *Novelties in the Heavens: Rhetoric and Science in the Copernican Controversy* (Chicago: University of Chicago Press, 1993), 76.

19. Also to secure suitable patronage. On the latter aspect of Galileo's career, see Westfall, "Science and Patronage." Also Mario Biagioli's provocative *Galileo, Courtier: The Practice of Science in the Culture of Absolutism* (Chicago: University of Chicago Press, 1993). However, for serious reservations regarding some of Biagioli's claims, see the essay review by Michael Shank in *Journal for the History of Astronomy,* 25 (1994), 236–42; Biagioli's reply, "Playing with the Evidence," *Early Science and Medicine,* 1 (1996), 70–105; and Shank's rejoinder, "How Shall We Practice History? The Case of Mario Biagioli's *Galileo Courtier,*" in ibid., 106–50.

20. Although the seeds of reform can be perceived as far back as the thirteenth and fourteenth centuries, the symbolic date for the beginning of the Protestant reformation is 1517, when Martin Luther nailed his ninety-five theses to the church door in Wittenberg.

21. Decree of the Council of Trent, 8 April 1546, trans. Richard J. Blackwell, *Galileo, Bellarmine, and the Bible* (Notre Dame: University of Notre Dame Press, 1991), 183. Blackwell's

book is an excellent source on the Council of Trent and the changes in the Catholic church that emanated from it.

22. Accounts of Galileo's relations with the church are innumerable. The best, most reliable general treatment is Fantoli, *Galileo: For Copernicanism and for the Church.* For an excellent, brief account, which concentrates on Galileo's arguments about the interpretation of Scripture, see Ernan McMullin, "Galileo on Science and Scripture," in Peter Machamer, ed., *The Cambridge Companion to Galileo* (Cambridge: Cambridge University Press, 1998), 271–347. Blackwell's *Galileo, Bellarmine, and the Bible* is fundamental for its treatment of issues of biblical interpretation; see also Blackwell's short, authoritative account, "Galileo Galilei," in Gary Ferngren et al., eds., *The History of Science and Religion in the Western Tradition: An Encyclopedia* (New York: Garland, 2000), 85–89. Langford's *Galileo, Science, and the Church,* though somewhat dated, is still useful. An interesting recent interpretation of the Galileo affair against the background of theological controversy within the Catholic church is Rivka Feldhay, *Galileo and the Church: Political Inquisition or Critical Dialogue?* (Cambridge: Cambridge University Press, 1995). An old standby, which adopts a harsh, judgmental stance toward the Catholic church, is de Santillana, *The Crime of Galileo.* Finally, Pietro Redondi's sensationalistic *Galileo Heretic,* trans. Raymond Rosenthal (Princeton: Princeton University Press, 1987), should not, in my judgment, be taken seriously.

23. By this time, the Jesuit order (or Society of Jesus), founded by Ignatius Loyola in 1540, had come to dominate higher education in Catholic Europe. The Collegio Romano was the flagship university of the Jesuit educational system.

24. Bruno, a radical utopian thinker, was burnt at the stake in Rome in 1600. Although heliocentrism was a prominent aspect of Bruno's worldview, it was not the reason for his execution. However, his cosmological ideas may have suffered from guilt by association with the heretical theological claims and dangerous political ambitions that doomed him. For a brief account of Bruno's trial and execution, see Frances A. Yates, *Giordano Bruno and the Hermetic Tradition* (Chicago: University of Chicago Press, 1964), 348–56.

25. Maurice A. Finocchiaro, *The Galileo Affair: A Documentary History* (Berkeley and Los Angeles: University of California Press, 1989), 51, 50. Finocchiaro provides the full text of this letter.

26. The *Letter to the Grand Duchess* is translated in ibid., 87–118. On these issues, see Blackwell, *Galileo, Bellarmine, and the Bible,* 53–85; McMullin, "Galileo on Science and Scripture."

27. De Santillana, *Crime of Galileo,* 42.

28. Finocchiaro, *Galileo Affair,* 134–35, for the text of the letter.

29. Trans. de Santillana, *Crime of Galileo,* 112–13, heavily edited.

30. Quoted by Langford, *Galileo, Science, and the Church,* 86–87, slightly edited.

31. Finocchiaro, *Galileo Affair,* 146, 149.

32. Ibid., 153.

33. Ibid., 147.

34. Fantoli, *Galileo,* 221–22. For debate on the subject, see de Santillana, *Crime of Galileo,* 261–74; Stillman Drake, "The Galileo-Bellarmine Meeting: A Historical Speculation," appendix A to Ludovico Geymonat, *Galileo Galilei: A Biography and Inquiry into His Philosophy*

of Science, trans. Stillman Drake, with a foreword by Giorgio de Santillana (New York: McGraw-Hill, 1965), 205–20; de Santillana's reply to Drake, appendix B to ibid., 221–25; and Langford, *Galileo, Science, and the Church,* 92–97.

35. Drake, *Discoveries and Opinions,* 227, judges this "the greatest polemic ever written in physical science." For a complete translation of *The Assayer,* see Stillman Drake and C. D. O'Malley, trans., *The Controversy on the Comets of 1618* (Philadelphia: University of Pennsylvania Press, 1960).

36. Fantoli, *Galileo,* 286.

37. de Santillana, *Crime of Galileo,* 166.

38. For examples, see Galileo's *Dialogue Concerning the Two Chief World Systems,* trans. Stillman Drake, 2d rev. ed. (Berkeley and Los Angeles: University of California Press, 1967), 53–54, 203. For more, including the judgment that "Galileo claims for the new science [which clearly includes heliocentric cosmology among its most notable achievements] an absolute monopoly on truth," see Brian Vickers, "Apodeictic Rhetoric in Galileo's *Dialogo,*" in Paolo Galluzzi, ed., *Novità celesti e crisi del sapere, Atti del Convegno internazionale di studi Galileiani* (supplement to the *Annali dell'Istituto e Museo di Storia della Scienza,* no. 8 [1983]), 86. In a letter to a friend Galileo promised that his forthcoming *Dialogue* would contain "a most ample confirmation of the Copernican system"; see Fantoli, *Galileo,* 333.

39. Maurice A. Finocchiaro, *Galileo on the World Systems: A New Abridged Translation and Guide* (Berkeley and Los Angeles: University of California Press, 1997), 305–6.

40. Ibid., 306–7, slightly edited.

41. Fantoli, *Galileo,* 335–43.

42. Ibid., 389–91.

43. Finocchiaro, *Galileo Affair,* 229 (slightly edited), 231, 232.

44. Ibid., 230, with revisions.

45. Fantoli, *Galileo,* 413–21.

46. On Galileo's trial, see Fantoli, *Galileo;* Westfall, *Essays on the Trial of Galileo;* Biagioli, *Galileo Courtier;* Langford, *Galileo, Science, and the Church;* Blackwell, *Galileo, Bellarmine, and the Bible;* and de Santillana, *Crime of Galileo.*

47. Blackwell, *Galileo, Bellarmine, and the Bible,* 131, 133–34. On this issue, see also Maurice A. Finocchiaro, "Science, Religion, and the Historiography of the Galileo Affair: On the Undesirability of Oversimplification," *Osiris,* 2d ser., vol. 16 (2001), 114–32 (at 125–28).

48. Langford, *Galileo, Science, and the Church,* 143–44, slightly edited.

49. Finocchiaro, *Galileo Affair,* 280.

50. Ibid., 286, slightly edited.

51. Ibid., 291, slightly edited.

52. Ibid., 292, slightly edited.

53. On Dominican-Jesuit politics, see especially Feldhay, *Galileo and the Church.*

CHAPTER 3

1. There is a voluminous literature on the origins and development of the mechanical philosophy; the best introduction and survey is still E. J. Dijksterhuis, *The Mechanization of the*

World Picture, trans. C. Dikshoorn (New York: Oxford University Press, 1961). A brief summary of seventeenth-century mechanical philosophy can be found in Richard S. Westfall, *The Construction of Modern Science: Mechanisms and Mechanics* (Cambridge: Cambridge University Press, 1977), 25–42.

2. Other studies of the interaction between Christianity and the mechanical philosophy include Gary B. Deason, "Reformation Theology and the Mechanistic Conception of Nature," in David C. Lindberg and Ronald L. Numbers, eds., *God and Nature: Historical Essays on the Encounter between Christianity and Science* (Berkeley and Los Angeles: University of California Press, 1986), 167–91; Richard S. Westfall, *Science and Religion in Seventeenth-Century England* (New Haven: Yale University Press, 1958), esp. 70–105; E. A. Burtt, *The Metaphysical Foundations of Modern Physical Science* (1924; rev. ed. 1932; reprinted Garden City: Doubleday, 1954).

3. The principal source on the skeptical crisis is Richard H. Popkin, *The History of Scepticism from Erasmus to Spinoza* (Berkeley and Los Angeles: University of California Press, 1979).

4. There are several good intellectual biographies of Descartes; one is Stephen Gaukroger, *Descartes: An Intellectual Biography* (Oxford: Clarendon Press, 1995); see also William R. Shea, *The Magic of Numbers and Motion: The Scientific Career of René Descartes* (Canton, Mass.: Science History Publications, 1991). A useful collection of essays on various aspects of Descartes's thought is John Cottingham, ed., *The Cambridge Companion to Descartes* (Cambridge: Cambridge University Press, 1992).

5. René Descartes, *Second Meditation;* translation based on *Philosophical Works of Descartes,* 2 vols., trans. Elizabeth S. Haldane and G. R. T. Ross (Cambridge: Cambridge University Press, 1911), 1:154; and Descartes, *Philosophical Writings,* trans. Elizabeth Anscombe and Peter T. Geach, rev. ed. (London: Thomas Nelson, 1970), 72.

6. The standard work on Descartes's cosmology is E. J. Aiton, *The Vortex Theory of Planetary Motions* (New York: American Elsevier, 1972); see esp. 43–64.

7. A good introduction to Aristotle's theory of substance is G. E. R. Lloyd, *Aristotle: The Growth and Structure of His Thought* (Cambridge: Cambridge University Press, 1968), 47–57. For a brief but quite clear discussion see David C. Lindberg, *The Beginnings of Western Science: The European Scientific Tradition in Philosophical, Religious, and Institutional Context, 600 B.C. to A.D. 1450* (Chicago: University of Chicago Press, 1992), 48–51. Norma A. Emerton, *The Scientific Reinterpretation of Form* (Ithaca: Cornell University Press, 1984), provides an entire history of forms in Western thought; although focused on mineralogy, it contains many useful insights.

8. On Aristotle's elements, see Lloyd, *Aristotle,* 164–75.

9. On natural magic in the Renaissance, see Brian P. Copenhaver, "Natural magic, Hermetism, and Occultism in Early Modern Science," in David C. Lindberg and Robert S. Westman, eds., *Reappraisals of the Scientific Revolution* (Cambridge: Cambridge University Press, 1990), 261–302; John Henry, "Magic and Science in the Sixteenth and Seventeenth Centuries," in R. C. Olby et al., eds., *Companion to the History of Modern Science* (London: Routledge, 1990), 583–96; Ingrid Merkel and Allen G. Debus, eds., *Hermeticism and the Renaissance: Intellectual History and the Occult in Early Modern Europe* (Washington: Folger Shakespeare Library, 1988); Brian Vickers, ed., *Occult and Scientific Mentalities in the Renaissance* (Cambridge: Cambridge University Press, 1984); D. P. Walker, *Spiritual and Demonic Magic from Ficino to Campanella* (Nedeln: Kraus, 1969).

10. An English translation, *Natural Magick* (1658), has been reprinted in facsimile (New York: Basic Books, 1957).

11. On the church and natural magic, see William B. Ashworth Jr., "Catholicism and Early Modern Science," in Lindberg and Numbers, *God and Nature,* 136–66 (esp. 148–49).

12. Most of Fludd's diagrams of correspondences are reproduced in Joscelyn Godwin, *Robert Fludd: Hermetic Philosopher and Surveyor of Two Worlds* (London: Thames and Hudson, 1979). See also Eberhard Knobloch, "Harmony and Cosmos: Mathematics Serving a Teleological Understanding of the World," *Physis,* 32 (1995), 55–89.

13. There is an enormous literature on Descartes, God, and religion; see Gary Hatfield, "Reason, Nature, and God in Descartes," *Science in Context,* 3 (1989), 175–201; Jean-Marie Beyssade, "The Idea of God and Proofs of His Existence," in Cottingham, *Cambridge Companion to Descartes,* 174–99; Margaret J. Osler, "Eternal Truths and the Laws of Nature: The Theological Foundations of Descartes' Philosophy of Nature," *Journal of the History of Ideas,* 46 (1985), 349–62; Richard S. Westfall, "The Rise of Science and the Decline of Orthodox Christianity: A Study of Kepler, Descartes, and Newton," in Lindberg and Numbers, *God and Nature,* 218–37.

14. On ancient Greek atomism, see Dijksterhuis, *Mechanization,* 8–13; Stephen Toulmin and June Goodfield, *The Architecture of Matter* (New York: Harper and Row, 1962), 63–72. For a complete history of atomism, see Andrew Pyle, *Atomism and Its Critics: From Democritus to Newton* (Bristol: Thoemmes Press, 1997), which is thorough and valuable, though possibly difficult for the nonspecialist, owing to its use of the technical language of the historian of philosophy.

15. On Epicurus, see Lindberg, *Beginnings of Western Science,* 77–80.

16. Book-length studies of Gassendi include Barry Brundell, *Pierre Gassendi: From Aristotelianism to a New Natural Philosophy* (Dordrecht: Reidel, 1987); and Lynn Sumida Joy, *Gassendi the Atomist: Advocate of History in an Age of Science* (Cambridge: Cambridge University Press, 1987).

17. On Gassendi's atomism, see Dijksterhuis, *Mechanization,* 425–31; Daniel Garber, "Apples, Oranges, and the Role of Gassendi's Atomism in Seventeenth-Century Science," *Perspectives on Science,* 3 (1995), 425–28; and Margaret J. Osler, "Baptizing Epicurean Atomism: Pierre Gassendi on the Immortality of the Soul," in Margaret J. Osler and Paul L. Farber, eds., *Religion, Science, and Worldview: Essays in Honor of Richard S. Westfall* (Cambridge: Cambridge University Press, 1985), 163–83.

18. On the contrasting views of Descartes and Gassendi, see Margaret J. Osler, *Divine Will and the Mechanical Philosophy: Gassendi and Descartes on Contingency and Necessity in the Created World* (Cambridge: Cambridge University Press, 1994).

19. On Hobbes and the mechanical philosophy, see Robert H. Kargon, *Atomism in England from Hariot to Newton* (Oxford: Clarendon Press, 1966), 54–62; Steven Shapin and Simon Schaffer, *Leviathan and the Air-Pump: Hobbes, Boyle, and the Experimental Life* (Princeton: Princeton University Press, 1985); Simon Schaffer, "Wallifaction: Thomas Hobbes on School Divinity and Experimental Pneumatics," *Studies in History and Philosophy of Science,* 19 (1988), 275–98. On Hobbes and materialism, see Samuel I. Mintz, *The Hunting of Leviathan: Seventeenth-Century Reactions to the Materialism and Moral Philosophy of Thomas Hobbes* (Cambridge:

Cambridge University Press, 1962), esp. 63–79. On Hobbes and religion, see Patricia Springborg, "Hobbes on Religion," in Tom Sorell, ed., *The Cambridge Companion to Hobbes* (Cambridge: Cambridge University Press, 1996), 346–80.

20. On More and the mechanical philosophy, see A. Rupert Hall, "Henry More and the Scientific Revolution," in Sarah Hutton, ed., *Henry More (1614–1687): Tercentenary Studies* (Dordrecht: Kluwer, 1990), 37–54; Alan Gabbey, "Henry More and the Limits of Mechanism," in ibid., 19–35; John Henry, "Henry More versus Robert Boyle: The Spirit of Nature and the Nature of Providence," in ibid., 55–76; Burtt, *Metaphysical Foundations,* 135–50.

21. For more detail on the attempts of More and Cudworth to refute Hobbesian materialism, see Mintz, *The Hunting of Leviathan,* 80–109.

22. On Cudworth, see J. A. Passmore, *Ralph Cudworth: An Interpretation* (Cambridge: Cambridge University Press, 1951); Alexander Jacob, "The Neoplatonic Conception of Nature in More, Cudworth, and Berkeley," in Stephen Gaukroger, ed., *The Uses of Antiquity: The Scientific Revolution and the Classical Tradition* (Dordrecht: Kluwer, 1991), 101–21; Alan Gabbey, "Cudworth, More, and the Mechanical Analogy," in Richard Kroll et al., eds., *Philosophy, Science, and Religion in England, 1640–1700* (Cambridge: Cambridge University Press, 1992), 109–27.

23. On Boyle's experimental science, see Shapin and Schaffer, *Leviathan and the Air-Pump,* esp. 22–79; Antonio Clericuzio, "The Mechanical Philosophy and the Spring of Air: New Light on Robert Boyle and Robert Hooke," *Nuncius,* 13 (1) (1998), 69–75.

24. Dijksterhuis, *Mechanization,* 433–44.

25. On Boyle and the mechanical philosophy, see Margaret J. Osler, "The Intellectual Sources of Robert Boyle's Philosophy of Nature: Gassendi's Voluntarism and Boyle's Physico-Theological Project," in Kroll, *Philosophy, Science, and Religion in England,* 178–98; Kargon, *Atomism in England,* 93–105; Peter Alexander, *Ideas, Qualities, and Corpuscles: Locke and Boyle on the External World* (Cambridge: Cambridge University Press, 1985); J. E. McGuire, "Boyle's Conception of Nature," *Journal of the History of Ideas,* 33 (1972), 523–42.

26. Boyle, "Notion of Nature," quoted in McGuire, "Boyle's Conception of Nature," 533.

27. Boyle, "Free Inquiry," quoted in Timothy Shanahan, "God and Nature in the Thought of Robert Boyle, *Journal of the History of Philosophy,* 26 (1988), 547–69 (on 558).

28. For more on Boyle's view of God and nature, see J. J. MacIntosh, "Robert Boyle's Epistemology: The Interaction between Scientific and Religious Knowledge," *International Studies in the Philosophy of Science,* 6 (1992), 91–121; Jan W. Wojcik, *Robert Boyle and the Limits of Reason* (Cambridge: Cambridge University Press, 1997); Shanahan, "God and Nature"; Osler, "Intellectual Sources."

29. On Boyle and natural theology, see Westfall, *Science and Religion in Seventeenth-Century England.*

30. On Boyle and miracles, see J. J. MacIntosh, "Locke and Boyle on Miracles and God's Existence," in Michael Hunter, ed., *Robert Boyle Reconsidered* (Cambridge: Cambridge University Press, 1994), 193–214.

31. The best introduction to all aspects of Newton's life and thought is the magisterial Richard S. Westfall, *Never at Rest: A Biography of Isaac Newton* (Cambridge: Cambridge University Press, 1980).

32. Walter Charleton was the English physician primarily responsible for introducing

Gassendi to England, through the publication of his *Physiologia Epicuro-Gassendi-Charltoniana* (1654), which, in spite of its formidable Latin title, was written in English. Charleton attempted to show that atomism does not necessarily lead to Hobbes's materialism, and Charleton was one of the first to argue that the mechanical philosophy can actually promote Christianity. For a good discussion of Charleton, see Kargon, *Atomism in England,* 84–89.

33. Westfall, *Never at Rest,* 83–104.

34. Christiaan Huygens, a brilliant Dutch mathematician, invented the first accurate pendulum clock, discovered the law of the pendulum, and worked out, before Newton, the law of centrifugal force. He took his mechanical philosophy from Descartes, and although he had many differences with Descartes, he believed that impact was the source of all motion. He worked out his own vortex theory of gravity, published in 1690. For detail on Huygens's opposition to Newtonian gravity, see H. A. M. Snelders, "Christiaan Huygens and Newton's Theory of Gravitation," *Notes and Records of the Royal Society of London,* 43 (1989), 202–22; Roberto de A. Martins, "Huygens's Reaction to Newton's Gravitational Theory," in J. V. Field and Frank A. J. L. James, eds., *Renaissance and Revolution: Humanists, Scholars, Craftsmen, and Natural Philosophers in Early Modern Europe* (Cambridge: Cambridge University Press, 1993), 203–13.

35. See John Hedley Brooke, "The God of Isaac Newton," in John Fauvel et al., eds., *Let Newton Be!* (Oxford: Oxford University Press, 1988), 169–83; Frank E. Manuel, *The Religion of Isaac Newton: The Fremantle Lectures, 1973* (Oxford: Clarendon Press, 1974); E. A. Burtt, *Metaphysical Foundations,* 283–302.

36. James E. Force, "Newton's God of Dominion: The Unity of Newton's Theological, Scientific, and Political Thought," in James E. Force and Richard H. Popkin, *Essays on the Context, Nature, and Influence of Isaac Newton's Theology* (Dordrecht: Kluwer, 1990), 75–102.

37. On Newton and active principles, see J. E. McGuire, "Force, Active Principles, and Newton's Invisible Realm," *Ambix,* 15 (1968), 154–208; Ernan McMullin, *Newton on Matter and Activity* (Notre Dame: University of Notre Dame Press, 1978); John Henry, "Newton, Matter, and Magic," in Fauvel et al., *Let Newton Be!,* 127–46; Alan Gabbey, "Newton and Natural Philosophy," in Olby et al., *Companion to the History of Modern Science,* 243–63 (esp. 258–62).

38. In 1715–16, Leibniz exchanged a series of letters with Samuel Clarke (who spoke for Newton), in which they disputed such matters as the nature of space, time, and God's relationship to his Creation. This correspondence, published in 1717, can be consulted in H. G. Alexander, ed., *The Leibniz-Clarke Correspondence, Together with Extracts from Newton's Principia and Opticks* (Manchester: Manchester University Press, 1956); see p. 11 for Leibniz's comment about Newton's God having to wind up his Creation.

CHAPTER 4

1. Peter Gay, *The Enlightenment: An Interpretation,* vol. 1, "The Rise of Modern Paganism" (New York: Norton, 1977), 339.

2. Gay, *The Enlightenment: An Interpretation,* vol. 2, "The Science of Freedom" (New York: Norton, 1977), 8.

3. Gay, *The Enlightenment,* 1:338.

4. Voltaire, *Letters on England,* trans. with an intro. by Leonard Tancock (London: Penguin, 1980), 70.

5. Edward Gibbon, *The History of the Decline and Fall of the Roman Empire,* with notes by the Rev. H. H. Milman,(Philadelphia: Porter & Coates, 1845), 1:550 n. 96. Beyond the attractions offered by Gibbon's sparkling prose, this particular edition also features a large number of apoplectic and outraged notes appended to the text by Milman.

6. Jean-Antoine-Nicolas de Condorcet, *Sketch for a Historical Picture of the Progress of the Human Mind,* trans. June Barraclough with an intro. by Stuart Hampshire (London: Weidenfeld and Nicolson, 1955), 72.

7. Immanuel Kant, *Critique of Pure Reason,* trans. Norman Kemp Smith (New York: St. Martin's, 1965), 9 n.

8. We would do well to remember that Christian belief was not alone in coming under critical scrutiny in the Enlightenment. Judaism too experienced a critical reform movement, known as the Haskalah, during the period. See David Sorkin, *Moses Mendelssohn and the Religious Enlightenment* (Berkeley and Los Angeles: University of California Press, 1996).

9. On the Leibniz-Clarke debate, see Steven Shapin, "Of Gods and Kings: Natural Philosophy and Politics in the Leibniz-Clarke Disputes," *Isis,* 72 (1981), 187–215; and Larry Stewart, *The Rise of Public Science: Rhetoric, Technology, and Natural Philosophy in Newtonian Britain, 1660–1750* (Cambridge: Cambridge University Press, 1992), 87ff.

10. Quoted in Gary B. Deason, "Reformation Theology and the Mechanistic Conception of Nature," in David C. Lindberg and Ronald L. Numbers, eds., *God and Nature: Historical Essays on the Encounter between Christianity and Science* (Berkeley and Los Angeles: University of California Press, 1986), 184. Deason's article is an excellent summary of the theological issues involved with the new natural philosophy of the seventeenth century.

11. John Locke, *An Essay Concerning Human Understanding,* bk. 2, chap. 21.

12. Jean d'Alembert, *Traité de dynamique,* quoted in Mary Terrall, "The Culture of Science in Frederick the Great's Berlin," *History of Science* (1990), 28:333–64, quoted on 355.

13. Terrall, "The Culture of Science in Frederick the Great's Berlin," esp. 353–58.

14. Quoted in Virginia Dawson, *Nature's Enigma: The Problem of the Polyp in the Letters of Bonnet, Trembley, and Réamur* (Philadelphia: American Philosophical Society, 1987), 96–97.

15. See Elizabeth Gasking, *Investigations into Generation, 1651–1828* (Baltimore: Johns Hopkins University Press, 1967), and John Farley, *The Spontaneous Generation Controversy from Descartes to Oparin* (Baltimore: Johns Hopkins University Press, 1977).

16. Friedrich Hoffmann, *Fundamenta medicinae,* trans. and intro. Lester S. King (London: Macdonald, 1971), 5, 7.

17. Hermann Boerhaave, *Institutiones medicae* (Leiden, 1730), sec. 40, pp. 12–13. The translation is taken from *Dr. Boerhaave's Academical Lectures on the Theory of Physick,* 6 vols. (London, 1742–46), 1:81.

18. Quoted in Shirley A. Roe, *Matter, Life, and Generation: Eighteenth-Century Embryology and the Haller-Wolff Debate* (Cambridge: Cambridge University Press, 1981), 1.

19. Charles Bonnet, "Expériences sur la régénération de la tête du limaçon terrestre," *Observations sur la physique, sur l'histoire naturelle et sur les arts* (1777), 11:165–79.

20. Pierre-Louis de Maupertuis, *The Earthly Venus,* trans. Simone Brangier Boas, with notes and an intro. by George Boas (New York: Johnson Reprint, 1966), 56.

21. Roe, *Matter, Life, and Generation,* 15–20. See also Jacques Roger, *Buffon,* trans. Sarah Lucille Bonnefoi, ed. L. Pearce Williams (Ithaca: Cornell University Press, 1997), 139–50.

22. For one example of ancient Greek ideas on this, see the treatise known in English as *Breaths,* ascribed to the ancient physician Hippocrates, in *Hippocrates,* vol. 2, trans. W. H. S. Jones (Cambridge: Loeb Classical Library, 1981), 219–53.

23. Richard Toellner, *Albrecht von Haller: Über die Einheit im Denken des letzten Universalgelehrten,* Sudhoffs Archiv Beihefte 10 (Wiesbaden: Franz Steiner, 1971), 49–51.

24. Roger, *Buffon,* 126–38.

25. Quoted in Roe, *Matter, Life, and Generation,* 29.

26. Albrecht von Haller, "A Dissertation on the Sensible and Irritable Parts of Animals," reprinted in Shirley A. Roe, ed., *The Natural Philosophy of Albrecht von Haller* (New York: Arno, 1981), 658–59. We note in passing the change from "corpori humani" in the original Latin to "animals" in the English translation.

27. On the controversies between Whytt and Haller, see Roger French, *Robert Whytt, the Soul, and Medicine* (London: Wellcome Institute of the History of Medicine, 1969), 63–76.

28. Julien Offray de La Mettrie, *Machine Man and Other Writings,* trans. and ed. Ann Thomson (Cambridge: Cambridge University Press, 1996), 26.

29. On La Mettrie's theories and the medical context of his work, see Kathleen Wellman, *La Mettrie: Medicine, Philosophy, and Enlightenment* (Durham: Duke University Press, 1992).

30. Joseph Priestley, *The History and Present State of Electricity* (1775; reprint, New York: Johnson Reprint, 1966), 2:134.

31. Quoted in Marcello Pera, *The Ambiguous Frog: The Galvani-Volta Controversy on Animal Electricity,* trans. Jonathan Mandelbaum (Princeton: Princeton University Press, 1992), 8.

32. Quoted in Philip C. Ritterbush, *Overtures to Biology: The Speculations of Eighteenth-Century Naturalists* (New Haven: Yale University Press, 1964), 20.

33. On the history of mesmerism, see Robert Darnton, *Mesmerism and the End of the Enlightenment in France* (Cambridge: Harvard University Press, 1968); and Alison Winter, *Mesmerized: Powers of Mind in Victorian Britain* (Chicago: University of Chicago Press, 1998). For a sample of Mesmer's own writings, see *Mesmerism: A Translation of the Original Medical and Scientific Writings of F. A. Mesmer, M.D.,* compiled and trans. by George J. Bloch, intro. by Ernest R. Hilgard (Los Altos, Calif.: William Kaufmann, 1980).

34. David Hume, *An Inquiry Concerning Human Understanding,* edit. with an intro. by Charles W. Hendel (Indianapolis: Bobbs-Merrill, 1979), 140–41.

CHAPTER 5

1. See especially Norman Cohn, *Noah's Flood: The Genesis Story in Western Thought* (New Haven: Yale University Press, 1996); and Don Cameron Allen, *The Legend of Noah* (Urbana: University of Illinois Press, 1949).

2. This is well documented in Cohn, *Noah's Flood,* 23–31.

3. Documented most conveniently in James Frazer, *Folklore in the Old Testament,* 3 vols. (London: Macmillan, 1918); D. B. Vitaliano, *Legends of the Earth: Their Geologic Origin* (Bloomington: Indiana University Press, 1973); and A. Dundes, ed., *The Flood Myth* (Berkeley and Los Angeles: University of California Press, 1988). The *Encyclopaedia Britannica,* 11th ed, 1910–12, has an informative article on the Deluge by T. K. Cheyne, a maverick biblical scholar. See also E. Zangger, *The Flood from Heaven: Deciphering the Atlantis Legend* (London: Sidgwick and Jackson, 1992).

4. See, for instance, L. R. Bailey, *Where Is Noah's Ark?* (Nashville: Abingdon, 1978); F. Navarra, *The Noah's Ark Expedition,* ed. D. Balsiger (London: Coverdale, 1974); R. L. Raikes, "The Physical Evidence for Noah's Flood," *Iraq,* 28 (1966): 52–63; and W. Ryan and W. C. Pittman, *Noah's Flood: The New Scientific Discoveries about the Event that Changed History* (New York: Simon and Schuster, 1999). There is much about the topic also in L. R. Bailey, *Noah: The Person and the Story in History and Tradition* (Columbia: University of South Carolina Press, 1989).

5. See especially Ronald Numbers, *The Creationists* (New York: Knopf, 1992). A Web site for the Christian Geology Ministry (www.kjvbible.org) lists many of the key issues of interpretation.

6. J. J. Scaliger, *De Emendatione Temporum* (Frankfurt, 1583). The date of the Deluge was often calculated: for example, the Masoretic text assigns it to the year 1656 after the Creation, the Samaritan to 1307, the Septuagint to 2242, Flavius Josephus to 2256. Another set of dates, based on modern interpretations, counts backward from Christ: the Masoretic, 2350 B.C.; the Samaritan, 2903 B.C.; and the Septuagint, 3134 B.C.

7. James Ussher, *Annals of the World Deduced from the Origin of Time* (London, 1658) (translation of *Annales Veteris Testamenti* [London, 1650]), 1. See Martin J. S. Rudwick, *The Meaning of Fossils: Episodes in the History of Palaeontology,* 2d ed. (New York: Science History, 1976; reprint, Chicago: University of Chicago Press, 1985), 70; and Stephen Jay Gould, "Fall in the House of Ussher," *Natural History,* November 1991, 12–21, reprinted in *Eight Little Piggies* (New York: Norton, 1993), 181–93.

8. See C. Reilly, *Athanasius Kircher: A Master of One Thousand Arts, 1610–1680* (Wiesbaden: Edizioni del Mondo, 1974). I have drawn on Allen's *Legend of Noah* for this section, and on Janet Browne, *The Secular Ark: Studies in the History of Biogeography* (New Haven: Yale University Press, 1983). See also Keith Thomas, *Man and the Natural World: Changing Attitudes in England, 1500–1800* (London: Allen Lane, 1983), and Cohn, *Noah's Flood.* For entry into the literature about European knowledge of exotic animals, see Wilma George, "Sources and Background to Discoveries of New Animals in the Sixteenth and Seventeenth Centuries," *History of Science,* 18 (1980), 79–104.

9. William B. Ashworth, "Emblematic Natural History of the Renaissance," in Nicholas Jardine, James A. Secord, and Emma Spary, eds., *The Cultures of Natural History* (Cambridge: Cambridge University Press, 1995), 17–37.

10. E. Tyson, *Orang-outang, sive Homo Sylvestris, or the Anatomy of a Pygmie compared with that of a Monkey, an Ape, and a Man* (London, 1699). See particularly Paula Findlen, *Possessing Nature: Museums, Collecting, and Scientific Culture in Early Modern Italy* (Berkeley and Los Angeles: University of California Press, 1994), and John Prest, *The Garden of Eden: The Botanic Garden and the Recreation of Paradise* (reprint, New Haven: Yale University Press, 1981).

11. Arthur MacGregor, ed., *Tradescant's Rarities: Essays on the Foundation of the Ashmolean Museum* (Oxford: Clarendon Press, 1983). An account of zoos and menageries is given in Doris Rybot, *It Began before Noah* (London: Michael Joseph, 1972).

12. David N. Livingstone, *The Preadamite Theory and the Marriage of Science and Religion* (Philadelphia: American Philosophical Society, 1992).

13. Rudwick, *The Meaning of Fossils,* 1–48. See also Michel Foucault, *The Order of Things: An Archaeology of the Human Sciences* (London: Tavistock, 1970).

14. Arthur O. Lovejoy, *The Great Chain of Being: A Study of the History of an Idea* (Cambridge:

Harvard University Press, 1936), and Charles Raven, *English Naturalists from Neckam to Ray: A Study of the Making of the Modern World* (Cambridge: Cambridge University Press, 1947).

15. See especially John Hedley Brooke, *Science and Religion: Some Historical Perspectives* (Cambridge: Cambridge University Press, 1991).

16. Out of a rich and varied analytical literature, see especially Richard S. Westfall, *Science and Religion in Seventeenth-Century England* (New Haven: Yale University Press, 1958); Margaret C. Jacob, *The Newtonians and the English Revolution, 1689–1720* (Hassocks, Sussex: Harvester Press, 1976); Steven Shapin, *The Scientific Revolution* (Chicago: University of Chicago Press, 1996), which includes a useful bibliographical essay; and Peter Harrison, *The Bible, Protestantism, and the Rise of Natural Science* (Cambridge: Cambridge University Press, 1998).

17. For the individuals discussed here and their views on the age and changing nature of the earth, see Gordon L. Davies, *The Earth in Decay: A History of British Geomorphology, 1578–1878* (London: Macdonald, 1968); Paolo Rossi, *The Dark Abyss of Time: The History of the Earth and the History of Nations from Hooke to Vico,* trans. Lydia G. Cochrane (Chicago: University of Chicago Press, 1984); Stephen Toulmin and June Goodfield, *The Discovery of Time* (New York: Harper and Row, 1965); R. Huggett, *Cataclysms and Earth History: The Development of Diluvialism* (Oxford: Clarendon Press, 1989); Francis Haber, *The Age of the World: Moses to Darwin* (Baltimore: Johns Hopkins University Press, 1959); and Rudwick, *The Meaning of Fossils.*

18. Frans A. Stafleu, *Linnaeus and the Linnaeans: The Spreading of Their Ideas in Systematic Botany, 1735–1789* (Utrecht: Oosthoek, 1971); and Browne, *The Secular Ark,* 27–31.

19. Findlen, *Possessing Nature;* Katie Whitaker, "The Culture of Curiosity," in Jardine, Secord, and Spary, *The Cultures of Natural History,* 75–90; David E. Allen, *The Naturalist in Britain: A Social History* (Harmondsworth: Penguin Books, 1978); and Janet Browne, "A Science of Empire: British Biogeography before Darwin," *Revue d'histoire des sciences,* 42 (1992), 453–75.

20. Rhoda Rappaport, "Geology and Orthodoxy: The Case of Noah's Flood in Eighteenth Century Thought," *British Journal for the History of Science,* 11 (1978), 1–18.

21. Jacques Roger, *The Life Sciences in Eighteenth-Century French Thought,* ed. Keith R. Benson, trans. Robert Ellrich (Stanford: Stanford University Press, 1997). See also the introduction to J. Lyon and P. Sloan, eds., *From Natural History to the History of Nature: Readings from Buffon and His Critics* (Notre Dame: University of Notre Dame Press, 1981).

22. James Larson, "Not without a Plan: Geography and Natural History in the Late Eighteenth Century," *Journal of the History of Biology,* 19 (1986), 447–88.

CHAPTER 6

1. William Buckland, *Vindiciae Geologicae; or, The Connexion of Geology with Religion Explained* (Oxford, 1820).

2. Charles Coulston Gillispie, *Genesis and Geology: A Study in the Relations of Scientific Thought, Natural Theology, and Social Opinion in Great Britain, 1790–1850* (Cambridge: Harvard University Press, 1951); Nicolaas A. Rupke, *The Great Chain of History: William Buckland and the English School of Geology (1814–1849)* (Oxford: Clarendon Press, 1983).

3. Rupke, *Great Chain of History,* 32–33; Gillispie, *Genesis and Geology,* 108. In Gillispie's version Buckland actually acquired the hyena.

4. Mott T. Greene, *Geology in the Nineteenth Century: Changing Views of a Changing World* (Ithaca: Cornell University Press, 1982), 85–87.

5. James Hutton, "On the Theory of the Earth; or, An Investigation of the Laws Observable in the Composition, Dissolution, and Restoration of the Globe," *Transactions of the Royal Society of Edinburgh*, 1 (1788), 209–304.

6. Georges Cuvier, *Recherches sur les ossemens fossiles des quadrupèdes: Discours préliminaire: Discours sur les révolutions de la globe* (Paris, 1812).

7. Rupke, *Great Chain of History*, 51. I agree with Rupke's contention that this was Buckland's real concern.

8. This is the principal argument of Gillispie, *Genesis and Geology*.

9. Ibid., 33.

10. Gordon L. Davies, *The Earth in Decay: A History of British Geomorphology, 1578–1878* (New York: American Elsevier, 1969); Hutton, "On the Theory of the Earth"; Greene, *Geology in the Nineteenth Century*, chap. 1; Erasmus Darwin, *Zoonomia; or, The Laws of Organic Life*, 2 vols. (London, 1794–96).

11. Rupke, *Great Chain of History;* William Buckland, "Account of an Assemblage of Fossil Teeth and Bones of Elephant, Rhinoceros, Hippopotamus, Bear, Tiger, and Hyaena, and Sixteen other Animals, Discovered in a Cave at Kirkdale, Yorkshire, in the Year 1821," *Philosophical Transactions of the Royal Society of London*, 112 (1822), 171–236.

12. Rupke, *Great Chain of History*, 36.

13. Gillispie, *Genesis and Geology*, 108.

14. Rupke, *Great Chain of History*, 51.

15. William Buckland, *Reliquae Diluvianae; or, Observations on the Organic Remains Contained in Caves, Fissures, and Diluvial Gravel, and on Other Geological Phenomena, Attesting to the Action of a Universal Deluge* (London, 1823).

16. Rupke, *Great Chain of History*, 39–41.

17. Hans W. Frei, *The Eclipse of Biblical Narrative* (New Haven: Yale University Press, 1974), chap. 2.

18. Gillispie, *Genesis and Geology*, 226.

19. Buckland, *Vindiciae Geologicae*, 19.

20. Quoted in Gillispie, *Genesis and Geology*, 200.

21. Ibid., 222.

22. Rupke, *Great Chain of History*, 49, 194.

23. Leonard G. Wilson, "Lyell, Charles," in *Dictionary of Scientific Biography*, 8:563–76, esp. 564.

24. Charles Lyell, *Principles of Geology*, 3 vols. (London: John Murray, 1830–33; reprinted in facsimile with an introduction by Martin Rudwick, Chicago: University of Chicago Press, 1990).

25. Ibid., 1:xvii.

26. Ibid., 3:270–72.

27. Ibid., 3:384–85.

28. Walter F. Cannon, "Buckland, William," in *Dictionary of Scientific Biography*, 2:566–71, esp. 567.

29. Greene, *Geology in the Nineteenth Century,* 69ff.

30. William Buckland, *Geology and Mineralogy Considered with Reference to Natural Theology,* 2 vols. (London: William Pickering, 1836).

31. Ibid., 1:x–xii.

32. Ibid., 1:11.

33. Ibid., 1:11–15.

34. Ibid., 1:16.

35. Ibid., 1:17.

36. Ibid., 1:18.

CHAPTER 7

1. L[ouis] A[gassiz], "The Diversity of Origin of the Human Races," *Christian Examiner,* 49 (1850), 110–45; Charles Lyell, *Geological Evidences of the Antiquity of Man* (London: Murray, 1863).

2. On the Protestant response to Darwinism and the timing of the debate in America, see Jon H. Roberts, *Darwinism and the Divine in America: Protestant Intellectuals and Organic Evolution, 1859–1900* (Madison: University of Wisconsin Press, 1988).

3. James Ussher, *Annals of the World Deduced from the Origin of Time* (London, 1658) (translation of *Annales Veteris Testamenti* [London, 1650]), 1. See William R. Brice, "Bishop Ussher, John Lightfoot, and the Age of Creation," *Journal of Geological Education,* 30 (1982), 18–24.

4. See Paolo Rossi, *The Dark Abyss of Time: The History of the Earth and the History of the Nations from Hooke to Vico,* trans. Lydia G. Cochrane (Chicago: University of Chicago Press, 1984), 107–8.

5. See Ronald L. Numbers, *Creation by Natural Law: Laplace's Nebular Hypothesis in American Thought* (Seattle: University of Washington Press, 1977), 88–104; and Numbers, *The Creationists* (New York: Knopf, 1992), xii–xiii, which contains a helpful chart comparing the gap and day-age theories.

6. The complete title of the English translation is *Men Before Adam; or, A Discourse upon the Twelfth, Thirteenth, and Fourteenth Verses of the Fifth Chapter of the Epistle of the Apostle Paul to the Romans, by Which Are Prov'd, That Men Were Created before Adam* (London, 1656). Richard Popkin has written the most on La Peyrère and preadamism. See his "Pre-Adamism in 19th Century American Thought: 'Speculative Biology' and Racism," *Philosophia,* 8 (1978), 205–39, and his biography of La Peyrère, *Isaac La Peyrère (1596–1676): His Life, Work, and Influence* (Leiden: Brill, 1987). See also David N. Livingstone, *The Preadamite Theory and the Marriage of Science and Religion* (Philadelphia: American Philosophical Society, 1992).

7. Thomas Jefferson, *Notes on the State of Virginia* (London, 1787; reprint, Chapel Hill: University of North Carolina Press, 1955), 138.

8. Henry Home, Lord Kames, *Sketches of the History of Man,* 2d ed., 4 vols. (Edinburgh, 1778; reprint, Hildesheim: Georg Olms Verlagsbuchhandlung, 1968), 1:3–84, esp. 76–79.

9. Samuel S. Smith, *An Essay on the Causes of the Variety of Complexion and Figure in the Human Species,* 2d ed. (New York, 1810; reprint, Cambridge: Harvard University Press, 1965), esp. 71–72, 89–90, 155, 184–85. I am grateful to David N. Livingstone for pointing out the political dimension of Smith's work. See also Winthrop D. Jordan's introduction to Smith's *Essay,* xv–xvi.

10. The most complete introduction to the American school of ethnology remains William Stanton, *The Leopard's Spots: Scientific Attitudes toward Race in America, 1815–59* (Chicago: University of Chicago Press, 1960). On geology and astronomy, respectively, see Rodney L. Stiling, "The Diminishing Deluge: Noah's Flood in Nineteenth-Century American Thought," Ph.D. diss., University of Wisconsin-Madison, 1991; and Numbers, *Creation by Natural Law.*

11. Samuel G. Morton, *Crania Americana; or, A Comparative View of the Skulls of Various Aboriginal Nations of North and South America, to Which Is Prefixed an Essay on the Varieties of the Human Species* (Philadelphia and London, 1839), 260. See Stephen Jay Gould, *The Mismeasure of Man* (New York: Norton, 1981), for an exposé of the unconscious racial bias in Morton's work.

12. Morton, *Crania Americana,* 260.

13. Samuel G. Morton, *Crania Ægyptiaca; or, Observations on Egyptian Ethnography, Derived from Anatomy, History and the Monuments* (Philadelphia, 1844), 66.

14. Stanton, *Leopard's Spots,* 146–47.

15. Reginald Horsman, *Josiah Nott of Mobile: Southerner, Physician, and Racial Theorist* (Baton Rouge: Louisiana State University Press, 1987), is the standard biography.

16. *American Journal of the Medical Sciences,* 6 (1843), 252–56. On the census, see Albert Deutsch, "The First U.S. Census of the Insane (1840) and Its Use as Pro-slavery Propaganda," *Bulletin of the History of Medicine,* 15 (1944), 469–82.

17. Stanton, *Leopard's Spots,* 66–72, 113–18.

18. Horsman, *Josiah Nott,* 5–26, 82.

19. Edward Lurie, *Louis Agassiz: A Life in Science* (Chicago: University of Chicago Press, 1960), is the standard biography.

20. Quoted ibid., 261.

21. Gould, *Mismeasure of Man,* 44–45. See also Lurie, *Louis Agassiz,* 256–57; Stanton, *Leopard's Spots,* 104–5.

22. Louis Agassiz, "Geographical Distribution of Animals," *Christian Examiner,* 48 (1850), 184–85.

23. A[gassiz], "Diversity of Origin of the Human Races"; Agassiz, "Contemplations of God in the Kosmos," *Christian Examiner,* 50 (1851), 1–17, quotation on p. 4.

24. Josiah Nott and George Gliddon, *Types of Mankind* (Philadelphia, 1854).

25. John Bachman, *The Doctrine of the Unity of the Human Race Examined on the Principles of Science* (Charleston, 1850). For Stanton's jaundiced view of Bachman, see *Leopard's Spots,* 123–36, 155. Lester D. Stephens, *Science, Race, and Religion in the American South: John Bachman and the Charleston Circle of Naturalists, 1815–1895* (Chapel Hill: University of North Carolina Press, 2000), provides an important corrective to Stanton.

26. Charles Hodge, "An Examination of Some Reasonings against the Unity of Mankind," *Biblical Repository and Princeton Review,* 34 (1862), 435–64; Hodge, "The Unity of Mankind," ibid., 31 (1859), 131; Gould, *Mismeasure of Man,* 54–69. On Hodge and science, see also Ronald L. Numbers, "Charles Hodge and the Beauties and Deformities of Science," in John W. Stewart and James H. Moorhead, eds., *Charles Hodge Revisited: A Critical Appraisal of His Life and Work* (Grand Rapids, Mich.: Eerdmans, 2002), 77–101.

27. Review of *Types of Mankind,* by Josiah Nott and George Gliddon, in *Presbyterian Magazine,* 4 (1854), 289.

28. S. R., review of *Types of Mankind,* by Josiah Nott and George Gliddon, in *Christian Examiner,* 57 (1854), 340–64; "Egyptian Antiquities," *[American] Church Review,* 3 (1850), 15.

29. Stanton, *Leopard's Spots,* 174.

30. Review of *Types of Mankind,* by Josiah Nott and George Gliddon, *Presbyterian Quarterly Review,* 3 (1854), 184.

31. Roberts, *Darwinism and the Divine,* 37.

32. "Chapters on Ethnology," *Christian Recorder,* 23 February to 15 June, 1861, esp. 2 March (doctrine) and 16 March (Gliddon); "The Origin of Species," ibid., 9 May 1863 (Darwinism). Eric D. Anderson has documented the relative indifference of African-American intellectuals to evolution and their continuing concern with polygenism into the twentieth century; see Anderson, "Black Responses to Darwinism, 1859–1915," in Ronald L. Numbers and John Stenhouse, eds., *Disseminating Darwinism: The Role of Place, Race, Religion, and Gender* (Cambridge: Cambridge University Press, 1999), 247–66.

33. Stanton, *Leopard's Spots,* 169–70.

34. Facsimiles of Buckner Payne's pamphlet and others from the "Ariel" controversy are found in John David Smith, ed., *Anti-Black Thought, 1863–1925,* 11 vols. (New York: Garland, 1993), vols. 5, 6. The latest contribution that I have located is Henry Parker Eastman, *The Negro, His Origin, History and Destiny, Containing a Reply to "The Negro a Beast"* (Boston: Eastern, 1905).

35. Buckner Payne [Ariel, pseud.], *The Negro: What Is His Ethnological Status?,* in Smith, *Anti-Black Thought,* 5: 30, 47–48.

36. Louis Agassiz to B. Payne, 29 August 1877, Agassiz Copybook, Houghton Library, Harvard University. The Christian press largely ignored Payne's pamphlets; but for an elegant demolition of his argument, see the review by B. W. Whilden in *Baptist Quarterly,* 2 (1868), 382–84.

37. Robert A. Young, *The Negro: A Reply to Ariel,* in Smith, *Anti-Black Thought,* 5:77.

38. See M.S., *The Adamic Race,* in Smith, *Anti-Black Thought,* 5:161, 166–67.

39. A. Bowdoin Van Riper, *Men among the Mammoths: Victorian Science and the Discovery of Human Prehistory* (Chicago: University of Chicago Press, 1993), describes the Brixham discovery and its effect on science and the public. See also Donald K. Grayson, *The Establishment of Human Antiquity* (New York: Academic Press, 1983).

40. See, e.g., "The Failures and Fallacies of the Pre-Historic Archaeology," *Southern Presbyterian Review,* 29 (1878), 672–98; "Science and Religion," *Nashville Christian Advocate,* 8 February 1873 (dating methods); review of *Studies in Science and Religion,* by G. Frederick Wright, *Methodist Quarterly Review,* 43 (1883), 168–71 (long biblical chronology); "The Antiquity of Man," ibid., 24 (1864), 41–57; review of *The Primeval Man,* by the duke of Argyll, ibid., 33 (1873), 169–71; review of *The Origin of the World,* by J. W. Dawson, ibid., 38 (1878), 367–72 (preadamism a possibility); and "Is Christianity to Be Modernized?" *Southwestern Presbyterian,* 3 July 1873. See also Ronald L. Numbers, "'The Most Important Biblical Discovery of Our Time': William Henry Green and the Demise of Ussher's Chronology," *Church History,* 69 (2000), 257–76.

41. Alexander Winchell, *Preadamites; or, A Demonstration of the Existence of Men before Adam* (Chicago, 1880); Livingstone, *Preadamite Theory,* 40–52.

42. Winchell, *Preadamites,* 295–96.

43. See Leonard Alberstadt, "Alexander Winchell's Preadamites: A Case for Dismissal from Vanderbilt University," *Earth Sciences History,* 13 (1994), 97–112, for an account of the affair.

44. Livingstone, *Preadamite Theory,* 52–53, 60–64. See also David N. Livingstone and Mark A. Noll, "B. B. Warfield (1851–1921): A Biblical Inerrantist as Evolutionist," *Isis,* 91 (2000), 283–304.

CHAPTER 8

1. Richard Dawkins, *The Blind Watchmaker* (London: Penguin, 1988), 6.

2. Michael Ruse, *Taking Darwin Seriously* (Oxford: Blackwell, 1984), 30.

3. *Hutchinson 20th Century Encyclopedia* (London: Hutchinson, 1999), s.v. "Darwin."

4. Adrian Desmond and James Moore, *Darwin* (London: Michael Joseph, 1991).

5. Frank Burch Brown, *The Evolution of Darwin's Religious Views* (Macon, Ga.: Mercer University Press, 1986), 3.

6. So, for instance, Maurice Mandelbaum, "Darwin's Religious Views," *Journal of the History of Ideas,* 19 (1958), 363–78. Similarly Jacques Roger, "Darwinism Today," in David Kohn, ed., *The Darwinian Heritage* (Princeton: Princeton University Press, 1985), 813–23.

7. James R. Moore, "Darwin of Down: The Evolutionist as Squarson Naturalist," in Kohn, *Darwinian Heritage,* 435–81.

8. Sylvan S. Schweber, "The Origin of the *Origin* Revisited," *Journal of the History of Biology,* 10 (1977), 229–316; Michael T. Ghiselin, "The Individual in the Darwinian Revolution," *New Literary History,* 3 (1972), 113–34.

9. Neal C. Gillespie, *Charles Darwin and the Problem of Creation* (Chicago: University of Chicago Press, 1979), 139. Similar is Dov Ospovat, *The Development of Darwin's Theory: Natural History, Natural Theology, and Natural Selection, 1838–1859* (Cambridge: Cambridge University Press, 1981).

10. James R. Moore, "Of Love and Death: Why Darwin 'Gave Up' Christianity," in Moore, ed., *History, Humanity, and Evolution* (Cambridge: Cambridge University Press, 1989), 195–229. Janet Browne concurs, observing that the death of Annie "was the formal beginning of Darwin's conscious dissociation from believing in the traditional figure of God. . . . The godless world of natural selection he was even then still creating came implacably face to face with the emptiness of bereavement." Janet Browne, *Charles Darwin: Voyaging* (New York: Knopf, 1995), 503.

11. Brown, *Evolution of Darwin's Religious Views,* 25, 28, 31.

12. Ibid., 27.

13. Thomas Söderqvist, "Existential Projects and Existential Choice in Science: Science Biography as an Edifying Genre," in Michael Shortland and Richard Yeo, eds., *Telling Lives in Science: Essays on Scientific Biography* (Cambridge: Cambridge University Press, 1996), 45–84.

14. Patrick Geddes, for example, spoke of the "substitution of Darwin for Paley." See Patrick Geddes, "Biology," in *Chambers's Encyclopaedia* (Edinburgh: W. & R. Chambers, 1925), 164.

15. Charles Darwin, *The Origin of Species by Charles Darwin: A Variorum Text,* ed. Morse Peckham (Philadelphia: University of Pennsylvania Press, 1959), 168–69.

16. Donald Fleming, "Charles Darwin, the Anaesthetic Man," *Victorian Studies,* 4 (1961), 219–36, on 231.

17. Some of these are explored in John Hedley Brooke, "Darwin's Science and His Religion," in John Durant, ed., *Darwinism and Divinity: Essays on Evolution and Religious Belief* (Oxford: Blackwell, 1985), 40–75.

18. William Irvine, *Apes, Angels, and Victorians: A Joint Biography of Darwin and Huxley* (London: Weidenfeld and Nicolson, 1956), 6.

19. These, and other advocates of similar persuasion, are discussed in J. Vernon Jensen, "Return to the Wilberforce-Huxley Debate," *British Journal for the History of Science,* 21 (1988), 161–79.

20. Various revisionist proposals have been advanced by J. R. Lucas, "Wilberforce and Huxley: A Legendary Encounter," *Historical Journal,* 22 (1979), 313–30; Sheridan Gilley, "The Huxley-Wilberforce Debate: A Reconstruction," in Keith Robbins, ed., *Religion and Humanism* (Oxford: Blackwell, 1981), 325–40; Jensen, "Return to the Wilberforce-Huxley Debate."

21. Quoted in Jensen, "Return to the Wilberforce-Huxley Debate," 166, 168, 170, 171.

22. Frank Miller Turner, "The Victorian Conflict between Science and Religion: A Professional Dimension," *Isis,* 69 (1978), 356–76.

23. Robert M. Young, *Darwin's Metaphor: Nature's Place in Victorian Culture* (Cambridge: Cambridge University Press, 1985).

24. Andrew D. White, "The Battle-Fields of Science," *New York Daily Tribune,* 18 December 1896, 4.

25. Quoted in Bruce Mazlish, preface to Andrew D. White, *A History of the Warfare of Science with Theology in Christendom* (abridged ed., New York, 1965), 13. For biographical details, see Glenn C. Altschuler, *Andrew D. White: Educator, Historian, Diplomat* (Ithaca: Cornell University Press, 1979).

26. John Hedley Brooke, *Science and Religion: Some Historical Perspectives* (Cambridge: Cambridge University Press, 1991), 35.

27. John William Draper, *History of the Conflict between Religion and Science* (New York: D. Appleton, 1874), 334.

28. Quoted in Donald Fleming, *John William Draper and the Religion of Science* (Philadelphia: University of Pennsylvania Press, 1950), 165. The following details rely on Fleming.

29. Draper, *History of the Conflict,* 323.

30. Fleming, *John William Draper,* 129, 133.

31. Colin A. Russell, "The Conflict Metaphor and Its Social Origins," *Science and Christian Belief,* 1 (1989), 3–26.

32. Desmond and Moore, *Darwin,* 526. On the X Club, see Ruth Barton, "'An Influential Set of Chaps': The X-Club and Royal Society Politics, 1864–85," *British Journal for the History of Science,* 23 (1990), 53–81.

33. Ruth Barton, "Evolution: The Whitworth Gun in Huxley's War for the Liberation of Science from Theology," in David Oldroyd and Ian Langham, eds., *The Wider Domain of Evolutionary Thought* (Dordrecht: Reidel, 1983), 262.

34. Cited in Brooke, *Science and Religion,* 308.

35. See Sophie Forgan, "The Architecture of Display: Museums, Universities, and Objects in Nineteenth-Century Britain," *History of Science,* 32 (1994), 139–62.

36. Benjamin B. Warfield, "Charles Darwin's Religious Life: A Sketch in Spiritual Biography," *Presbyterian Review,* 9 (1888), 569–601.

37. General Protestant responses are the subject of James R. Moore, *The Post-Darwinian Controversies: A Study of the Protestant Struggle to Come to Terms with Darwin in Great Britain and America, 1870–1900* (Cambridge: Cambridge University Press, 1979). For American Protestantism, the most exhaustive account is Jon H. Roberts, *Darwinism and the Divine in America: Protestant Intellectuals and Organic Evolution, 1859–1900* (Madison: University of Wisconsin Press, 1988). More specifically evangelical reactions are discussed in David N. Livingstone, *Darwin's Forgotten Defenders: The Encounter between Evangelical Theology and Evolutionary Thought* (Grand Rapids: Eerdmans; Edinburgh: Scottish Academic Press, 1987). Adventist and Pentecostal attitudes are reviewed in Ronald L. Numbers, *Darwinism Comes to America* (Cambridge: Harvard University Press, 1998), chaps. 5, 6. For Catholics, it has been suggested, polygenist anthropology was significantly more unnerving that either uniformitarian geology or Darwinian biology, because the unity of the human race had to be preserved at all theological costs. See William J. Astore, "Gentle Skeptics? American Catholic Encounters with Polygenism, Geology, and Evolutionary Theories from 1845 to 1875," *Catholic Historical Review,* 82 (1996), 40–76.

38. Charles Hodge, *What Is Darwinism?* (New York: Scribner's, 1874).

39. B. B. Warfield, "Personal Reflections of Princeton Undergraduate Life: IV—The Coming of Dr. McCosh," *Princeton Alumni Weekly,* 19 April 1916, 650–53.

40. B. B. Warfield, "Calvin's Doctrine of the Creation," *Princeton Theological Review,* 13 (1915), 190–255.

41. Quoted in Moore, *Post-Darwinian Controversies,* 196–97.

42. A. H. Strong, *Systematic Theology* (1907; reprint, London: Pickering and Inglis, 1956), 123.

43. See the discussion in James R. Moore, "Evangelicals and Evolution: Henry Drummond, Herbert Spencer, and the Naturalization of the Spiritual World," *Scottish Journal of Theology,* 38 (1985), 383–417.

44. The section that follows draws on my "Science, Region, and Religion: The Reception of Darwinism in Princeton, Belfast, and Edinburgh," in Ronald L. Numbers and John Stenhouse, eds., *Disseminating Darwinism: The Role of Place, Race, Religion, and Gender* (New York: Cambridge University Press, 1999), 7–38. Full referencing is provided there.

45. See George Elder Davie, "Scottish Philosophy and Robertson Smith," in *The Scottish Enlightenment and Other Essays* (Edinburgh: Polygon, 1991), 101–45.

Chapter 9

1. These are the Greek words used in the New Testament. In the Old Testament the Hebrew terms were *oth* (sign), and *mopheth* (wonder).

2. John Calvin, "Prefatory Address to King Francis," in *Institutes of the Christian Religion,* ed. John T. McNeill, trans. Ford Lewis Battles, 2 vols. (Philadelphia: Westminster Press, 1960), 1:116. Much of the following discussion is based upon my volume *Miracles and the Modern Religious Imagination* (New Haven: Yale University Press, 1996).

3. John Locke, "A Discourse on Miracles," in *The Reasonableness of Christianity,* ed. and abridged by I. T. Ramsey (London: A. & C. Black, 1958), 80.

4. Contrast Paley's *Natural Theology,* which elaborated his famous defense of the order

of nature, with his appeal to the authority of the Bible found in his *View of the Evidences of Christianity*.

5. Mozley, *Eight Lectures on Miracles Preached Before the University of Oxford* (London: Rivingtons, 1867), 107, 54ff; *Recent Inquiries in Theology . . . Being Essays and Reviews*, 2d ed. (Boston: Walker, Wise and Co., 1862), 122–23.

6. On the issue of science and class, see (among many) Frank M. Turner, "The Victorian Conflict between Science and Religion: A Professional Dimension," *Isis*, 69 (1978), 356–76; Martin Fichman, "Ideological Factors in the Dissemination of Darwinism in England: 1860–1900," in Everett Mendelsohn, ed., *Transformation and Tradition in the Sciences* (Cambridge: Harvard University Press, 1984), 471–85; and Adrian Desmond and James Moore, *Darwin* (New York: Norton, 1991).

7. Leonard Huxley, ed., *Life and Letters of Thomas Henry Huxley*, 2 vols. (New York: Appleton, 1900), 1:83–85.

8. John Tyndall, "Miracles and Special Providences," in *Fragments of Science*, 2 vols. (New York: P. F. Collier, 1897), 2: 31, 24, 30. This essay was originally published in the *Fortnightly Review* (1867).

9. Tyndall, *Fragments*, 2:19; T. H. Huxley, *Hume* (New York: Macmillan, [1874]), 130.

10. John Stuart Mill, *A System of Logic*, 8th ed. (New York: Harper and Bros., 1900), 440; Frederick Temple, *The Relation between Religion and Science* (New York: Macmillan, 1884), 31.

11. Tyndall, *Fragments*, 2: 12, 32.

12. Sidney Lee, *King Edward VII: A Biography* (London: Macmillan, 1925), 320–22; Philip Magnus, *King Edward the Seventh* (London: J. Murray, 1964), 113–14; John Tyndall, "The 'Prayer for the Sick': Hints towards a Serious Attempt to Estimate Its Value," *Contemporary Review*, 20 (1872), 205–6. For details of the controversy, see Frank M. Turner, "Rainfall, Plagues, and the Prince of Wales: A Chapter in the Conflict of Science and Religion," *Journal of British Studies*, 13 (1973), 46–65; and Stephen G. Brush, "The Prayer Test," *American Scientist*, 62 (1974), 561–63.

13. Tyndall, "Prayer," 208, 209, 210.

14. Tyndall, *Fragments*, 2:4. This is from an earlier essay "Reflections on Prayer and Natural Law" (1861). On the import of these presuppositions, see Frank Miller Turner, *Between Science and Religion: The Reaction to Scientific Naturalism in Late Victorian England* (New Haven: Yale University Press, 1974), 24–29.

15. Tyndall, *Fragments*, 2:15; Frank M. Turner, "John Tyndall and Victorian Scientific Naturalism," in W. H. Brock, N. D. McMillan, and R. C. Mollan, eds., *John Tyndall: Essays on a Natural Philosopher* (Dublin: Royal Dublin Society, 1981), 174ff.

16. J. W. Reynolds, *The Supernatural in Nature: A Verification of the Free Use of Science* (London: C. K. Paul, 1878), 353. The second quotation is from "Captain Galton Criticized," *Spectator*, 3 August 1872. This article and many of the other responses (pro and con) were collected and published in John O. Means, ed., *The Prayer-Gauge Debate: By Prof. Tyndall, Francis Galton, and Others, against Dr. Littledale, President McCosh, the Duke of Argyll, Canon Liddon, and "The Spectator"* (Boston: Congregational Publishing Society, 1876). On the prayer-gauge debate in America, see Rick Ostrander, *The Life of Prayer in a World of Science: Protestants, Prayer, and American Culture, 1870–1930* (New York: Oxford University Press, 2000).

17. James McCosh, untitled letter (originally published in *The Contemporary Review*, October 1872), in Means, *Prayer-Gauge Debate*, 136ff.

18. Untitled piece from *The Spectator*, in Means, *Prayer-Gauge Debate*, 26.

19. On Galton, see Francis Galton, *Memoirs of My Life* (London: Methuen, 1910); Karl Pearson, *The Life, Letters, and Labours of Francis Galton*, 3 vols. in 4 (Cambridge: Cambridge University Press, 1914–30); and D. W. Forrest, *Francis Galton: The Life and Work of a Victorian Genius* (New York: P. Elek, 1974).

20. Francis Galton, "Statistical Inqueries into the Efficacy of Prayer," in Means, *Prayer-Gauge Debate*, 87.

21. Ibid., 89, 99, 102.

22. Ibid., 105.

23. Means, *Prayer-Gauge Debate*, 158, 224, 225. The first of these contributors was an anonymous writer, "Protagorus." The second was the Scottish writer William Knight, who received ecclesiastical censure because of his essay. For a sophisticated critique of Knight, see George J. Romanes, *Christian Prayers and General Laws* (London: Macmillan, 1874), 203–68.

24. Article from *The Spectator*, in Means, *Prayer-Gauge Debate*, 174–75, 246. The former citation is from Charles W. Stubbs, who was later made Anglican bishop of Truro.

25. Richard Frederick Littledale, "The Rationale of Prayer," ibid., 49, 48. Buxton (1707–82) was a great mathematical prodigy, but was legendary because his mind was so occupied in mental calculations that he was unable to absorb a conventional education, and so remained a farm laborer all his life.

26. Ibid., 49.

27. Ibid., 75.

28. The duke of Argyll, "The Two Spheres: Are They Two?" in Means, *Prayer-Gauge Debate*, 267 (emphasis added). On Argyll, see Neal C. Gillespie, "The Duke of Argyll, Evolutionary Anthropology, and the Art of Scientific Controversy," *Isis*, 68 (1977), 40, 54.

29. Littledale, "The Rationale of Prayer," 76, 61.

30. Ibid., 61.

31. *The Ladies Repository*, n.s., 8 (1871), 313. The article referred to Tyndall's earlier essay.

32. *Proceedings at the Farewell Banquet to Professor Tyndall* (New York: D. Appleton, l873), 29. For an account of this banquet, see Katherine R. Sopka, "John Tyndall: International Populariser of Science," in Brock, McMillan, and Mollan, *Tyndall*, 193–203.

33. Charles Cullis, *Faith Cures* (Boston: Willard Trust, 1879), 5.

34. D. D. Smith, quoted in *Record of the International Conference on Divine Healing and True Holiness, Held at the Agricultural Hall, London, June 1 to 5, 1885* (London: J. Snow and Co., 1885), 92.

35. Daniel Hack Tuke, *Illustrations of the Influence of the Mind on the Body in Health and Disease Designed to Elucidate the Action of the Imagination* (London: J & A Churchill, 1872), ix.

36. William James, "What Psychical Research Has Accomplished," in *The Will to Believe and Other Essays in Popular Philosophy* (London: Longmans Green, 1910), 301. On the American movement, see R. Laurence Moore, *In Search of White Crows: Spiritualism, Parapsychology, and American Culture* (New York: Oxford University Press, 1977); on the British society, see

John T. Cerullo, *The Secularization of the Soul: Psychical Research in Modern Britain* (Philadelphia: Institute for the Study of Human Issues, 1982).

37. See "The Anglican Church and Faith Healing," *British Medical Journal,* 16 December 1911, 1609; *Spiritual Healing: Report of a Clerical and Medical Committee of Inquiry into Spiritual, Faith, and Mental Healing* (London: Macmillan, 1914); and W. F. Cobb, *Spiritual Healing* (London: G. Bell and Sons, 1914).

38. The current interest in scientific examinations of the efficacy of prayer was perhaps inaugurated by Randolph C. Byrd, "Positive Therapeutic Effects of Intercessory Prayer in a Coronary Care Unit Population," *Southern Medical Journal,* 81 (1988): 826–29. For a summary of the discussion, see Gary Thomas, "Doctors Who Pray: How the Medical Community Is Rediscovering the Healing Power of Prayer," *Christianity Today,* 6 January 1997, 20–30.

CHAPTER 10

1. Sigmund Freud to Oskar Pfister, 9 October 1918, in *Psychoanalysis and Faith: The Letters of Sigmund Freud and Oskar Pfister,* ed. Heinrich Meng and Ernst L. Freud, trans. Eric Mosbacher (London: Hogarth Press and the Institute of Psycho-Analysis, 1963), 63; Allison Stokes, *Ministry after Freud* (New York: Pilgrim Press, 1985), xi–xv, 4–7 (quotation on xv). See also David Shakow and David Rapaport, "The Influence of Freud on American Psychology," *Psychological Issues,* 13 (1964), 21; Sonya Michel, "American Conscience and the Unconscious: Psychoanalysis and the Rise of Personal Religion, 1906–1963," *Psychoanalysis and Contemporary Thought,* 7 (1984), 387–88.

2. My discussion of the prevailing view of mental processes in the United States and Europe is drawn from Nathan G. Hale Jr., *Freud and the Americans: The Beginnings of Psychoanalysis in the United States, 1876–1917* (New York: Oxford University Press, 1971); John Chynoweth Burnham, *Psychoanalysis and American Medicine, 1894–1918: Medicine, Science, and Culture,* Psychological Issues 5, monograph 20 (New York: International Universities Press, 1967); Eric Caplan, *Mind Games: American Culture and the Birth of Psychotherapy* (Berkeley and Los Angeles: University of California Press, 1998); and Barbara Sicherman, *The Quest for Mental Health in America, 1870–1917* (New York: Arno Press, 1980).

3. Among the useful secondary sources on the subject of American nervousness, see especially F. G. Gosling, *Before Freud: Neurasthenia and the American Medical Community, 1870–1910* (Urbana: University of Illinois Press, 1987); Tom Lutz, *American Nervousness, 1903: An Anecdotal History* (Ithaca: Cornell University Press, 1991); Anita Clair Fellman and Michael Fellman, *Making Sense of Self: Medical Advice Literature in Late Nineteenth-Century America* (Philadelphia: University of Pennsylvania Press, 1981); Hale, *Freud and the Americans;* and Sicherman, *Quest for Mental Health.*

4. Charles G. Hill, quoted in Hale, *Freud and the Americans,* 17; Hale, *Freud and the Americans,* 147; Sicherman, *Quest for Mental Health,* 196–98.

5. Hale, *Freud and the Americans,* 147–48.

6. Sigmund Freud, "The Question of Lay Analysis" (1926), in *The Standard Edition of the Complete Psychological Works of Sigmund Freud,* ed. James Strachey (London: Hogarth Press and the Institute of Psycho-Analysis, 1959), 20:253 (hereafter referred to as *Standard Edition*). My discussion of Freud's life in this and the following paragraphs draws on Peter Gay, *Freud: A Life for Our Time* (New York: Norton, 1988); Richard Wollheim, *Sigmund Freud* (New York:

Viking Press, 1971); and Roger Smith, *The Norton History of the Human Sciences* (New York: Norton, 1997).

7. Sigmund Freud, "Project for a Scientific Psychology" [1895], in *Standard Edition,* 1:295; Shakow and Rapaport, "Influence of Freud," 45 n. 21; Gay, *Freud,* 78.

8. The description of Freud's theory in the following paragraphs is a close paraphrase of my discussion in Jon H. Roberts, "The Human Mind and Personality," in Stanley I. Kutler, ed., *Encyclopedia of the United States in the Twentieth Century,* 4 vols. (New York: Scribner's, 1996), 2:882. I have also relied on the discussion in Gay, *Freud,* and Smith, *Norton History.*

9. Sigmund Freud, "The Psychotherapy of Hysteria," in Sigmund Freud and Josef Breuer, *Studies on Hysteria* [1895], in *Standard Edition,* 2:305.

10. Useful treatments of psychotherapy include Caplan, *Mind Games;* Burnham, *Psychoanalysis and American Medicine,* 73–81; Hale, *Freud and the Americans,* 147–51.

11. James J. Putnam, "Remarks on the Psychical Treatment of Neurasthenia," *Boston Medical and Surgical Journal,* 132 (1895), 505–6; James J. Putnam, quoted in Hale, *Freud and the Americans,* 133. See also Sicherman, *Quest for Mental Health,* 189–91.

12. Hale, *Freud and the Americans,* 138.

13. Caplan, *Mind Games,* 149; William Newton Clarke, *An Outline of Christian Theology* (1898; reprint, New York: Scribner's, 1922), 188.

14. William Rupp, "The Theory of Evolution and the Christian Faith," *Reformed Quarterly Review,* 35 (1888), 166. See also Clarke, *An Outline,* 183–84, 191. For a more extensive discussion of the importance of the mind in Christian theology, see Jon H. Roberts, *Darwinism and the Divine in America: Protestant Intellectuals and Organic Evolution, 1859–1900* (Madison: University of Wisconsin Press, 1988), 104–5, 176–77, 212.

15. George Harris, "The Function of the Christian Consciousness," *Andover Review,* 2 (1884), 351, 349; Newman Smyth, *The Religious Feeling: A Study for Faith* (New York: Scribner, Armstrong & Co., 1877), 15. See also Lewis French Stearns, *The Evidence of Christian Experience* (New York: Charles Scribner's Sons, 1890), 257; D. H. Meyer, "American Intellectuals and the Victorian Crisis of Faith," in Daniel Walker Howe, ed., *Victorian America* (Philadelphia: University of Pennsylvania Press, 1976), 69.

16. Useful discussions of Christian Science and New Thought include Donald B. Meyer, *The Positive Thinkers: Popular Religious Psychology from Mary Baker Eddy to Norman Vincent Peale and Ronald Reagan,* rev. ed. (Middletown, Conn.: Wesleyan University Press, 1988); Stephen Gottschalk, *The Emergence of Christian Science in American Religious Life* (Berkeley and Los Angeles: University of California Press, 1973); and Gail Thain Parker, *Mind Cure in New England: From the Civil War to World War I* (Hanover, N.H.: University Press of New England, 1973). For valuable discussions of mind cure within the context of psychotherapy, see Caplan, *Mind Games,* 61–88; and Hale, *Freud and the Americans,* 121–22.

16. William James, *The Varieties of Religious Experience: A Study in Human Nature* (1902; reprint, New York: Longmans, Green, and Co., 1914), 95, 13. See also Caplan, *Mind Games,* 62–63.

18. Elwood Worcester, Samuel McComb, and Isador H. Coriat, *Religion and Medicine: The Moral Control of Nervous Disorders* (New York: Moffat, Yard and Company, 1908), 59, 14–15, 52, 342, 354–56, 368, 376–81, 385–88; and Elwood Worcester and Samuel McComb, *The Christian Religion as a Healing Power: A Defense and Exposition of the Emmanuel Movement* (New

York: Moffat, Yard and Company, 1909), 7, 84–97. Among the numerous historical treatments of the Emmanuel movement, I have found the following most illuminating: Caplan, *Mind Games,* 117–48; Raymond J. Cunningham, "The Emmanuel Movement: A Variety of American Religious Experience," *American Quarterly,* 14 (1962), 48–63; John Gardner Greene, "The Emmanuel Movement 1906–1929," *New England Quarterly,* 7 (1934), 494–532; E. Brooks Holifield, *A History of Pastoral Care in America: From Salvation to Self-Realization* (Nashville: Abingdon Press, 1983), 201–9; Hale, *Freud and the Americans,* 228–49. The standard account of the movement is Elwood Worcester's autobiography, *Life's Adventure: The Story of a Varied Career* (New York: Charles Scribner's Sons, 1932), 218–73, passim.

19. For discussion of the opposition, see Caplan, *Mind Games,* 131–148; Hale, *Freud and the Americans,* 228. Freud denounced the Emmanuel movement as an exercise in mysticism. Hale, *Freud and the Americans,* 226. Predictably, Freud found the alliance between psychotherapy and the churches unsavory. Nathan G. Hale Jr., "Freud's Reich, the Psychiatric Establishment, and the Founding of the American Psychoanalytic Association: Professional Styles in Conflict," *Journal of the History of the Behavioral Sciences,* 15 (1979), 138.

20. In 1910, for example, H. Addington Bruce, a freelance journalist who frequently discussed topics relating to psychology in popular magazines, informed readers of *American Magazine* that "no other leading psychopathologist has accepted" Freud's "sweeping, audacious theory." H. Addington Bruce, "Masters of the Mind," *American Magazine,* 71 (1910–11), 81. During the next decade Freud's views remained enormously controversial. An insightful analysis of psychologists' reception of Freudianism is Gail A. Hornstein, "The Return of the Repressed: Psychology's Problematic Relations with Psychoanalysis, 1909–1960," *American Psychologist,* 47 (1992), 254–63. My discussion of this issue is greatly indebted to her work. See also F. H. Matthews, "The Americanization of Sigmund Freud: Adaptations of Psychoanalysis before 1917," *Journal of American Studies,* 1 (1967), 42–43; Hale, *Freud and the Americans,* 114.

21. Trigant Burrow, "Psychoanalysis and Society," *Journal of Abnormal Psychology,* 7 (1912), 342; Theodore Schroeder, "The Erotogenetic Interpretation of Religion: Its Opponents Review," *Journal of Religious Psychology,* 7 (1914–15), 23–44 (quotation on 43). For other examples of the use of psychoanalysis to assail religion, see Leonard T. Troland, "The Freudian Psychology and Psychical Research," *Journal of Abnormal Psychology,* 8 (1913–14), 405–28; Alfred Kuttner, "Revivalism," *New Republic,* 4 (1915), 15–16; and Preserved Smith, "Luther's Early Development in the Light of Psycho-analysis," *American Journal of Psychology,* 24 (1913), 360–77.

22. Arthur Bardwell Patten, *Can We Find God? The New Mysticism* (New York: George H. Doran, 1924), 52–53. See also Walter Marshall Horton, *A Psychological Approach to Theology* (New York: Harper and Brothers, 1931), 4; C. K. Mahoney, *The Religious Mind: A Psychological Study of Religious Experience* (New York: Macmillan, 1927), 26–27; Selden Peabody Delany, "Religion and the New Psychology," *American Church Monthly,* 14 (1923–24), 247; S. Parkes Cadman, *Imagination and Religion* (New York: Macmillan, 1926), 88–89; D. Maurice Allan, "Mind in the Remaking," *Union Seminary Review,* 38 (1926–27), 321; Edgar Sheffield Brightman, *A Philosophy of Religion* (New York: Prentice-Hall, 1940), 73.

23. [Leander S. Keyser], review of *Psychoanalysis Explained and Criticised,* by A. E. Baker, *Bible Champion,* 33 (1927), 304; Cecil V. Crabb, *Psychology's Challenge to Christianity* (Richmond,

Va.: Presbyterian Committee of Publication, 1923), 38–39, 111–12, 147–48, 145. See also Frazer Hood, "The New Psychology," *Union Seminary Review,* 38 (1927), 179; Mahoney, *Religious Mind,* 35–36; J. C. Faw, "The Newer Psychology," *Union Seminary Review,* 40 (1928–29), 322; [Frank M.] T[homas], "The Psychology of the Christian Life," *Methodist Review Quarterly,* 69 (1920), 184; Jared S. Moore, "Psychoanalysis: Its Values and Its Dangers," *American Church Monthly,* 11 (1922), 869; Pryor McN. Grant, "The Moral and Religious Life of the Individual in the Light of the New Psychology: Religious Values in Mental Hygiene," *Mental Hygiene,* 12 (1928), 456; [James M. Gillis], "Sigmund Freud," *Catholic World,* 118 (1923–24), 787; Allan, "Mind in the Remaking," 321; Brightman, *Philosophy of Religion,* 74; Cadman, *Imagination and Religion,* 89–90; Sv. Norborg, *Varieties of Christian Experience* (Minneapolis: Augsburg Publishing House, 1937), 64; Nels F. S. Ferre, "Christianity and the Freudian Fever," *Christian Century,* 56 (1939), 900.

24. Delany, "Religion and the New Psychology," 244; T[homas], "Psychology of the Christian Life," 184; and Crabb, *Psychology's Challenge,* 124–28 (quotations on 128 and 125), 112, 132–33. See also James Bissett Pratt, "The Subconscious and Religion," *Harvard Theological Review,* 6 (1913), 222; [Gillis], "Sigmund Freud," 794; Jules Bois, "A New Psychoanalysis: The Superconscious," *Catholic World,* 119 (1924), 582–83; Leander S. Keyser, *A Handbook of Christian Psychology* (1928; reprint, Burlington, Iowa: Lutheran Literary Board, 1945), 71–72, 130–31; Granville Mercer Williams, "Psychology and the Confessional," *American Church Monthly,* 22 (1927–28), 197–98; Charles T. Holman, *The Religion of a Healthy Mind* (New York: Round Table Press, 1939), 84; Patten, *Can We Find God?,* 42–44, 53; Edwin A. Burtt, *Types of Religious Philosophy* (New York: Harper and Brothers, 1939), 333–39.

25. Allan, "Mind in the Remaking," 321; Keyser, *Handbook of Christian Psychology,* 72; William Hallock Johnson, review of *The Freudian Wish and Its Place in Ethics,* by Edwin B. Holt, and *The Will in Ethics,* by Theophilus B. Stork, *Princeton Theological Review,* 15 (1917), 327; Holman, *Religion of a Healthy Mind,* 69, 72, 84–87 (quotations on 69). See also Norborg, *Varieties,* 52–53, 64–65; Faw, "Newer Psychology," 322; Ferre, "Christianity," 900; Patten, *Can We Find God?,* 54; Earl S. Rudisill, "Psychology and Religion," *Lutheran Church Quarterly,* 1 (1928), 20. Christians in this period illustrated the persistent tendency Hale identified among Americans to make the unconscious beneficent. Hale, *Freud and the Americans,* 408–9.

26. John Wright Buckham, "What Has Psychology Done to Us?" *Christian Century,* 57 (1940), 1171; Charles Bruehl, "Psychoanalysis," *Catholic World,* 116 (1922–23), 578, 586–88.

27. T[homas], "Psychology of the Christian Life," 184; Patten, *Can We Find God?,* 43–47, 55–56 (quotation on 44); Pratt, "Subconscious," 224; "How Far the Church Can Indorse Psychoanalysis," *Current Opinion,* 71 (1921), 623. See also Pratt, "Subconscious," 215, 221–22; Mahoney, *Religious Mind,* 34; Keyser, *Handbook of Christian Psychology,* 71. Most Christians joined the fundamentalist clergyman John Roach Straton in believing that "the conscious mind can send down, as it were, into the subconscious mind whatever suggestions it may desire, and the subconscious mind then will seize upon those suggestions and tend to work them out in reality." John Roach Straton, *Divine Healing in Scripture and Life* (New York: Christian Alliance Publishing, 1927), 18.

28. Scotford, "Psychoanalysis," 366–69 (quotation on 366). See also Grant, "Moral and Religious Life," 455–56; Moore, "Psychoanalysis," 870–71; Karl Ruf Stolz, "The Church and Psychoanalysis," *Methodist Review,* 5th ser., 44 (1928), 176, 181–82; John Moore,

"Psycho-analysis and the Ministry," *Homiletic Review,* 83 (1922), 15; Williams, "Psychology and the Confessional," 200.

29. Allan, "Mind in the Remaking," 321. See also "Editorial Comment," *Catholic World,* 125 (1927), 337; "How Far the Church," 623–24; Holman, *Religion of a Healthy Mind,* 84–86; Norborg, *Varieties,* 250.

30. Russell Henry Stafford, *Religion Meets the Modern Mind* (New York: Round Table Press, 1934), 126–29; Holman, *Religion of a Healthy Mind,* 74–77; Williams, "Psychology and the Confessional," 205.

31. Moore, "Psycho-analysis and the Ministry," 16; Crabb, *Psychology's Challenge,* 195, 191. See also "Editorial Comment," *Catholic World,* 120 (1924–25), 551–52; Norborg, *Varieties,* 138, 142; Cadman, *Imagination and Religion,* 89–90; "Editorial Comment," *Catholic World,* 132 (1930–31), 228.

32. Cadman, *Imagination and Religion,* 88–90; Norborg, *Varieties,* 250–51. See also Allan, "Mind in the Remaking," 321; Straton, *Divine Healing,* 16; [Gillis], "Sigmund Freud," 795; Shakow and Rapaport, "Influence of Freud," 25–26, 115; Sanford Gifford, "The American Reception of Psychoanalysis, 1908–1922," in Adele Heller and Lois Rudnick, eds., *1915: The Cultural Moment: The New Politics, the New Woman, the New Psychology, the New Art, and the New Theatre in America* (New Brunswick: Rutgers University Press, 1991), 137; Hale, *Freud and the Americans,* 298.

33. Norborg, *Varieties,* 269–70. See also Norborg, *Varieties,* 65, 139, 143; Holman, *Religion of a Healthy Mind,* 114–15, 120–21; Stolz, "Church and Psychoanalysis," 177–80; Crabb, *Psychology's Challenge,* 189; Williams, "Psychology and the Confessional," 198–200, 205; Delany, "Religion and the New Psychology," 246–47; Albert Edward Day, *Jesus and Human Personality* (New York: Abingdon Press, 1934), 69–70; Keyser, *Handbook of Christian Psychology,* 141–42; Moore, "Psychoanalysis," 872.

34. Norborg, *Varieties,* 254–56, 270–72 (quotation on 254); Richard Roberts, "Building Tomorrow's World: Beyond the Four Walls," *World Tomorrow,* 10 (1927), 396; Burtt, *Types of Religious Philosophy,* 333–34; Moore, "Psychoanalysis," 871; Harry Emerson Fosdick, quoted in Robert Moats Miller, *Harry Emerson Fosdick: Preacher, Pastor, Prophet* (New York: Oxford University Press, 1985), 261; Crabb, *Psychology's Challenge,* 138–39, 130–35, 118–21, 123, 152; Holman, *Religion of a Healthy Mind,* 135–41, 92–95 (quotations on 140, 93); Samuel Nowell Stevens, *Religion in Life Adjustments* (New York: Abingdon Press, 1930), 40–41, 124, 33–34, 87–106; W. A. Cameron, *The Clinic of a Cleric* (New York: Ray Long & Richard R. Smith, 1931), 23. See also Clarence H. Gaines, "Psychoanalysis and Faith," *North American Review,* 219 (1924), 265–66; Day, *Jesus and Human Personality,* 68; Horton, *Psychological Approach,* 2; Delany, "Religion and the New Psychology," 247; Manfred A. Carter, "What Has the New Psychology Done to Religion?," *Methodist Review,* 5th ser., 44 (1928), 891; Straton, *Divine Healing,* 16–19, 23; Moore, "Psycho-analysis and the Ministry," 16, 17; and Georgia Harkness, *Conflicts in Religious Thought* (New York: Henry Holt and Company, 1929), 16–17.

35. John Sutherland Bonnell, *Pastoral Psychiatry* (New York: Harper & Brothers, 1938), 201, 52. See also Roberts, "Building Tomorrow's World," 394–96; Norborg, *Varieties,* 270–71; Stolz, "Church and Psychoanalysis," 185; Harold Leonard Bowman, "Mental Hygiene in Relation to Religion," *Mental Hygiene,* 20 (1936), 183–84.

36. Holman, *Religion of a Healthy Mind,* 23–25, 27, 91–93; Hugh Hartshorne, "Religion and Psychology," *World Tomorrow,* 13 (1930), 311.

37. Williams, "Psychology and the Confessional," 209; Holman, *Religion of a Healthy Mind*, 136–37, 87, 93; L. L. Leh, *Christianity Reborn* (New York: Macmillan, 1928), 157; Bonnell, *Pastoral Psychiatry*, 73. See also Norborg, *Varieties*, 268; Bowman, "Mental Hygiene," 186–87.

38. Sigmund Freud, *The Future of an Illusion*, trans. James Strachey (1927; reprint, New York: Norton, 1961), esp. 22, 30, 39, 70–71. An illuminating discussion of *The Future of an Illusion* and Freud's religious views can be found in Gay, *Freud*, esp. 523–37. My discussion draws on Gay's work.

39. Gay, *Freud*, 530; J. E. Turner, "Freud and the Illusion of Religion," *Journal of Religion*, 11 (1931), 213; Norborg, *Varieties*, 24–25, 64–65; Holman, *Religion of a Healthy Mind*, 185, 209–10 (quotation on 185); Brightman, *Philosophy of Religion*, 73–74; Gaines, "Psychoanalysis and Faith," 266; Stevens, *Religion in Life Adjustments*, 34. See also Buckham, "What Has Psychology," 1171; Horton, *Psychological Approach*, 26; Crabb, *Psychology's Challenge*, 152–53; Bowman, "Mental Hygiene," 182. For a psychoanalytically oriented interpretation of atheism, see Paul E. Johnson, "Psychoanalyzing the Atheist," *Christian Century*, 53 (1936), 292–94. For his part, Freud always assumed that belief rather than unbelief was what needed to be explained. Gay, *Freud*, 526.

40. Stevens, *Religion in Life Adjustments*, 62–63. See also Grant, "Moral and Religious Life," 469–70; Holman, *Religion of a Healthy Mind*, 6–9; Goodwin Watson, "What Does Psychology Do to Religion?" *Christian Century*, 48 (1931), 605; Bowman, "Mental Hygiene," 180–81; Hartshorne, "Religion and Psychology," 311.

41. Norborg, *Varieties*, 274; H. Richard Niebuhr, "Theology and Psychology: A Sterile Union," *Christian Century*, 44 (1927), 47–48; Buckham, "What Has Psychology," 1171–72; Helmut Richard Niebuhr, "Religious Realism in the Twentieth Century," in D.C. Macintosh, ed., *Religious Realism* (New York: Macmillan, 1931), 413–28; E. G. Homrighausen, *Christianity in America: A Crisis* (New York: Abingdon Press, 1936), 52–53.

CHAPTER 11

1. Jerome Lawrence and Robert E. Lee, *Inherit the Wind* (New York: Bantam, 1960), 58–59. For recent accounts of the Scopes trial, see Edward J. Larson, *Summer for the Gods: The Scopes Trial and America's Continuing Debate over Science and Religion* (New York: Basic Books, 1997); and Ronald L. Numbers, *Darwinism Comes to America* (Cambridge: Harvard University Press, 1998), chap. 4, "The Scopes Trial: History and Legend."

2. Lawrence and Lee, *Inherit the Wind*, vii, 3, 7, 64. To avoid unnecessary confusion, the real names of persons are used in this article even when discussing actions of their characters in *Inherit the Wind*. Where those names appear in quotations from the play's script, the bracketed real name replaces the stage name.

3. Joseph Wood Krutch, "The Monkey Trial," *Commentary*, May 1967, 84.

4. 1925 Tenn. House Bill 185.

5. "Plan Assault on State Law on Evolution," *Chattanooga Times*, 4 May 1925, 5.

6. *Why Dayton, of All Places?* (Chattanooga, Tenn.: Andrews, 1925), 3.

7. On the changing population of Dayton, see U.S. Census Bureau, *1880 Census: Population* (Washington: Government Printing Office, 1883), I, 338; and U.S. Census Bureau, *1900 Census: Population* (Washington: Government Printing Office, 1901), I, pt. 1, 373.

8. "Was Converted through Science," *Chattanooga Times*, 21 May 1925, 2; and G. W. Rappleyea to editor, ibid., 19 May 1925, 5.

9. F. E. Robinson, quoted in Warren Allem, "Backgrounds of the Scopes Trial at Dayton, Tennessee," M.A. thesis, University of Tennessee, 1959, 58.

10. John T. Scopes, *The Center of the Storm: Memoirs of John T. Scopes* (New York: Holt, 1967), 58–59; Juanita Glenn, "Judge Still Recalls 'Monkey Trial'—50 Years Later," *Knoxville Journal*, 11 July 1975, 17; Allem, "Backgrounds," 60–61.

11. T. W. Callaway, "Father of Scopes Renounced Church," *Chattanooga Times*, 10 July 1925, 1.

12. Arthur Garfield Hays, *Let Freedom Ring* (New York: Boni and Liveright, 1928), 33.

13. Lawrence and Lee, *Inherit the Wind*, 9.

14. Ibid., 108.

15. William Jennings Bryan, "Prince of Peace," in *Speeches of William Jennings Bryan*, 2 vols. (New York: Funk and Wagnalls, 1909), 2:266–68.

16. Lawrence and Lee, *Inherit the Wind*, 114.

17. Lawrence W. Levine, *Defender of the Faith: William Jennings Bryan, the Last Decade, 1915–1925* (New York: Oxford University Press, 1965), vii.

18. Lawrence and Lee, *Inherit the Wind*, 23–24.

19. David E. Lilienthal, "Clarence Darrow," *Nation*, 20 April 1927, 417.

20. For Darrow's views on free will, see, e.g., Clarence Darrow and Will Durant, *Is Man a Machine?* (New York: League for Public Discussion, 1927), 51.

21. Will Herberg, *Protestant, Catholic, Jew* (Garden City, N.Y.: Doubleday, 1960), 259–60.

22. Kevin Tierney, *Darrow: A Biography* (New York: Croswell, 1979), 85.

23. Robert G. Ingersoll, "Reply to Dr. Lymann Abbott," in Robert G. Ingersoll, *The Works of Robert G. Ingersoll*, 12 vols. (New York: Dresden, 1903), 4:463.

24. "Darrow Asks W. J. Bryan to Answer These," *Chicago Daily Tribune*, 4 July 1923, 1.

25. See W. B. Norton, "Bryan's Ailment is Intolerance, Pastors Assert," *Chicago Daily Tribune*, 28 June 1923, 3.

26. Clarence Darrow, *The Story of My Life* (New York: Grosset, 1932), 249.

27. Ibid.

28. Jerome Lawrence and Robert E. Lee, "Inherit the Wind: The Genesis and Exodus of the Play," *Theatre Arts*, August 1957, 33; and Elizabeth J. Haybe, "A Comparison Study of *Inherit the Wind* and the Scopes 'Monkey Trial,'" M.A. thesis, University of Tennessee, 1964, 66.

29. "Trial July 10 Suits Bryan," *Nashville Banner*, 26 May 1925, 1.

30. Arthur Garfield Hays, "The Strategy of the Scopes Defense," *Nation*, 5 August 1925, 158.

31. *The World's Most Famous Court Case: Tennessee Evolution Case* (Dayton, Tenn.: Bryan College, 1990), 66–68. This reprinted edition of the trial transcript, originally published in 1925, is hereafter cited as "Transcript."

32. Lawrence and Lee, *Inherit the Wind*, 84.

33. H. L. Mencken, "The Monkey Trial: A Reporter's Account," in Jerry D. Tompkins,

ed., *D-Days at Dayton: Reflections on the Scopes Trial* (Baton Rouge: Louisiana State University Press, 1965), 39 (reprint of 11 July 1925 syndicated column for the *Baltimore Sun*).

34. Lawrence and Lee, *Inherit the Wind*, 65.

35. Transcript, 180.

36. Scopes, *Center of the Storm*, 270.

37. Lawrence and Lee, *Inherit the Wind*, 73–74.

38. "Defense Counsel Make Ready for Final Battle," *Nashville Tennessean*, 19 July 19 1925, 1.

39. Scopes, *Center of the Storm*, 166.

40. Transcript, 288.

41. Ralph Perry, "Added Thrill Given Dayton," *Nashville Banner*, 21 July 1925, 2.

42. Darrow, *My Life*, 267.

43. Transcript, 285.

44. Ibid., 302.

45. Hays, *Let Freedom Ring*, 77. On the limits of biblical inerrancy, see David N. Livingstone and Mark A. Noll, "B. B. Warfield (1851–1921): A Biblical Inerrantist as Evolutionist," *Isis*, 91 (2000), 283–304.

46. Scopes, *Center of the Storm*, 178. On the overwhelmingly positive response of fundamentalists to Bryan's performance in Dayton, see Numbers, *Darwinism Comes to America*, 81–84.

47. Transcript, 302.

48. "Bryan, Made Witness in Open Air Court, Shakes His Fist at Darrow Amid Cheers," *New York Times*, 21 July 1925, 1.

49. Transcript, 299.

50. Ibid., 304.

51. Lawrence and Lee, *Inherit the Wind*, 7, 30, 63.

52. Ibid., 85, 91, 103.

53. Ibid., 32, 42.

54. Ibid., 112–15.

55. Andrew Sarris, "Movie Guide," *Village Voice*, 10 November 1960, 11.

56. Lawrence and Lee, *Inherit the Wind*, 109.

57. H. L. Mencken, editorial, *American Mercury*, 6 (1925), 159.

58. Vincent Canby, "Of Monkeys, Reason, and the Creation," *New York Times*, 5 April 1996, C1.

CHAPTER 12

1. Opinion of Judge William R. Overton in *McLean v. Arkansas Board of Education*, in Marcel C. La Follette, ed., *Creationism, Science, and the Law: The Arkansas Case* (Cambridge: MIT Press, 1983), 60. See also Michael Ruse, "Creation-Science Is Not Science," ibid., 151–54; Stephen Jay Gould, "Impeaching a Self-Appointed Judge," *Scientific American*, July 1992, 118–20; *Science and Creationism: A View from the National Academy of Sciences* (Washington: National Academy Press, 1984), 8; *Teaching about Evolution and the Nature of Science* (Washington: National Academy Press, 1998), chap. 3.

2. Bernard Lightman, "'Fighting Even with Death': Balfour, Scientific Naturalism,

and Thomas Henry Huxley's Final Battle," in Alan P. Barr, ed., *Thomas Henry Huxley's Place in Science and Letters: Centenary Essays* (Athens: University of Georgia Press, 1997), 338. Historians have also used naturalism to identify a philosophical point of view, common centuries ago but rare today, that attributes intelligence to matter. The phrase "methodological naturalism" seems to have been coined by the philosopher Paul de Vries, then at Wheaton College, who introduced it at a conference in 1983 in a paper subsequently published as "Naturalism in the Natural Sciences," *Christian Scholar's Review,* 15 (1986), 388–96. De Vries distinguished between what he called "methodological naturalism," a disciplinary method that says nothing about God's existence, and "metaphysical naturalism," which "denies the existence of a transcendent God."

3. G. S. Kirk and J. E. Raven, *The Presocratic Philosophers* (Cambridge: Cambridge University Press, 1957), 73; G. E. R. Lloyd, *Magic, Reason, and Experience: Studies in the Origin and Development of Greek Science* (Cambridge: Cambridge University Press, 1979), 11, 15–29; John Clarke, *Physical Science in the Time of Nero* (London: Macmillan, 1910), 228; Edward B. Davis and Robin Collins, "Scientific Naturalism," in Gary B. Ferngren, ed., *The History of Science and Religion in the Western Tradition: An Encyclopedia* (New York: Garland, 2000), 201–7.

4. David C. Lindberg, *The Beginnings of Western Science: The European Scientific Tradition in Philosophical, Religious, and Institutional Context, 600 B.C. to A.D. 1450* (Chicago: University of Chicago Press, 1992), 198–202, 228–33. In recent years the history of medieval natural philosophy has generated heated debate. See Andrew Cunningham, "How the *Principia* Got Its Name; or, Taking Natural Philosophy Seriously," *History of Science,* 29 (1991), 377–92; Roger French and Andrew Cunningham, *Before Science: The Invention of the Friars' Natural Philosophy* (Aldershot: Scolar Press, 1996); Edward Grant, "God, Science, and Natural Philosophy in the Late Middle Ages," in Lodi Nauta and Arjo Vanderjagt, eds., *Between Demonstration and Imagination: Essays in the History of Science and Philosophy Presented to John D. North* (Leiden: Brill, 1999), 243–67.

5. Bert Hansen, *Nicole Oresme and the Marvels of Nature* (Toronto: Pontifical Institute of Mediaeval Studies, 1985), 59–60, 137. On Buridan's reputation, see Edward Grant, *The Foundations of Modern Science in the Middle Ages: Their Religious, Institutional, and Intellectual Contexts* (Cambridge: Cambridge University Press, 1996), 144.

6. Maurice A. Finocchiaro, ed., *The Galileo Affair: A Documentary History* (Berkeley and Los Angeles: University of California Press, 1989), 93.

7. Perez Zagorin, *Francis Bacon* (Princeton: Princeton University Press, 1998), 48–51; Paul H. Kocher, *Science and Religion in Elizabethan England* (San Marino, Calif.: Huntington Library, 1953), 27–28, 75.

8. Ronald L. Numbers, *Creation by Natural Law: Laplace's Nebular Hypothesis in American Thought* (Seattle: University of Washington Press, 1977), 3–4; Aram Vartanian, *Diderot and Descartes: A Study of Scientific Naturalism in the Enlightenment* (Princeton: Princeton University Press, 1953), 52 (quoting Pascal). See also Stephen Gaukroger, *Descartes: An Intellectual Biography* (Oxford: Clarendon Press, 1995), 146–292; Margaret J. Osler, *Divine Will and the Mechanical Philosophy: Gassendi and Descartes on Contingency and Necessity in the Created World* (Cambridge: Cambridge University Press, 1994); and Margaret J. Osler, "Mixing Metaphors: Science and Religion or Natural Philosophy and Theology in Early Modern Europe," *History of Science,* 36 (1998), 91–113. On nature as "a law-bound system of matter in motion," see John

C. Greene, *The Death of Adam: Evolution and Its Impact on Western Thought* (Ames: Iowa State University Press, 1959).

9. Robert Boyle, *A Free Enquiry into the Vulgarly Received Notion of Nature,* ed. Edward B. Davis and Michael Hunter (Cambridge: Cambridge University Press, 1996), ix–x, xv, 38–39. See also Rose-Mary Sargent, *The Diffident Naturalist: Robert Boyle and the Philosophy of Experiment* (Chicago: University of Chicago Press, 1995), 93–103. On Christianity and the mechanical philosophy, see Osler, "Mixing Metaphors"; and John Hedley Brooke, *Science and Religion: Some Historical Perspectives* (Cambridge: Cambridge University Press, 1991), chap. 4, "Divine Activity in a Mechanical Universe."

10. Isaac Newton, *Mathematical Principles of Natural Philosophy and His System of the World,* trans. Andrew Motte, revised by Florian Cajori (Berkeley and Los Angeles: University of California Press, 1960), 543–44; H. W. Turnbull, ed., *The Correspondence of Isaac Newton,* 7 vols. (Cambridge: Cambridge University Press, 1959–77), 2:331–34. For a statement that the solar system could not have been produced by natural causes alone, see Isaac Newton, *Papers and Letters on Natural Philosophy and Related Documents,* ed. I. Bernard Cohen (Cambridge: Cambridge University Press, 1958), 282. See also David Kubrin, "Newton and the Cyclical Cosmos: Providence and the Mechanical Philosophy," *Journal of the History of Ideas,* 28 (1967), 325–46; Richard S. Westfall, *Never at Rest: A Biography of Isaac Newton* (Cambridge: Cambridge University Press, 1980); and Cunningham, "How the *Principia* Got Its Name."

11. Kocher, *Science and Religion,* 97; Keith Thomas, *Religion and the Decline of Magic: Studies in Popular Beliefs in Sixteenth and Seventeenth Century England* (London: Weidenfeld and Nicolson, 1971).

12. This paragraph is based largely on Ronald L. Numbers and Ronald C. Sawyer, "Medicine and Christianity in the Modern World," in Martin E. Marty and Kenneth L. Vaux, eds., *Health/Medicine and the Faith Traditions: An Inquiry into Religion and Medicine* (Philadelphia: Fortress Press, 1982), 133–60, esp. 138–39. On the naturalization of mental illness, see Michael MacDonald, "Insanity and the Realities of History in Early Modern England," *Psychological Medicine,* 11 (1981), 11–15; Michael MacDonald, "Religion, Social Change, and Psychological Healing in England, 1600–1800," in W. J. Sheils, ed., *The Church and Healing,* Studies in Church History 19 (Oxford: Basil Blackwell, 1982), 101–25; Ronald L. Numbers and Janet S. Numbers, "Millerism and Madness: A Study of 'Religious Insanity' in Nineteenth-Century America," in Ronald L. Numbers and Jonathan M. Butler, eds., *The Disappointed: Millerism and Millenarianism in the Nineteenth Century* (Bloomington: Indiana University Press, 1987), 92–117.

13. John Duffy, *Epidemics in Colonial America* (Baton Rouge: Louisiana State University Press, 1953), 30–32; Otho T. Beall Jr. and Richard H. Shryock, *Cotton Mather: First Significant Figure in American Medicine* (Baltimore: Johns Hopkins Press, 1954), 104–5; Perry Miller, *The New England Mind: From Colony to Province* (Cambridge: Harvard University Press, 1953), 345–66; Maxine van de Wetering, "A Reconsideration of the Inoculation Controversy," *New England Quarterly,* 58 (1985), 46–67. On the transformation of cholera from divine punishment to public-health problem, see Charles E. Rosenberg, *The Cholera Years: The United States in 1832, 1849, and 1866* (Chicago: University of Chicago Press, 1962).

14. I. Bernard Cohen, *Benjamin Franklin's Science* (Cambridge: Harvard University Press, 1990), chap. 8, "Prejudice against the Introduction of Lightning Rods."

15. Theodore Hornberger, "The Science of Thomas Prince," *New England Quarterly,* 9 (1936), 31; William D. Andrews, "The Literature of the 1727 New England Earthquake," *Early American Literature,* 7 (1973), 283.

16. Thomas, *Religion and the Decline of Magic,* 643; Theodore Hornberger, *Scientific Thought in the American Colleges, 1638–1800* (Austin: University of Texas Press, 1945), 82.

17. Andrew Cunningham, "Getting the Game Right: Some Plain Words on the Identity and Invention of Science," *Studies in History and Philosophy of Science,* 19 (1988), 365–89. See also Sydney Ross, "Scientist: The Story of a Word," *Annals of Science,* 18 (1962), 65–85, which dates the earliest use of the term "scientist" to 1834.

18. Vartanian, *Diderot and Descartes,* 206 (quoting La Mettrie); Kathleen Wellman, *La Mettrie: Medicine, Philosophy, and Enlightenment* (Durham: Duke University Press, 1992), 175, 186–88. See also Aram Vartanian, ed., *La Mettrie's L'Homme Machine: A Study in the Origins of an Idea* (Princeton: Princeton University Press, 1960); and Jacques Roger, *The Life Sciences in Eighteenth-Century French Thought,* ed. Keith R. Benson, trans. Robert Ellrich (Stanford: Stanford University Press, 1997).

19. Georges-Louis Leclerc de Buffon, *Natural History: General and Particular,* trans. William Smellie, 7 vols. (London: W. Strahan & T. Cadell, 1781), 1: 34, 63–82; Numbers, *Creation by Natural Law,* 6–8. On Buffon's religious beliefs, see Jacques Roger, *Buffon: A Life in Natural History,* trans. Sarah Lucille Bonnefoi, ed. L. Pearce Williams (Ithaca: Cornell University Press, 1997), 431. On the naturalism of Buffon and his contemporaries, see Kenneth L. Taylor, "Volcanoes as Accidents: How 'Natural' Were Volcanoes to 18th-Century Naturalists?" in Nicoletta Morello, ed., *Volcanoes and History* (Genova: Brigati, 1998), 595–618.

20. Roger Hahn, "Laplace and the Vanishing Role of God in the Physical Universe," in Harry Woolf, ed., *The Analytic Spirit: Essays in the History of Science in Honor of Henry Guerlac* (Ithaca: Cornell University Press, 1981), 85–95; Roger Hahn, "Laplace and the Mechanistic Universe," in David C. Lindberg and Ronald L. Numbers, eds., *God and Nature: Historical Essays on the Encounter between Christianity and Science* (Berkeley and Los Angeles: University of California Press, 1986), 256–76.

21. Thomas Chalmers, *On the Power, Wisdom, and Goodness of God as Manifested in the Adaptation of External Nature to the Moral and Intellectual Constitution of Man,* 2 vols. (London: William Pickering, 1835), 1:30–32. On English attitudes toward French science, see Adrian Desmond, *The Politics of Evolution: Morphology, Medicine, and Reform in Radical London* (Chicago: University of Chicago press, 1989), chap. 2. This and the following five paragraphs are extracted in large part from Numbers, *Creation by Natural Law,* esp. 79–83, 93.

22. J. P. N[ichol], "State of Discovery and Speculation Concerning the Nebulae," *Westminster Review,* 25 (1836), 406–8; J. P. Nichol, *Views of the Architecture of the Heavens,* 2d ed. (New York: Dayton & Newman, 1842), 103–5. On Nichol, see also Simon Schaffer, "The Nebular Hypothesis and the Science of Progress," in James R. Moore, ed., *History, Humanity, and Evolution: Essays for John C. Greene* (Cambridge: Cambridge University Press, 1989), 131–64.

23. John Le Conte, "The Nebular Hypothesis," *Popular Science Monthly,* 2 (1873), 655; "The Nebular Hypothesis," *Southern Quarterly Review,* 10 (1846), 228 (Great Architect). It is unclear from the transcript of his lecture whether Le Conte was expressing his own view or quoting the words of someone else.

24. [Daniel Kirkwood], "The Nebular Hypothesis," *Presbyterian Quarterly Review,* 2 (1854), 544.

25. "The Nebular Hypothesis," *Southern Quarterly Review,* n.s., 1 (1856), 110–11.

26. "The Nebular Hypothesis," *Southern Quarterly Review,* 10, (1846), 240.

27. Arnold Guyot, *Creation; or, The Biblical Cosmogony in the Light of Modern Science* (Edinburgh: T. and T. Clark, [1883]), 135. For an early version of Guyot's harmonizing scheme, see Arnold Guyot, "The Mosaic Cosmogony and Modern Science Reconciled," (New York) *Evening Post,* 6, 12, 15, and 23 March 1852 (no pagination). See also James D. Dana, "Memoir of Arnold Guyot, 1807–1884," National Academy of Sciences, *Biographical Memoirs,* 2 (1886), 309–47.

28. Charles Hodge, *Systematic Theology,* 3 vols. (New York: Scribner, 1871–73), 1:573. See also Ronald L. Numbers, "Charles Hodge and the Beauties and Deformities of Science," in John W. Stewart and James H. Moorhead, eds., *Charles Hodge Revisited: A Critical Appraisal of His Life and Work* (Grand Rapids, Mich.: Eerdmans, 2002), 77–101.

29. John Frederick William Herschel, *A Preliminary Discourse on the Study of Natural Philosophy,* new ed. (Philadelphia: Lea & Blanchard, 1839), 6, 27–28, 59, 108; Walter F. Cannon, "John Herschel and the Idea of Science," *Journal of the History of Ideas,* 22 (1961), 215–39; William Minto, *Logic: Inductive and Deductive* (New York: C. Scribner's Sons, 1904), 157, quoted in Laurens Laudan, "Theories of Scientific Method from Plato to Mach: A Bibliographical Review," *History of Science,* 7 (1968), 30. On the meaning of science, see Richard Yeo, *Defining Science: William Whewell, Natural Knowledge, and Public Debate in Early Victorian Britain* (Cambridge: Cambridge University Press, 1993).

30. Martin Rudwick, "Uniformity and Progression: Reflections on the Structure of Geological Theory in the Age of Lyell," in Duane H. D. Roller, ed., *Perspectives in the History of Science and Technology* (Norman: University of Oklahoma Press, 1971), 209–27; Roy Porter, *The Making of Geology: Earth Science in Britain, 1660–1815* (Cambridge: Cambridge University Press, 1977), 2; James R. Moore, "Charles Lyell and the Noachian Deluge," *Journal of the American Scientific Affiliation,* 22 (1970), 107–15; Leonard G. Wilson, *Charles Lyell: The Years to 1841: The Revolution in Geology* (New Haven: Yale University Press, 1972), 256. On Lyell's indebtedness to Herschel, see Rachel Laudan, *From Mineralogy to Geology: The Foundations of a Science, 1650–1830* (Chicago: University of Chicago Press, 1987), 203–4. See also Martin J. S. Rudwick, "The Shape and Meaning of Earth History," in Lindberg and Numbers, eds., *God and Nature,* 296–321; and James R. Moore, "Geologists and Interpreters of Genesis in the Nineteenth Century," ibid., 322–50.

31. Charles Lyell, *Principles of Geology, Being an Attempt to Explain the Former Changes of the Earth's Surface, by Reference to Causes Now in Operation,* 3 vols. (London: John Murray, 1830–1833), 1: 75–76. On the role of Christian ministers in naturalizing earth history, see, e.g., Nicolaas A. Rupke, *The Great Chain of History: William Buckland and the English School of Geology (1814–1849)* (Oxford: Clarendon Press, 1983); and Rodney L. Stiling, "The Diminishing Deluge: Noah's Flood in Nineteenth-Century American Thought," Ph.D. diss., University of Wisconsin-Madison, 1991.

32. James A. Secord, introduction to *Principles of Geology,* by Charles Lyell (London: Penguin, 1997), xxxii–xxxiii. On the naturalization of physiology, see Alison Winter, "The Construction of Orthodoxies and Heterodoxies in the Early Victorian Life Sciences," in

Bernard Lightman, ed., *Victorian Science in Context* (Chicago: University of Chicago Press, 1997), 24–50.

33. Howard E. Gruber, *Darwin on Man: A Psychological Study of Scientific Creativity,* together with *Darwin's Early and Unpublished Notebooks,* transcribed and annotated by Paul H. Barrett (New York: Dutton, 1974), 417–18; Leonard Huxley, ed., *The Life and Letters of Thomas Henry Huxley,* 2 vols. (New York: Appleton, 1900), 2:320; Mario A. di Gregorio, *T. H. Huxley's Place in Natural Science* (New Haven: Yale University Press, 1984), 51; Charles Darwin, *On the Origin of Species,* with an introduction by Ernst Mayr (Cambridge: Harvard University Press, 1966), 483; Adrian Desmond, *Huxley: From Devil's Disciple to Evolution's High Priest* (Reading, Mass.: Addison-Wesley, 1997), 256. According to Desmond, Huxley first described atoms flashing into elephants in his "Lectures," *Medical Times and Gazette,* 12 (1856), 482–83, which includes the statement about lack of evidence for creation. For American references to atomic elephants, see Ronald L. Numbers, *Darwinism Comes to America* (Cambridge: Harvard University Press, 1998), 52. As Neal C. Gillespie has shown, natural theology "had virtually ceased to be a significant part of the day-to-day practical explanatory structure of natural history" well before 1859. Gillespie, "Preparing for Darwin: Conchology and Natural Theology in Anglo-American Natural History," *Studies in History of Biology,* 7 (1983), 95.

34. Darwin, *On the Origin of Species,* 466; S. R. Calthrop, "Religion and Evolution," *Religious Magazine and Monthly Review,* 50 (1873), 205; [W. N. Rice], "The Darwinian Theory of the Origin of Species," *New Englander,* 26 (1867), 608.

35. Asa Gray, *Darwiniana: Essays and Reviews Pertaining to Darwinism* (New York: D. Appleton, 1876), 78–79, from a review first published in 1860 (sufficient answer); Asa Gray, *Natural Science and Religion: Two Lectures Delivered to the Theology School of Yale College* (New York: Charles Scribner's Sons, 1880), 77 (business of science); George F. Wright, "Recent Works Bearing on the Relation of Science to Religion: No. II," *Bibliotheca Sacra,* 33 (1876), 480.

36. Duke of Argyll, *The Reign of Law,* 5th ed. (New York: John W. Lovell, 1868), 34; B. B. Warfield, review of *Darwinism To-Day,* by Vernon L. Kellogg, in *Princeton Theological Review,* 6 (1908), 640–50. I am indebted to David N. Livingstone for bringing the Warfield statement to my attention. For more on Warfield, see David N. Livingstone and Mark A. Noll, "B. B. Warfield (1851–1921): A Biblical Inerrantist as Evolutionist," *Isis,* 91 (2000), 283–304. See also Desmond, *Huxley,* 555–56. On the scientific context of Argyll's views, see Nicolaas A. Rupke, *Richard Owen: Victorian Naturalist* (New Haven: Yale University Press, 1994).

37. Jon H. Roberts, *Darwinism and the Divine in America: Protestant Intellectuals and Organic Evolution, 1859–1900* (Madison: University of Wisconsin Press, 1988), 136; Numbers, *Darwinism Comes to America,* 40–41; Joseph Le Conte, "Evolution in Relation to Materialism, *Princeton Review,* n.s. 7 (1881), 166, 174. Jon H. Roberts and James Turner, *The Sacred and the Secular University* (Princeton: Princeton University Press, 2000), 28–30, discusses "the triumph of methodological naturalism." Of the eighty anthropologists, botanists, geologists, and zoologists elected to the National Academy of Sciences between its creation in 1863 and 1900, only thirteen were known agnostics or atheists; ibid., 41. James H. Leuba's pioneering survey of the beliefs of American scientists, in 1916, turned up only 41.8 percent who affirmed belief "in a God in intellectual and affective communication with humankind, i.e. a God to whom one may pray in expectation of receiving an answer." Fifty years later the

figure had declined to 39.3 percent of respondents. See Edward J. Larson and Larry Witham, "Scientists Are Still Keeping the Faith," *Nature,* 386 (1997), 435–36.

38. Frank M. Turner, *Between Science and Religion: The Reaction to Scientific Naturalism in Late Victorian England* (New Haven: Yale University Press, 1974), 16. See also Frank M. Turner, *Contesting Cultural Authority: Essays in Victorian Intellectual Life* (Cambridge: Cambridge University Press, 1993); and Bernard Lightman, *The Origins of Agnosticism: Victorian Unbelief and the Limits of Knowledge* (Baltimore: Johns Hopkins University Press, 1987).

39. Frank M. Turner, "John Tyndall and Victorian Scientific Naturalism," in W. H. Brock, N. D. McMillan, and R. C. Mollan, eds., *John Tyndall: Essays on a Natural Philosopher* (Dublin: Royal Dublin Society, 1981), 172; Turner, *Between Science and Religion,* 11–12; Lightman, "'Fighting Even with Death,'" 323–50. See also Ruth Barton, "John Tyndall, Pantheist: A Rereading of the Belfast Address," *Osiris,* 2d ser., 3 (1987), 111–34.

40. John C. Burnham, "The Encounter of Christian Theology with Deterministic Psychology and Psychoanalysis," *Bulletin of the Menninger Clinic,* 49 (1985), 321–52, quotations on 333, 337; Turner, *Between Science and Religion,* 14–16. For a comprehensive discussion of naturalism and the rise of the social sciences, see Roberts and Turner, *The Sacred and the Secular University,* 43–60. On earlier efforts to naturalize psychology, see Roger Cooter, *The Cultural Meaning of Popular Science: Phrenology and the Organization of Consent in Nineteenth-Century Britain* (Cambridge: Cambridge University Press, 1984).

41. Phillip E. Johnson, foreword to *The Creation Hypothesis: Scientific Evidence for an Intelligent Designer,* ed. J. P. Moreland (Downers Grove, Ill.: InterVarsity Press, 1994), 7–8; Larry Vardiman, "Scientific Naturalism as Science," *Impact* #293, inserted in *Acts & Facts,* 26 (November 1997); Paul Nelson at a conference on "Design and Its Critics," Concordia University Wisconsin, 24 June 2000. On scientific creationists and naturalism, see Ronald L. Numbers, *The Creationists* (New York: Knopf, 1982), 90, 96, 207–8.

42. Michael J. Behe, *Darwin's Black Box: The Biochemical Challenge to Evolution* (New York: Free Press, 1996), 15, 33, 193, 232–233; "The Evolution of a Skeptic: An Interview with Dr. Michael Behe, Biochemist and Author of Recent Best-Seller, *Darwin's Black Box,*" *Real Issue,* November/December 1996, 1, 6–8. For a brief historical overview of the intelligent design movement, see Numbers, *Darwinism Comes to America,* 15–21.

43. William A. Dembski, "What Every Theologian Should Know about Creation, Evolution, and Design," Center for Interdisciplinary Studies, *Transactions,* May/June 1995, 1–8. See also William A. Dembski, *Intelligent Design: The Bridge between Science and Theology* (Downers Grove, Ill.: InterVarsity Press, 1999). For a philosophical critique of methodological naturalism from a Christian perspective, see Alvin Plantinga, "Methodological Naturalism?" *Perspectives on Science and Christian Faith,* September 1997, 143–54. For a philosophical critique of intelligent design, see Robert T. Pennock, *Tower of Babel: The Evidence against the New Creationism* (Cambridge: MIT Press, 1999).

44. David K. Webb, letter to the editor, *Origins & Design,* spring 1996, 5; J. W. Haas Jr., "On Intelligent Design, Irreducible Complexity, and Theistic Science," *Perspectives on Science and Christian Faith,* March 1997, 1, who quotes Dawkins.

45. John Hedley Brooke has similarly challenged the common notion "that the sciences have eroded religious belief by explaining physical phenomena in terms of natural

law"; see his "Natural Law in the Natural Sciences: The Origins of Modern Atheism?" *Science and Christian Belief,* 4 (1992), 83.

46. Larson and Witham, "Scientists Are Still Keeping the Faith," 435–36; Eugenie C. Scott, "Gallup Reports High Level of Belief in Creationism" *NCSE Reports,* fall 1993, 9; John Cole, "Gallup Poll Again Shows Confusion," ibid., spring 1996, 9; Claudia Wallis, "Faith and Healing," *Time,* 24 June 1996, 63.

A GUIDE TO FURTHER READING

GENERAL

Brooke, John Hedley. *Science and Religion: Some Historical Perspectives.* Cambridge: Cambridge University Press, 1991. A survey of the relationship between science and Christianity since the sixteenth century, with emphasis on the complexity of the interaction.

Brooke, John Hedley, and Geoffrey Cantor. *Reconstructing Nature: The Engagement of Science and Religion.* Edinburgh: T & T Clark, 1998. More examples of the complex relationship between science and Christianity.

Brooke, John Hedley, Margaret J. Osler, and Jitse M. van der Meer, eds. *Science in Theistic Contexts: Cognitive Dimensions.* Special number of *Osiris,* 2d ser., vol. 16 (2001). A collection of articles, chronologically broad and reflecting a post-positivist historiography of science.

Cohn, Norman. *Noah's Flood: The Genesis Story in Western Thought.* New Haven: Yale University Press, 1996. Traces the flood story from Mesopotamian mythology to twentieth-century fundamentalism.

Ferngren, Gary B., ed. *The History of Science and Religion in the Western Tradition: An Encyclopedia.* New York: Garland, 2000. A collection of more than one hundred essays: historiographical, biographical, and topical.

Lindberg, David C., and Ronald L. Numbers, eds. *God and Nature: Historical Essays on the Encounter between Christianity and Science.* Berkeley and Los Angeles: University of California Press, 1986. Eighteen essays surveying the relationship between science and Christianity from the early church through the twentieth century.

Livingstone, David N. *The Preadamite Theory and the Marriage of Science and Religion.* Philadelphia: American Philosophical Society, 1992. Sketches the history of preadamism from antiquity into the twentieth century.

Livingstone, David N., D. G. Hart, and Mark A. Noll, eds. *Evangelicals and Science in Historical Perspective.* New York: Oxford University Press, 1999. Essays on topics ranging from Puritanism to fundamentalism.

Numbers, Ronald L., and Darrel W. Amundsen, eds. *Caring and Curing: Health and Medicine in the Western Religious Traditions.* New York: Macmillan, 1986. Histories of various religious traditions, from Judaism and Catholicism to Mormonism and Adventism.

Olson, Richard. *Science Deified and Science Defied: The Historical Significance of Science in Western Culture.* 2 vols. Berkeley and Los Angeles: University of California Press, 1982, 1990. Looks at episodes of interaction between science and other aspects of Western culture, including religion, from the Bronze Age to the early nineteenth century.

EARLY CHRISTIANITY AND THE MIDDLE AGES

Amundsen, Darrel W. *Medicine, Society, and Faith in the Ancient and Medieval Worlds.* Baltimore: Johns Hopkins University Press, 1996. A collection of essays on medicine and Christianity before about 1500.

Cochrane, Charles N. *Christianity and Classical Culture: A Study of Thought and Action from Augustus to Augustine.* Oxford: Clarendon Press, 1940. A brilliant investigation of the relationship between early Christianity and the classical tradition.

Ferngren, Gary B. "Early Christianity as a Religion of Healing," *Bulletin of the History of Medicine,* 66 (1992), 1–15. The best introduction to the topic.

French, Roger, and Andrew Cunningham. *Before Science: The Invention of the Friars' Natural Philosophy.* Aldershot: Scolar Press, 1996. A provocative, if tendentious and unreliable, account of Dominican and Franciscan natural philosophy.

Grant, Edward. *The Foundations of Modern Science in the Middle Ages: Their Religious, Institutional, and Intellectual Contexts.* Cambridge: Cambridge University Press, 1996. Examines medieval science in its medieval context.

———. *God and Reason in the Middle Ages.* Cambridge: Cambridge University Press, 2001. Argues that the medieval period was forerunner (in theology as well as in science and law) of seventeenth-century rationalism.

Hackett, Jeremiah, ed. *Roger Bacon and the Sciences: Commemorative Essays.* Leiden: Brill, 1997. The best source for Bacon's life and scholarly career.

Lindberg, David C. *The Beginnings of Western Science: The European Scientific Tradition in Philosophical, Religious, and Institutional Context, 600 B.C. to A.D. 1450.* Chicago: University of Chicago Press, 1992. A history of ancient and medieval science, in which theology and religion are treated unapologetically, as major players.

———. "Medieval Science and Its Religious Context," *Osiris,* 2d ser., 10 (1995), 61–79. Offers a series of proposals about the ways in which science and Christianity interacted during the medieval period.

———. "Science as Handmaiden: Roger Bacon and the Patristic Tradition," *Isis,* 78 (1987), 518–36. Bacon's scientific program in the context of the Augustinian tradition.

McCluskey, Stephen C. *Astronomies and Cultures in Early Medieval Europe.* Cambridge: Cambridge University Press, 1998. A study of astronomy in the monastic culture of the early Middle Ages.

Rist, John M. *Augustine: Ancient Thought Baptized.* Cambridge: Cambridge University Press, 1994. An excellent account of Augustine's relationship to the classical tradition.

Russell, Jeffrey Burton. *Inventing the Flat Earth: Columbus and Modern Historians.* New York: Praeger, 1991. Deconstructs the myth that medieval thinkers believed in a flat earth.

Southern, Richard W. *Robert Grosseteste: The Growth of an English Mind in Medieval Europe.* Oxford: Clarendon Press, 1986. Treats one of the outstanding theologians and natural philosophers of the Middle Ages, including the relationship between his theology and his scientific efforts.

THE EARLY MODERN PERIOD

Allen, Don Cameron. *The Legend of Noah: Renaissance Rationalism in Art, Science, and Letters.* Urbana: University of Illinois Press, 1949. Centers on efforts to prove that a universal flood actually happened.

Blackwell, Richard J. *Galileo, Bellarmine, and the Bible.* Notre Dame: University of Notre Dame Press, 1991. Analyzes events leading up to Galileo's trial from the vantage point of attitudes toward biblical exegesis and biblical authority.

Blair, Ann. *The Theater of Nature: Jean Bodin and Renaissance Science.* Princeton: Princeton University Press, 1997. Religion and natural philosophy in the late Renaissance.

Burns, William E. *An Age of Wonders: Prodigies, Politics, and Providence in England, 1627–1727.* Manchester: University of Manchester Press, 2002. From divine providence to natural philosophy.

Cipolla, Carlo M. *Faith, Reason, and the Plague in Seventeenth-Century Tuscany.* Ithaca: Cornell University Press, 1979. A case study of conflict between religious leaders and public health authorities.

Cohen, H. Floris. *The Scientific Revolution: A Historiographical Inquiry.* Chicago: University of Chicago Press, 1994. Discusses interpretations of the Scientific Revolution, including attempts to bring religion into the picture.

Cohen, I. Bernard, ed. *Puritanism and the Rise of Modern Science: The Merton Thesis.* New Brunswick: Rutgers University Press, 1990. A useful collection.

Cunningham, Andrew. *The Anatomical Renaissance: The Resurrection of the Anatomical Projects of the Ancients.* Aldershot, England: Scolar Press, 1997. An "attempt to put the religion back into sixteenth century anatomizing."

Daston, Lorraine, and Katharine Park. *Wonders and the Order of Nature, 1150–1750.* New York: Zone Books, 1998. Traces the changing fortunes of marvels and monsters as objects of curiosity and signs from God.

Dear, Peter. *Mersenne and the Learning of the Schools.* Ithaca: Cornell University Press, 1988. Looks at the contributions of a seventeenth-century Catholic monk to the birth of modern science.

Dick, Steven J. *Plurality of Worlds: The Origins of the Extraterrestrial Life Debate from Democritus to Kant.* Cambridge: Cambridge University Press, 1982. Shows how the concept of other worlds was transformed from heresy to orthodoxy in Western thought.

Dobbs, Betty Jo T. *The Janus Faces of Genius: The Role of Alchemy in Newton's Thought.* Cambridge: Cambridge University Press, 1991. Treats Newton's views on divine activity in the world.

Fantoli, Annibale. *Galileo: For Copernicanism and for the Church.* Trans. George V. Coyne. 2d ed. Vatican City: Vatican Observatory, 1996. The best full account of the Galileo affair.

Feldhay, Rivka. *Galileo and the Church: Political Inquisition or Critical Dialogue?* Cambridge: Cambridge University Press, 1995. Situates the Galileo affair in the context of Counter-Reformation Catholicism.

Finocchiaro, Maurice A. *The Galileo Affair: A Documentary History.* Berkeley and Los Angeles: University of California Press, 1989. A valuable collection of primary documents.

Genuth, Sara Schechner. *Comets, Popular Culture, and the Birth of Modern Cosmology.* Princeton: Princeton University Press, 1997. Focuses on the role of comets as divine signs.

Grell, Ole Peter, and Andrew Cunningham, eds. *Medicine and the Reformation.* London: Routledge, 1993. Essays on medicine and religion in the sixteenth century.

———, eds. *Religio Medici: Medicine and Religion in Seventeenth-Century England.* Aldershot, England: Scolar Press, 1996. More essays on medicine and religion.

Hannaway, Owen. *The Chemists and the Word: The Didactic Origins of Chemistry.* Baltimore: Johns Hopkins University Press, 1975. Examines the role of religion in the emergence of chemistry as a scientific discipline.

Harrison, Peter. *The Bible, Protestantism, and the Rise of Natural Science.* Cambridge: Cambridge University Press, 1998. Argues that radical new principles of biblical interpretation developed by the Protestant reformers, when transferred to the reading of the book of nature, gave rise to the new natural history of the seventeenth century.

———. "Original Sin and the Problem of Knowledge in Early Modern Europe," *Journal of the History of Ideas,* 63 (2002), 239–59. Argues for the pervasive influence of the doctrine of original sin on seventeenth-century theories of knowledge.

———. *"Religion" and the Religions in the English Enlightenment.* Cambridge: Cambridge University Press, 1990. A study of the emergence of comparative religion from sacred history to natural history.

———. "Voluntarism and Early Modern Science," *History of Science,* 40 (2002), 63–89. An analysis and searching critique of arguments that link the development of empirical science with theological voluntarism.

Heilbron, John L. *Electricity in the 17th and 18th Centuries: A Study of Early Modern Physics.* Berkeley and Los Angeles: University of California Press, 1979. Includes extensive discussions of Jesuit contributions.

———. *The Sun in the Church: Cathedrals as Solar Observatories.* Cambridge: Harvard University Press, 1999. Argues that Catholic churches "were the best solar observatories in the world."

Heyd, Michael. *Between Orthodoxy and the Enlightenment: Jean-Robert Chouet and the Introduction of Cartesian Science in the Academy of Geneva.* The Hague: Martinus Nijhoff, 1982. Calvinism and Cartesianism in Geneva.

Hodgen, Margaret T. *Early Anthropology in the Sixteenth and Seventeenth Centuries.* Philadelphia: University of Pennsylvania Press, 1964. Stresses the biblical context in which discussions of human history took place.

Hooykaas, Reijer. *Religion and the Rise of Modern Science.* Grand Rapids, Mich.: Eerdmans, 1972. A partisan defense of the proposition that the biblical worldview (especially that of English Puritans) created conditions that made possible the emergence of modern science in the sixteenth and seventeenth centuries.

Howell, Kenneth J. *God's Two Books: Copernican Cosmology and Biblical Interpretation in Early Modern Science.* Notre Dame: University of Notre Dame Press, 2002. Focuses on the writings of Lutheran and Calvinist astronomers, natural philosophers, and theologians.

Hunter, Michael, ed. *Robert Boyle Reconsidered.* Cambridge: Cambridge University Press, 1994. A sampler of recent scholarly opinion.

———. *Science and the Shape of Orthodoxy: Intellectual Change in Late Seventeenth-Century Britain.* Woodbridge, England: Boydell Press, 1995. Essays on science and religion.

Kusukawa, Sachiko. *The Transformation of Natural Philosophy: The Case of Philip Melanchthon.* Cambridge: Cambridge University Press, 1995. Focuses on the creation of what the author sees as a distinctive Lutheran natural philosophy.

Langford, Jerome J. *Galileo, Science, and the Church.* Rev. ed. Ann Arbor: University of Michigan Press, 1966. A reconsideration of the events surrounding Galileo's defense of heliocentrism that softens traditional criticism of the Catholic church.

MacDonald, Michael. *Mystical Bedlam: Madness, Anxiety, and Healing in Seventeenth-Century England.* Cambridge: Cambridge University Press, 1981. Illustrates the fusion of natural and supernatural theories of insanity.

Machamer, Peter, ed. *The Cambridge Companion to Galileo.* Cambridge: Cambridge University Press, 1998. A collection of historical essays.

Merton, Robert K. "Science, Technology, and Society in Seventeenth-Century England," *Osiris,* 4 (1938), 360–632. Occasionally reprinted as a book, this classic monograph attempted to establish a close connection between English Puritanism and the emergence of modern science.

Morgan, John. "Puritanism and Science: A Reinterpretation," *The Historical Journal,* 22 (1979), 535–60. A critical analysis of Merton's thesis.

O'Malley, John W., et al., eds. *The Jesuits: Cultures, Sciences, and the Arts, 1540–1773.* Toronto: University of Toronto Press, 1999. Includes a number of important essays on Jesuit science.

Osler, Margaret J. *Divine Will and the Mechanical Philosophy: Gassendi and Descartes on Contingency and Necessity in the Created World.* Cambridge: Cambridge University Press, 1994. Places the mechanical philosophy in theological context.

———. "Mixing Metaphors: Science and Religion or Natural Philosophy and Theology in Early Modern Europe," *History of Science,* 36 (1998), 91–113. Proposes appropriation and translation as new metaphors to replace the traditional conflict, harmony, and segregation.

———. ed. *Rethinking the Scientific Revolution.* Cambridge: Cambridge University Press, 2000. Contains half a dozen articles treating science and religion in the early modern period.

Park, Katharine. "The Criminal and the Saintly Body: Autopsy and Dissection in Renaissance Italy," *Renaissance Quarterly,* 47 (1994), 1–33. Challenges the notion that there was a widespread taboo against autopsies and dissections in medieval and Renaissance Europe.

Pedersen, Olaf. "Galileo and the Council of Trent: The Galileo Affair Revisited," *Journal for the History of Astronomy,* 14 (1983), 1–29. Looks at Galileo's troubles with the Roman Catholic church in the context of the theology that emerged from the Council of Trent.

Popkin, Richard H. *Isaac La Peyrère (1596–1676): His Life, Work, and Influence.* Leiden: Brill, 1987. A study of the French biblical scholar who argued that there had been men before Adam.

Thomas, Keith. *Religion and the Decline of Magic.* New York: Scribner, 1971. A much-cited social history of magic and witchcraft in sixteenth- and seventeenth-century England, which explores the influence of Reformation thought on supernatural belief and practice.

Webster, Charles. *The Great Instauration: Science, Medicine, and Reform, 1626–1660.* London: Duckworth, 1975. An examination of the impact of Puritan millennialism on English science, emphasizing practical achievements.

Westfall, Richard S. *Essays on the Trial of Galileo.* Vatican City: Vatican Observatory, 1989. Thoughtful essays on aspects of the Galileo affair.

———. *Never at Rest: A Biography of Isaac Newton.* Cambridge: Cambridge University Press, 1980. The best biography of Newton, covering both his religious and his scientific concerns.

———. *Science and Religion in Seventeenth-Century England.* New Haven: Yale University Press, 1958. A pioneering study of the opinions of scientific "virtuosi" regarding the relationship between science and Christianity.

Westman, Robert S., ed. *The Copernican Achievement.* Berkeley and Los Angeles: University of California Press, 1975. Includes several essays on religious responses to Copernican astronomy and cosmology.

Wojcik, Jan W. *Robert Boyle and the Limits of Reason.* Cambridge: Cambridge University Press, 1997. Places Boyle's thought on reason and scientific method in theological context.

THE EIGHTEENTH CENTURY

Browne, Janet. *The Secular Ark: Studies in the History of Biogeography.* New Haven: Yale University Press, 1983. Examines the influence of the Flood story.

Cohen, I. Bernard. *Benjamin Franklin's Science.* Cambridge: Harvard University Press, 1990. Includes a long chapter on "Prejudice against the Introduction of Lightning Rods."

Crowe, Michael J. *The Extraterrestrial Life Debate, 1750–1900: The Idea of a Plurality of Worlds from Kant to Lowell.* Cambridge: Cambridge University Press, 1986. Demonstrates the role of religion in the debates.

Force, James E. *William Whiston: Honest Newtonian.* Cambridge: Cambridge University Press, 1985. A biography of the preacher who succeeded Newton as Lucasian Professor of Mathematics at Cambridge in 1703.

Gascoigne, John. *Cambridge in the Age of the Enlightenment: Science, Religion, and Politics from the Restoration to the French Revolution.* Cambridge: Cambridge University Press, 1989. Science and religion in eighteenth-century Cambridge.

Gould, Stephen Jay. *Time's Arrow, Time's Cycle: Myth and Metaphor in the Discovery of Geological Time.* Cambridge: Harvard University Press, 1987. Reassesses the role of religion in the intellectual transition from thousands to billions of years.

Jacob, Margaret C. *The Radical Enlightenment: Pantheists, Freemasons, and Republicans.* London: George Allen & Unwin, 1981. Explores the relationship between modern science and the rise of pantheistic materialism.

Levine, Joseph M. *Dr. Woodward's Shield: History, Science, and Satire in Augustan England.* Berkeley and Los Angeles: University of California Press, 1977. A biography of the London physician who wrote an influential natural history of the earth.

Pinto-Correia, Clara. *The Ovary of Eve: Egg and Sperm and Preformation.* Chicago: University of Chicago Press, 1997. The debate over the origins of new life.

Rappaport, Rhoda. *When Geologists Were Historians, 1665–1750.* Ithaca: Cornell University Press, 1997. Examines the debates over diluvialism and the geological role of Noah's Flood.

Roe, Shirley A. *Matter, Life, and Generation: Eighteenth-Century Embryology and the Haller-Wolff Debate.* Cambridge: Cambridge University Press, 1981. "Preformationists," who believed that God had made all embryos at the Creation, versus "epigenesists," who stressed the gradual development of embryos.

Roger, Jacques. *The Life Sciences in Eighteenth-Century French Thought.* Ed. Keith R. Benson. Trans. Robert Ellrich. Stanford: Stanford University Press, 1997. The standard introduction.

Schofield, Robert E. *The Enlightenment of Joseph Priestley: A Study of His Life and Work from 1733 to 1773.* University Park: Pennsylvania State University Press, 1997. A biography of the Unitarian preacher who discovered oxygen.

Semonin, Paul. *American Monster: How the Nation's First Prehistoric Creature Became a Symbol of National Identity.* New York: New York University Press, 2000. Were the huge bones that Americans began to find in 1705 the remnants of antediluvian humans or of prehistoric monsters?

BEFORE DARWIN

Astore, William J. *Observing God: Thomas Dick, Evangelicalism, and Popular Science in Victorian Britain and America.* Aldershot, England: Ashgate, 2001. A biographical study of science, Christianity, and popular culture.

Bozeman, Theodore Dwight. *Protestants in an Age of Science: The Baconian Ideal and Antebellum American Religious Thought.* Chapel Hill: University of North Carolina Press, 1977. Shows how Old School Presbyterians used the Baconian philosophy to monitor science without opposing it.

Cantor, Geoffrey. *Michael Faraday, Sandemanian and Scientist: A Study of Science and Religion in the Nineteenth Century.* New York: St. Martin's, 1991. Explores the role of religion in the life of the great experimentalist who contributed to the science of electromagnetism.

Conser, Walter H., Jr. *God and the Natural World: Religion and Science in Antebellum America.* Columbia: University of South Carolina Press, 1993. Stresses accommodation rather than conflict.

Corsi, Pietro. *Science and Religion: Baden Powell and the Anglican Debate, 1800–1860.* Cambridge: Cambridge University Press, 1988. A study of the first prominent Anglican to embrace organic evolution.

Gillespie, Neal C. "Preparing for Darwin: Conchology and Natural Theology in Anglo-American Natural History," *Studies in History of Biology,* 7 (1983), 93–145. Assesses the extent to which working naturalists incorporated natural theology into their scientific practice.

Gillispie, Charles Coulston. *Genesis and Geology: A Study in the Relations of Scientific Thought, Natural Theology, and Social Opinion in Great Britain, 1790–1850.* Cambridge: Harvard University Press, 1951. A seminal study that emphasizes "religion in science" rather than "religion versus science."

Hovenkamp, Herbert. *Science and Religion in America, 1800–1860.* Philadelphia: University of Pennsylvania Press, 1978. A flawed but useful overview.

Klaver, J. M. I. *Geology and Religious Sentiment: The Effect of Geological Discoveries on English Society and Literature between 1829 and 1859.* Leiden: Brill, 1997. Focuses on the writings of Charles Lyell and of three clergymen-geologists: William Buckland, Adam Sedgwick, and William Whewell.

Numbers, Ronald L. "Charles Hodge and the Beauties and Deformities of Science." In *Charles Hodge Revisited: A Critical Appraisal of His Life and Work,* ed. John W. Stewart and James H. Moorhead (Grand Rapids, Mich.: Eerdmans, 2002), 77–101. A study of the most influential Presbyterian theologian in nineteenth-century America.

———. *Creation by Natural Law: Laplace's Nebular Hypothesis in American Thought.* Seattle: University of Washington Press, 1977. Explores the theological and scientific consequences of accepting the nebular cosmogony.

Ospovat, Dov. *The Development of Darwin's Theory: Natural History, Natural Theology, and Natural Selection, 1838–1859.* Cambridge: Cambridge University Press, 1981. Stresses the influence of natural theology in the development of Darwin's thought.

Rosenberg, Charles E. *The Cholera Years: The United States in 1832, 1849, and 1866.* Chicago: University of Chicago Press, 1962. A classic case study of the changing relationship between medical theory and religious belief.

Rudwick, Martin J. S. *The Meaning of Fossils: Episodes in the History of Palaeontology.* New York: Science History, 1976; reprint, Chicago: University of Chicago Press, 1985. Examines changing interpretations of fossils from the Renaissance to the late nineteenth century.

Rupke, Nicolaas A. *The Great Chain of History: William Buckland and the English School of Geology (1814–1849).* Oxford: Clarendon Press, 1983. Argues that Buckland tried to harmonize science and religion because of institutional constraints, not religious orthodoxy.

———. *Richard Owen: Victorian Naturalist.* New Haven: Yale University Press, 1994. A study of the foremost British natural historian before Darwin and a major critic of natural selection.

Secord, James A. *Victorian Sensation: The Extraordinary Publication, Reception, and Secret Authorship of "Vestiges of the Natural History of Creation."* Chicago: University of Chicago Press, 2000. A major study of popular science and religion in the years before Darwin.

Shortland, Michael, ed. *Hugh Miller and the Controversies of Victorian Science.* Oxford: Clarendon Press, 1996. Essays on an influential harmonizer of Genesis and geology.

Stanton, William. *The Leopard's Spots: Scientific Attitudes toward Race in America, 1815–59.* Chicago: University of Chicago Press, 1960. Although weak on religious issues, still the only extended treatment of the American school of ethnology, which discarded the story of Adam and Eve and lengthened human history.

Stephens, Lester D. *Science, Race, and Religion in the American South: John Bachman and the Charleston Circle of Naturalists, 1815–1895.* Chapel Hill: University of North Carolina Press, 2000. Supplies an important corrective to Stanton's treatment of Bachman.

Trautmann, Thomas R. *Lewis Henry Morgan and the Invention of Kinship.* Berkeley and Los Angeles: University of California Press, 1987. Examines the anthropologist's contribution to "the explosion of the traditional biblically based chronology" for human history.

Van Riper, A. Bowdoin. *Men among the Mammoths: Victorian Science and the Discovery of Human Prehistory.* Chicago: University of Chicago Press, 1993. Focuses on the British debate over human antiquity, which began in earnest in the late 1850s.

DARWIN AND THE DARWINIAN DEBATES

Browne, Janet. *Charles Darwin.* 2 vols. New York: Knopf, 1995, 2002. A major new biography.

Conkin, Paul K. *When All the Gods Trembled: Darwinism, Scopes, and American Intellectuals.* Lanham, Md.: Rowman & Littlefield, 1998. Provides a useful discussion of the theological context.

Desmond, Adrian. *Huxley: From Devil's Disciple to Evolution's High Priest.* Reading, Mass.: Addison-Wesley, 1997. A biography of the zoologist who became Darwin's "bulldog" and the father of agnosticism.

———. *The Politics of Evolution: Morphology, Medicine, and Reform in Radical London.* Chicago: University of Chicago Press, 1989. Looks at the atheists and deists who promoted evolution in the 1830s.

Desmond, Adrian, and James Moore. *Darwin.* London: Michael Joseph, 1991. A compelling biography of Darwin, which sets his scientific work in its wider social context.

Dupree, A. Hunter. *Asa Gray, 1810–1888.* Cambridge: Harvard University Press, 1959. A biography of the American botanist who sought to reconcile Darwinism and Christianity.

Durant, John, ed. *Darwinism and Divinity: Essays on Evolution and Religious Belief.* Oxford: Blackwell, 1985. A series of essays examining a range of interactions between Darwinism and Christianity.

Jensen, J. Vernon. "Return to the Wilberforce-Huxley Debate," *British Journal for the History of Science,* 21 (1988), 161–79. Provides a new interpretation of the famous encounter between Bishop Samuel Wilberforce and T. H. Huxley at the 1860 meeting of the British Association for the Advancement of Science.

Kelly, Alfred. *The Descent of Darwin: The Popularization of Darwinism in Germany, 1860–1914.* Chapel Hill: University of North Carolina Press, 1981. Shows how popular Darwinism was used as a weapon against Christianity.

Larson, Edward J. *Summer for the Gods: The Scopes Trial and America's Continuing Debate over Science and Religion.* New York: Basic Books, 1997. A Pulitzer Prize–winning history.

———. *Trial and Error: The American Controversy over Creation and Evolution.* Updated ed. New York: Oxford University Press, 1989. Focuses on the legal aspects of the controversy.

Livingstone, David N. *Darwin's Forgotten Defenders: The Encounter between Evangelical Theology and Evolutionary Thought.* Edinburgh: Scottish Academic Press, 1987. An examination of the responses of major intellectual leaders of evangelical Christianity to evolutionary theory.

Livingstone, David N., and Mark A. Noll. "B. B. Warfield (1851–1921): A Biblical Inerrantist as Evolutionist," *Isis,* 91 (2000), 283–304. A study of an influential conservative Presbyterian theologian who embraced Darwinism.

Moore, James R. *The Darwin Legend.* Grand Rapids, Mich.: Baker Books, 1994. A fascinating account of Darwin's legendary conversion to Christianity.

———, ed. *History, Humanity, and Evolution: Essays for John C. Greene.* Cambridge: Cambridge University Press, 1989. Includes several important essays on evolution and religion.

———. *The Post-Darwinian Controversies: A Study of the Protestant Struggle to Come to Terms with Darwin in Great Britain and America, 1870–1900.* Cambridge: Cambridge University Press, 1979. A contentious interpretation of Protestant responses to Darwinism in the English-speaking world, accompanied by a superb history of the warfare thesis.

Numbers, Ronald L. *The Creationists.* New York: Knopf, 1992. Looks at creationists who possessed (or who claimed to possess) scientific credentials.

———. *Darwinism Comes to America.* Cambridge: Harvard University Press, 1998. A collection of essays.

Numbers, Ronald L., and John Stenhouse, eds. *Disseminating Darwinism: The Role of Place, Race, Religion, and Gender.* Cambridge: Cambridge University Press, 1999. A collection of essays illustrating the role of various contextual factors that helped shape responses to Darwinian thinking.

Pancaldi, Giuliano. *Darwin in Italy: Science across Cultural Frontiers.* Updated and expanded ed. Trans. Ruey Brodine Morelli. Bloomington: Indiana University Press, 1991. Includes a discussion of Catholic responses.

Paul, Harry W. *The Edge of Contingency: French Catholic Reaction to Scientific Change from Darwin to Duhem.* Gainesville: University Presses of Florida, 1979. Describes a broad spectrum of Catholic responses to science.

Richards, Robert J. *Darwin and the Emergence of Evolutionary Theories of Mind and Behavior.* Chicago: University of Chicago Press, 1987. A magisterial, if controversial, interpretation.

Roberts, Jon H. *Darwinism and the Divine in America: Protestant Intellectuals and Organic Evolution, 1859–1900.* Madison: University of Wisconsin Press, 1988. The most thorough investigation of the reaction of American Protestants to evolutionary theory, emphasizing the biblical rather than philosophical concerns of the participants.

Smith, Crosbie. *The Science of Energy: A Cultural History of Energy Physics in Victorian Britain.* Chicago: University of Chicago Press, 1998. Looks at the group of north British physicists and engineers who constructed the "science of energy" and repudiated the scientific naturalism of Huxley and Tyndall.

THE LATE NINETEENTH AND TWENTIETH CENTURIES

Bowler, Peter J. *Reconciling Science and Religion: The Debate in Early Twentieth-Century Britain.* Chicago: University of Chicago Press, 2001. Science and religion in Britain from the late nineteenth century to World War II.

Burnham, John C. "The Encounter of Christian Theology with Deterministic Psychology and Psychoanalysis," *Bulletin of the Menninger Clinic,* 49 (1985), 321–52. Emphasizes the clash between theological and psychological thinkers.

Evans, John H. *Playing God? Human Genetic Engineering and the Rationalization of Public Bioethical Debate.* Chicago: University of Chicago Press, 2002. Explains how bioethicists secularized the debates over human genetic engineering.

Gay, Peter. *A Godless Jew: Freud, Atheism, and the Making of Psychoanalysis.* New Haven: Yale University Press, 1987. Includes both Christian and Jewish responses.

Gilbert, James. *Redeeming Culture: American Religion in an Age of Science.* Chicago: University of Chicago Press, 1997. A kaleidoscopic history of science and religion in America from the Scopes trial of 1925 to the Seattle World's Fair of 1962.

Gregory, Frederick. *Nature Lost? Natural Science and the German Theological Traditions of the Nineteenth Century.* Cambridge: Harvard University Press, 1992. Examines various theological responses to scientific developments.

Harris, Ruth. *Lourdes: Body and Spirit in the Secular Age.* New York: Viking, 1991. Uses the history of the Catholic healing shrine in France to explore the relationship between religion and medical science.

Holifield, E. Brooks. *A History of Pastoral Care in America: From Salvation to Self-Realization.* Nashville: Abingdon, 1983. Focuses on the turn to pastoral psychology.

Hollinger, David A. "Justification by Verification: The Scientific Challenge to the Moral Authority of Christianity in Modern America." In *Religion and Twentieth-Century American Intellectual Life,* ed. Michael J. Lacey (Cambridge: Cambridge University Press, 1989), 116–35. A "neo-conflictist" interpretation.

———. *Science, Jews, and Secular Culture: Studies in Mid-Twentieth-Century American Intellectual History.* Princeton: Princeton University Press, 1996. Explores the de-Christianization of elite American culture.

Jammer, Max. *Einstein and Religion: Physics and Theology.* Princeton: Princeton University Press, 1999. Includes Christian responses.

Lightman, Bernard. *The Origins of Agnosticism: Victorian Unbelief and the Limits of Knowledge.* Baltimore: Johns Hopkins University Press, 1987. Focuses on the founding generation of agnostics, which included prominent men of science.

Moore, R. Laurence. *In Search of White Crows: Spiritualism, Parapsychology, and American Culture.* New York: Oxford University Press, 1977. Science and religion on the margins of American culture.

Mullin, Robert Bruce. *Miracles and the Modern Religious Imagination.* New Haven: Yale University Press, 1996. Includes an excellent discussion of the prayer-gauge debate.

Numbers, Ronald L. "'The Most Important Biblical Discovery of Our Time': William Henry Green and the Demise of Ussher's Chronology," *Church History,* 69 (2000), 257–76. Biblical scholarship, not science, led the way.

Orel, Vítezslav. *Gregor Mendel: The First Geneticist.* Trans. Stephen Finn. Oxford: Oxford University Press, 1996. A biography of the Augustinian monk who laid the foundations of genetics.

Orsi, Robert A. *Thank You, St. Jude: Women's Devotion to the Patron Saint of Hopeless Causes.* New Haven: Yale University Press, 1996. A magnificent study of the persistence of religious faith in an age of science.

Ostrander, Rick. *The Life of Prayer in a World of Science: Protestants, Prayer, and American Culture, 1870–1930.* New York: Oxford University Press, 2000. Explores the historical relationship between prayer and healing.

Roberts, Jon H., and James Turner. *The Sacred and the Secular University.* Princeton: Princeton University Press, 2000. A brief, highly readable survey of the de-Christianization of the American academy.

Schoepflin, Rennie B. *Christian Science on Trial: Religious Healing in America.* Baltimore: Johns Hopkins University Press, 2002. Focuses on Christian Science as a healing sect.

Turner, Frank Miller. *Between Science and Religion: The Reaction to Scientific Naturalism in Later Victorian England.* New Haven: Yale University Press, 1974. A study of six Englishmen who rejected the dogmas of both religion and science.

———. *Contesting Cultural Authority: Essays in Victorian Intellectual Life.* Cambridge: Cambridge University Press, 1993. Contains important essays on the history of secularization and scientific naturalism.

Turner, James. *Without God, without Creed: The Origins of Unbelief in America.* Baltimore: Johns Hopkins University Press, 1985. Argues that religion, not science, caused unbelief.

CONTRIBUTORS

WILLIAM B. ASHWORTH JR. is associate professor of history at the University of Missouri–Kansas City and consultant for the history of science at the Linda Hall Library of Science and Technology in Kansas City. He has published numerous articles on seventeenth-century astronomy, natural history, and scientific illustration and is completing a book, "Allegorical Images of the Scientific Revolution," a thematic study of illustrated title pages.

THOMAS H. BROMAN is associate professor of the history of science and the history of medicine at the University of Wisconsin-Madison, where he teaches a course on science in the Enlightenment. He is author of *The Transformation of German Academic Medicine, 1750–1820* (1996) and coeditor, with Lynn K. Nyhart, of *Science and Civil Society*, vol. 17 of *Osiris* (2002).

JANET BROWNE is reader in the history of biology at the Wellcome Trust Centre for the History of Medicine at University College London. Formerly editor of the *British Journal for the History of Science* and associate editor of the *Correspondence of Charles Darwin,* she is the author of *The Secular Ark: Studies in the History of Biogeography* (1983), *Charles Darwin: Voyaging* (1995), and *Charles Darwin: The Power of Place* (2002).

MOTT T. GREENE is the John B. Magee Professor of Science and Values at the University of Puget Sound in Tacoma, Washington. He is the author of *Geology in the Nineteenth Century* (1982) and *Natural Knowledge in Preclassical Antiquity* (1992). A former MacArthur Foundation Fellow, he has a long-standing interest in the relationship between science and belief. He is currently completing a biography of Alfred Wegener.

EDWARD J. LARSON is the Richard B. Russell Professor of History and the Herman Talmadge Professor of Law at the University of Georgia. He is the author of *Trial and Error: The American Controversy over Creation and Evolution* (1985), *Sex, Race, and Science: Eugenics in the Deep South* (1995), *Summer for the Gods: The Scopes Trial and America's Continuing Debate over Science and Religion* (1997), which received the 1998 Pulitzer Prize in history, and *Evolution's Workshop: God and Science on the Galapagos Islands* (2001). In 2001, he held the Fulbright Program's John Adams Chair in American Studies at the University of Leiden.

DAVID C. LINDBERG is the Hilldale Professor Emeritus of the History of Science at the University of Wisconsin-Madison. He has written or edited a dozen books on topics in the history of medieval and early modern science, including *Theories of Vision from al-Kindi to Kepler* (1976), *The Beginnings of Western*

Science (1992), and *Roger Bacon and the Origins of Perspectiva in the Middle Ages* (1996). He and Ronald L. Numbers have edited *God and Nature: Historical Essays on the Encounter between Christianity and Science* (1986) and are editing the eight-volume Cambridge History of Science (the first volumes of which appeared in 2003). A fellow of the American Academy of Arts and Sciences, he has been a recipient of the Sarton Medal of the History of Science Society, of which he is also past president (1994–95).

DAVID N. LIVINGSTONE is professor of geography and intellectual history at the Queen's University of Belfast. He is the author of *Nathaniel Southgate Shaler and the Culture of American Science* (1987), *Darwin's Forgotten Defenders* (1987), *The Geographical Tradition: Episodes in the History of a Contested Enterprise* (1992), and, with Ronald A. Wells, *Ulster-American Religion* (1999). His *Spaces of Science* is currently forthcoming from the University of Chicago Press, and he is coediting, with Ronald L. Numbers, the volume *Modern Science in National and International Context* in the Cambridge History of Science. He is a fellow of the British Academy and a member of the Royal Irish Academy.

ROBERT BRUCE MULLIN is the Society for the Promotion of Religion and Learning Professor of History at the General Theological Seminary of the Episcopal Church, New York City. He has written or edited six books, including *Episcopal Vision/American Reality: High Church Theology and Social Thought in Evangelical America* (1986), *Miracles and the Modern Religious Imagination* (1996), and *The Puritan as Yankee: A Life of Horace Bushnell* (2002).

G. BLAIR NELSON is completing his Ph.D. in the history of science at the University of Wisconsin-Madison. "Men before Adam!" is based on his dissertation research.

RONALD L. NUMBERS is the Hilldale and William Coleman Professor of the History of Science and Medicine at the University of Wisconsin-Madison. He has written or edited some two dozen books, including *The Creationists* (1992), *Darwinism Comes to America* (1998), and *Disseminating Darwinism: The Role of Place, Race, Religion, and Gender* (1999), edited with John Stenhouse. With David C. Lindberg, he has edited *God and Nature: Historical Essays on the Encounter between Christianity and Science* (1986) and is editing the eight-volume Cambridge History of Science. He is currently writing a history of science in America. He is a fellow of the American Academy of Arts and Sciences and past president of the American Society of Church History and of the History of Science Society.

JON H. ROBERTS is professor of history at Boston University. He is the author of the Brewer Prize–winning book *Darwinism and the Divine in America: Protestant Intellectuals and Organic Evolution, 1859–1900* (1998) and coauthor, with James Turner, of *The Sacred and the Secular University* (2000). He is currently writing a history of psychology and Protestant thought in the United States from 1870 to 1940.

Index

References to figures are in boldface type.